AF361909

Disaster Risk Reduction and Climate Change Adaptation: An Interdisciplinary Approach

Disaster Risk Reduction and Climate Change Adaptation: An Interdisciplinary Approach

Editors

Mikio Ishiwatari

Daisuke Sasaki

MDPI • Basel • Beijing • Wuhan • Barcelona • Belgrade • Manchester • Tokyo • Cluj • Tianjin

Editors

Mikio Ishiwatari
University of Tokyo
Japan

Daisuke Sasaki
Tohoku University
Japan

Editorial Office
MDPI
St. Alban-Anlage 66
4052 Basel, Switzerland

This is a reprint of articles from the Special Issue published online in the open access journal *International Journal of Environmental Research and Public Health* (ISSN 1660-4601) (available at: https: //www.mdpi.com/journal/ijerph/special_issues/DRRCCA).

For citation purposes, cite each article independently as indicated on the article page online and as indicated below:

LastName, A.A.; LastName, B.B.; LastName, C.C. Article Title. *Journal Name* **Year**, *Volume Number*, Page Range.

ISBN 978-3-0365-6686-3 (Hbk)
ISBN 978-3-0365-6687-0 (PDF)

Contents

International Journal of
*Environmental Research
and Public Health*

Editorial

Special Issue "Disaster Risk Reduction and Climate Change Adaptation: An Interdisciplinary Approach"

Mikio Ishiwatari [1,*] and Daisuke Sasaki [2]

[1] Graduate School of Frontier Sciences, The University of Tokyo, Kashiwa 277-8563, Japan
[2] International Research Institute of Disaster Science, Tohoku University, Sendai 980-8577, Japan
*   Correspondence: ishiwatari.mikio@jica.go.jp

The UN member states adopted three international agreements for the post-2015 agenda: the Sendai Framework for Disaster Risk Reduction 2015–2030, the Paris Agreement of the United Nations Framework Convention on Climate Change, and the 2030 Agenda for Sustainable Development. Climate change is exacerbating disaster risks worldwide, forcing countries to enhance disaster reduction measures. Approaches geared toward adapting to climate change involve a wide range of measures that reduce disaster risks [1]. Interdisciplinary approaches to climate change adaptation (CCA) and disaster risk reduction (DRR) could help make society more resilient to various shocks and multi-hazards and help achieve the three global agendas mentioned above. Developing interdisciplinary approaches involves integrating multiple disciplines and concepts. This is because disaster risks vary by risk factors, people's perceptions, spatial scales, development stages, and region [2]. Integrating the DRR and CCA approaches is challenging because experts and researchers have engaged with them separately [3]. Informed policymaking requires climate and socio-economic data as well as evidence of approaches' effectiveness, something of which developing countries do not have enough [4].

This Special Issue has accepted 15 papers and the papers included cover a wide range of issues related to interdisciplinary approaches to DRR and CCA, such as methods of assessing risks and damage, people's risk perception, financing, and policies. The findings of these studies could help promote interdisciplinary approaches at central, local, and community levels as well as internationally. We hope that this Special Issue will help accelerate research associated with the global agendas mentioned above, especially the SFDRR, which is due to undergo a midterm review soon.

1. Overview of Natural Disaster Adaptation

Jia et al. [2] reviewed recent studies on natural disaster adaptation. They found that studies primarily cover socio-economic responses for farm-scale adaptation and that studies for evaluating adaptation focus on vulnerability and not on other areas, such as resilience and countermeasures. There are research gaps in adaptive governance, lifestyle and behavior changes, and innovative financing mechanisms. Moreover, some papers in this Special Issue cover people's perceptions leading to behavior change and financing of DRR, but not governance. Future studies should cover risk governance issues.

2. Risk and Damage Assessment to Mitigate Damage

Formulating evidence-based policies requires data-based analysis. The risks of extreme events provide fundamental information for formulating DRR and CCA policies. Two papers within this Special Issue assessed the risks caused by extreme temperatures in China. Shi and Ye [5] analyzed the temporal and spatial variation of extreme temperatures from 1970 to 2014 in the Yangtze River Basin, China. Indices show a decreasing trend for extreme cold temperatures and an increasing trend for extreme warm temperatures. In addition to climate change, rapid development and urbanization from the 1980s may contribute to abrupt changes in extreme temperature indices starting in the same decade. Ma et al. [6]

Citation: Ishiwatari, M.; Sasaki, D. Special Issue "Disaster Risk Reduction and Climate Change Adaptation: An Interdisciplinary Approach". *Int. J. Environ. Res. Public Health* **2023**, *20*, 2641. https://doi.org/10.3390/ijerph20032641

Received: 19 January 2023
Accepted: 29 January 2023
Published: 1 February 2023

assessed the risks of high-temperature disasters affecting kiwifruit in Shaanxi Province, China. They developed models that can identify suitable areas for producing kiwifruit and areas at risk of high-temperature disasters.

Guo et al. [7] analyzed drought vulnerability in China and revealed that the vulnerability of agriculture to drought has decreased since the 1970s. The northwest and southwest regions' vulnerability is more severe than that of other regions.

Improving the damage assessment process following disasters could strengthen recovery efforts. Providing accurate damage information could assist decision-makers to undertake scientifically based response and rehabilitation. Two papers in this Special Issue proposed methods of estimating damage following disasters. Li et al. [8] proposed a rapid estimation method of earthquake fatality by combining physical simulations and empirical statistics in China. Zheng et al. [9] studied excess mortality of indirect deaths caused by the Great East Japan Earthquake and Tsunami in 2011 and found that the government underreported indirect deaths. Indirect deaths are caused by factors indirectly related to disasters, such as illness deterioration due to difficult conditions while evacuating, increased stress due to drastic changes in living conditions, and suicides among evacuees. They estimated that the government had underreported 873 deaths.

3. People's Risk Perception for Changing Behavior

As risk perception determines people's protective behavior, understanding how people's risk perception affects their behavior is useful for formulating policies to encourage people to change their behavior to reduce risks. Regarding this, two studies included in this Special Issue reached different conclusions. Wu et al. [10] analyzed people's risk perception in at-risk areas in Sichuan Province, China. They found a positive correlation between people's risk perception and willingness to evacuate and a negative correlation with the population at risk. Lestari et al. [11] asserted the opposite conclusion. They examined the relationships between people's initial protection behavior, evacuation behavior, concern over the possibility of a tsunami, and natural-hazard-triggered technological (Natech) situations in an earthquake in Indonesia. The results of their study did not support the hypothesis that higher risk perception is associated with evacuation behavior or that immediate evacuation is related to foreseeing cascading sequential consequences, contrary to the existing literature.

Two studies examined people's perceptions related to the COVID-19 pandemic and they propose policies for managing the pandemic. These studies quantitatively clarified key factors toward realizing evidence-based policymaking for managing pandemics. Sasaki et al. [12] investigated people's perception of well-being during the COVID-19 pandemic in Japan and advocated that the government should pay more attention to single-person households affected by the COVID-19 pandemic to improve their well-being. Pelupessy et al. [13] analyzed people's perceptions of COVID-19 risk in Greater Jakarta, Indonesia. Individual-level perceptions affect protection behavior at a family level against COVID-19. Thus, the results suggested that improving individual-level perceptions could strengthen family-level responses to the pandemic.

4. Financing Investment and Policy Formulation Based on Evidence

Several studies within this Special Issue analyzed the DRR and CCA approaches, investigated capacities, and recommended policies based on evidence. Shimada [14] analyzed the impact of climate–natural disasters on economic and social variables and the impact of international aid in Africa. The study revealed that natural disasters affect economic growth, agriculture, poverty, and cause armed conflicts. In particular, droughts were the main cause of negative impacts. Although international aid had positive effects, these effects were insignificant compared to the negative impacts of natural disasters. Moreover, cereal food aid had a negative crowding-out effect on cereal production.

Ishiwatari and Sasaki [15] examined the factors affecting investments in flood protection by analyzing investment trends over a 150-year period in Japan and found investment cycles affected by damage. They proposed approaches to securing investments in DRR by enhancing policies, legislation, and institutions.

Guo et al. [16] analyzed the effects of agricultural productive services on farmers' climate-responsive behaviors in Jilin Province. It is common among maize farmers to change to appropriate varieties in accordance with the frost-free season. Agricultural productive services significantly affect climate-responsive behaviors by farmers.

Since local governments are responsible for responding to disasters and adapting to climate change on the ground, understanding their capacities and preparation is crucial for mitigating damage. Zhai and Lee [17] developed a model of evaluating disaster preparedness capability at a local level. They applied this model to a local government and for areas requiring improvement. Ramalho et al. [18] investigated adaptation processes by local governments in Portugal and found that most local governments have developed and are implementing CCA strategies. The local governments that were studied are familiar with nature-based solutions but underestimate community-based adaptation.

Author Contributions: M.I. wrote this paper. D.S. reviewed and modified the paper. All authors have read and agreed to the published version of the manuscript.

Funding: This study was supported by the Japan Society for the Promotion of Science KAKENHI, Grant Number 21H03680.

Institutional Review Board Statement: Not applicable.

Informed Consent Statement: Not applicable.

Data Availability Statement: Not applicable.

Conflicts of Interest: The authors declare no conflict of interest.

References

1. Kelman, I. Linking disaster risk reduction, climate change, and the sustainable development goals. *Disaster Prev. Manag.* **2017**, *26*, 254–258. [CrossRef]
2. Jia, H.; Chen, F.; Du, E. Adaptation to Disaster Risk—An Overview. *Int. J. Environ. Res. Public Health* **2021**, *18*, 11187. [CrossRef] [PubMed]
3. Valente, M.; Trentin, M.; Dell'Aringa, M.F.; Bahattab, A.; Lamine, H.; Linty, M.; Ragazzoni, L.; Della Corte, F.; Barone-Adesi, F. Dealing with a changing climate: The need for a whole-of-society integrated approach to climate-related disasters. *Int. J. Disaster Risk Reduct.* **2021**, *68*, 102718. [CrossRef]
4. United Nations Climate Change Secretariat. *Opportunities and Options for Integrating Climate Change Adaptation with the Sustainable Development Goals and the Sendai Framework for Disaster Risk Reduction 2015–2030*; United Nations Climate Change Secretariat: Bonn, Germany, 2017.
5. Shi, G.; Ye, P. Assessment on Temporal and Spatial Variation Analysis of Extreme Temperature Indices: A Case Study of the Yangtze River Basin. *Int. J. Environ. Res. Public Health* **2021**, *18*, 10936. [CrossRef] [PubMed]
6. Ma, Y.; Guga, S.; Xu, J.; Zhang, J.; Tong, Z.; Liu, X. Comprehensive Risk Assessment of High Temperature Disaster to Kiwifruit in Shaanxi Province, China. *Int. J. Environ. Res. Public Health* **2021**, *18*, 10437. [CrossRef] [PubMed]
7. Guo, H.; Chen, J.; Pan, C. Assessment on Agricultural Drought Vulnerability and Spatial Heterogeneity Study in China. *Int. J. Environ. Res. Public Health* **2021**, *18*, 4449. [CrossRef] [PubMed]
8. Li, Y.; Zhang, Z.; Wang, W.; Feng, X. Rapid Estimation of Earthquake Fatalities in Mainland China Based on Physical Simulation and Empirical Statistics—A Case Study of the 2021 Yangbi Earthquake. *Int. J. Environ. Res. Public Health* **2022**, *19*, 6820. [CrossRef] [PubMed]
9. Zheng, X.; Feng, C.; Ishiwatari, M. Examining the Indirect Death Surveillance System of The Great East Japan Earthquake and Tsunami. *Int. J. Environ. Res. Public Health* **2022**, *19*, 12351. [CrossRef] [PubMed]
10. Wu, S.; Lei, Y.; Jin, W. An Interdisciplinary Approach to Quantify the Human Disaster Risk Perception and Its Influence on the Population at Risk: A Case Study of Longchi Town, China. *Int. J. Environ. Res. Public Health* **2022**, *19*, 16393. [CrossRef] [PubMed]
11. Lestari, F.; Jibiki, Y.; Sasaki, D.; Pelupessy, D.; Zulys, A.; Imamura, F. People's Response to Potential Natural Hazard-Triggered Technological Threats after a Sudden-Onset Earthquake in Indonesia. *Int. J. Environ. Res. Public Health* **2021**, *18*, 3369. [CrossRef] [PubMed]
12. Sasaki, D.; Suppasri, A.; Tsukuda, H.; Nguyen, D.N.; Onoda, Y.; Imamura, F. People's Perception of Well-Being during the COVID-19 Pandemic: A Case Study in Japan. *Int. J. Environ. Res. Public Health* **2022**, *19*, 12146. [CrossRef] [PubMed]
13. Pelupessy, D.C.; Jibiki, Y.; Sasaki, D. Exploring People's Perception of COVID-19 Risk: A Case Study of Greater Jakarta, Indonesia. *Int. J. Environ. Res. Public Health* **2023**, *20*, 336. [CrossRef] [PubMed]

14. Shimada, G. The Impact of Climate-Change-Related Disasters on Africa's Economic Growth, Agriculture, and Conflicts: Can Humanitarian Aid and Food Assistance Offset the Damage? *Int. J. Environ. Res. Public Health* **2022**, *19*, 467. [CrossRef] [PubMed]

15. Ishiwatari, M.; Sasaki, D. Disaster Risk Reduction Funding: Investment Cycle for Flood Protection in Japan. *Int. J. Environ. Res. Public Health* **2022**, *19*, 3346. [CrossRef] [PubMed]

16. Guo, H.; Xia, Y.; Pan, C.; Lei, Q.; Pan, H. Analysis in the Influencing Factors of Climate-Responsive Behaviors of Maize Growers: Evidence from China. *Int. J. Environ. Res. Public Health* **2022**, *19*, 4274. [CrossRef] [PubMed]

17. Zhai, L.; Lee, J.E. Analyzing the Disaster Preparedness Capability of Local Government Using AHP: Zhengzhou 7.20 Rainstorm Disaster. *Int. J. Environ. Res. Public Health* **2023**, *20*, 952. [CrossRef] [PubMed]

18. Ramalho, M.; Ferreira, J.C.; Jóia Santos, C. Climate Change Adaptation Strategies at a Local Scale: The Portuguese Case Study. *Int. J. Environ. Res. Public Health* **2022**, *19*, 16687. [CrossRef] [PubMed]

International Journal of
***Environmental Research
and Public Health***

Review

Adaptation to Disaster Risk—An Overview

Huicong Jia [1], Fang Chen [1,2,3,4,*] and Enyu Du [2]

1   Key Laboratory of Digital Earth Science, Aerospace Information Research Institute,
    Chinese Academy of Sciences, Beijing 100094, China; jiahc@radi.ac.cn
2   College of Resources and Environment, University of Chinese Academy of Sciences, Beijing 100049, China;
    duenyu@cug.edu.cn
3   Hainan Key Laboratory of Earth Observation, Institute of Remote Sensing and Digital Earth,
    Chinese Academy of Sciences, Sanya 572029, China
4   State Key Laboratory of Remote Sensing Science, Aerospace Information Research Institute,
    Chinese Academy of Sciences, Beijing 100094, China
*   Correspondence: chenfang_group@radi.ac.cn; Tel.: +86-10-8217-8105

Abstract: The role of natural disaster adaptation is increasingly being considered in academic research. The Paris Agreement and Sustainable Development Goal 13 require measuring the progress made on this adaptation. This review summarizes the development stages of adaptation, the multiple attributes and analysis of adaptation definitions, the models and methods for adaptation analysis, and the research progress of natural disaster adaptation. Adaptation research methods are generally classified into two types: case analysis and mathematical models. The current adaptive research in the field of natural disasters focuses primarily on the response of the social economy, especially the adaptive decision making and risk perception at farm-level scales (farmer households). The evaluation cases of adaptation in the field of disasters exist mostly as a part of vulnerability evaluation. Adaptation and adaptive capacity should focus on four core issues: adaptation to what; who or what adapts; how does adaptation occur; what is adaptation; and how good is the adaptation. The main purpose of the "spatial scale–exposure–vulnerability" three-dimensional scales of adaptation assessment is to explore the differences in index system under different scenarios, the spatial pattern of adaptations, and the geographical explanation of its formation mechanism. The results of this study can help and guide future research on integrating climate change and disaster adaptations especially in regional sustainable development and risk reduction strategies.

Keywords: disaster risk; climate change; adaptation; method; digital disaster reduction

Citation: Jia, H.; Chen, F.; Du, E. Adaptation to Disaster Risk—An Overview. *Int. J. Environ. Res. Public Health* **2021**, *18*, 11187. https://doi.org/10.3390/ijerph182111187

Academic Editors: Mikio Ishiwatari and Daisuke Sasaki

Received: 30 September 2021
Accepted: 22 October 2021
Published: 25 October 2021

Publisher's Note: MDPI stays neutral with regard to jurisdictional claims in published maps and institutional affiliations.

1. Introduction

Since the proposal of climate change in the 1970s and its impact on human society, the international scientific community and governments began discussing how human society should respond to global changes and adopt corresponding countermeasures. The specific research direction proposed in the 1970s was prevention, with mitigation in the 1980s, and adaptation to date. Prevention, mitigation, and adaptation are all human response behaviors [1–5]. The term "adaptation" is currently used in the climate field. It originated from natural science in the field of population biology and evolutionary ecology [6]. It originally referred to the general characteristics that ensure the survival and reproduction of organic individuals in living environments. These characteristics result in sustainable survival and the development of species or ecosystems, while evolving to changes in an organism or species makes it more adapted to survive [7]. The IHDP (International Human Dimensions Program) launched the "Inter-vulnerability framework to assess interacting impacts of global processes" in January 2005, which proposed dynamic changes in time and space to form the vulnerability and process of climate change and that of globalization [2,8]. A vulnerability evaluation concept combined with a global change process proposes to transform the general index evaluation methods in most studies into the vulnerability evaluation of the adapter. The vulnerability of the adapter is not only a function of the exposure level, sensitivity, and adaptability but also includes the adaptors' cognitive process for changes, risks, trade-offs, and the selection of adaptation methods [9].

Through the 2015 Paris Agreement on Climate Change, 197 countries have committed to 47 ambitious efforts to combat climate change, adapt to its effects, and provide enhanced support to developing countries [10]. Enhancing understanding and management of risks, as well as the impacts that impinge upon individuals, households, communities, cities, countries, economies, or ecologies through time, is at the heart of the aspirations and goals of the Sendai Framework, which was adopted by Member States at the United Nations General Assembly in June 2015 [11]. In 2015, UN member countries also adopted the 2030 Agenda for Sustainable Development—a comprehensive global plan of action for "people, planet and prosperity" comprised of 17 Sustainable Development Goals (SDGs) and 169 targets to be achieved by 2030, including Goal 13 on climate action [12]. The goal of "Climate Action: Take urgent action to combat climate change and its impacts" (SDG 13) is to reduce the impact of climate change on people and improve the ability to respond to climate change. An important way to reinforce SDG 13.1 (Strengthen resilience and adaptive capacity to climate-related hazards and natural disasters in all countries) is to explore and analyze the connotation and assessment methods of adaptation. In this review, we fill this gap by providing an overview of adaptation and adaptive capacity evaluation methods and their applications in the fields of climate change and natural disaster risk reduction.

In a changing climate, disasters are inevitable due to the uncertainty and abnormality of hazards and the expanded exposure. Over the past three decades, evidence has mounted that the global climate is changing and that anthropogenic greenhouse gas emissions are largely to blame [12]. According to the Intergovernmental Panel on Climate Change (IPCC), climate change includes increasing temperatures, changing rainfall patterns, rising sea levels, saltwater intrusion, and a higher probability of extreme weather events that could lead to natural disasters [10,12]. The length, frequency, and/or intensity of heat waves will increase with higher temperatures when extremes occur. Heavy rainfalls associated with tropical cyclones are likely to increase with continued warming. The intensity and frequency of extreme precipitation events are very likely to increase over many areas, and the return period of extreme rainfall events is projected to decline, resulting in more numerous floods and landslides. Mid-continental areas will generally become dryer, which is likely to increase the risk of summer droughts and wild fires.

This paper is organized as follows: after the Introduction section, the definition of adaptation and development stages of adaptation are discussed in Sections 2 and 3. Sections 4 and 5 summarizes analysis methods of adaptations and regional applications for different fields of the natural sciences. Finally, Section 6 provides a summary and discussions.

2. Definition of Adaptation

2.1. Definition of Adaptation and Multiple Properties

Understanding the concept of adaptability has undergone a change process to include processing power and the response to adjustments. Although these definitions have respective focuses, they all emphasize the need to adjust the system and reduce its vulnerability to improve and strengthen its ability to adapt to climate change. The content of adaptation involves the process of natural and sudden disaster impact assessments, which includes countermeasures against climate change to enhance the process of designing and improving measures for sustainable regional development [13].

The definition of "adaptation" has many attributes, including the two most important points (Table 1). First is the spatial scale of adaptation, which depends on who is responsible. Second is the nature of adaptive behavior, whether it is spontaneous or conscious or it is planned or prescriptive. The former is usually short-term and tactical adaptation, which is directly related to specific climate change. The latter is more strategic, long-term, and proactive and is usually formulated by government departments and used as part of policy adaptation measures [14]. The adaptation to climate change in the literature is sometimes divergent at the temporal and spatial scales. Short-term adaptation is more of a reaction, and higher-scale adaptation is considered an expected adaptation through policies, projects, and recent plans and actions [15].

Table 1. Definition of adaptation.

Reference	Definition
	(1) Adaptation is a handling capability.
[16]	Handle the ability of short-term and long-term "possibilities"
[17]	The behavior and features of "system adjustment" can enhance the ability to process external pressure
	(2) Adaptation is a response.
[18]	Ecology–social–economic system: a response to actual and expected climate oscillation and its impact
[19]	Adaptation of climate change refers to a response to human or natural systems on existing or future climate stimuli or influence
[20]	A region or department's adaptability to climate change relies on many non-climate factors, such as its availability, social and economic policies, cultural and political considerations, individual and public property (economic development and investment levels), and markets or insurance; the adaptability analysis is an important part of the policy response of climate change
	(3) Adaptation is a change (adjustment) process.
[21]	Adaptation includes changes in processes, measures, or structures to reduce or offset potential hazards associated with climate change or to take advantage of the opportunities brought about by climate change, which include reducing the vulnerability of society, regions, or activities to climate change and variability adjustments
[22]	Climate adaptation is a process by which people reduce the negative impact of climate on health and welfare and take advantage of opportunities provided by climate and environmental changes
[23]	Any adjustment measures, whether passive or active, are aimed at reducing the expected adverse effects of climate change
[24]	Climate adaptation countermeasures are adjustment measures taken by individuals for short-term and long-term climate change and extreme weather disasters to enhance the viability of social and economic activities and reduce vulnerability
[25]	Climate change adaptation is defined as the degree to which the implementation, operation process, or structure of the system can be adjusted under possible or actual climate change conditions in the future or the system's adaptive capacity; adaptation behavior can be spontaneous or planned and can be put into practice in actual processes to handle climate change that has occurred or is expected to occur
[26]	Climate change adaptation includes all human actions or economic structural adjustment measures taken to reduce the vulnerability of all society
[27]	The adjustment of individual organizations and institutional behaviors to reduce the vulnerability of society to the climate change
[18]	The adjustment of the ecological–social–economic system responses to actual or predicted climate change
[28]	Adaptation is a policy option to reduce the negative impact of climate change
[29]	An adjustment of the socio-economic system response to actual or expected climate change
[30]	An adjustment to reduce the risks associated with climate change and vulnerability under its influence to a predetermined level without affecting the existing economic, social, and environmental sustainability
[31]	Adaptation includes both moderating harm and exploiting beneficial opportunities
[32]	Adaptation refers to the process of adjustment to actual or expected climate change and its effects to moderately harm or exploit beneficial opportunities
[33]	Adaptability is a manifestation of adaptation, which is the ability to absorb hazard impacts and to prepare for and recover from them; adaptation in most cases is a proactive action to the anticipated hazards so that potential negative effects or risks can be alleviated in advance
[11]	Incorporate disaster risk reduction measures into multilateral and bilateral development assistance programs within and across all sectors as appropriate, which is related to poverty reduction, sustainable development, natural resource management, environment, urban development, and adaptation to climate change
[34]	Adaptation is a process with varied and changing goals and risk context
[35]	The goal of adaptation is to reduce vulnerability and increase resilience

Smit and Skinner further defined several key characteristics of adaptability, including the purpose, time and duration, scale and responsibility, and form of adaptation [20]. At the same time, several adaptive approaches have been proposed, such as technological development, government projects and insurance, production reform, and financial management, which provide a useful framework for the development and selection of adaptive strategies toward human vulnerability. Fankhauser et al. proposed three elements of adaptation: necessity, motivation, and capability [36]. Bryant et al. gave four main components of adaptability: pressure characteristics, system characteristics (including cultural, economic, political, institutional, and biophysical environment), multi-scale, and adaptive response [37]. The different spatial scales (from local adjustments to regional and national resource reorganization strategies and policies) and temporal scale changes (from short-term changes to longer time scales) cause adaptations to vary widely [34]. Climate change is only one aspect of response and adaptation, which is closely related to other human and economic factors.

2.2. Analysis of Adaptation-Related Terms

As an important attribute of disaster systems, adaptation is closely related to the concepts of other disasters; however, they exhibit differences when used.

2.2.1. Adaptation, Adaptability, and Capacity of Response

Adaptability in ecology is the ability to adapt to certain environmental changes, while adaptation is the characteristic of structure, function, and organizational behaviors [38]. Adaptability is the external manifestation of adaptive ability and shows a way to reduce vulnerability [18]. A system's capability to better handle exposure and sensitivity reflects the capability to adapt [39] (Figure 1). Many adaptive forms and levels can be divided based on timing (anticipated, current), intention (automatic, planned), spatial scales (local, wide-area), and form (technical, behavioral, financial, economic, institutional, and information) [40]. The adaptation of the original system can be distinguished from the degree of adjustment [41].

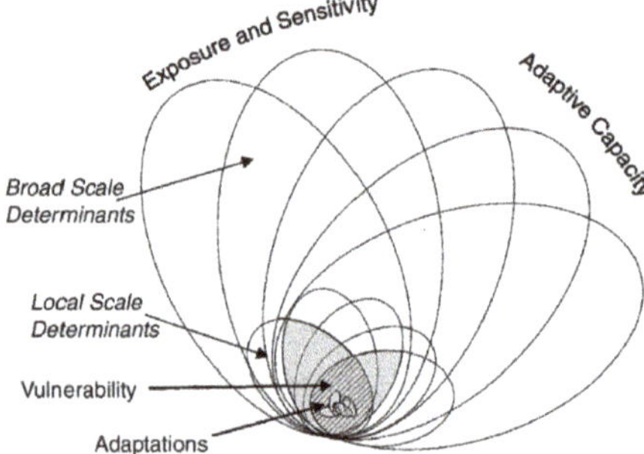

Figure 1. Nested hierarchy model of adaptation, adaptive capacity, and vulnerability.

Local adaptive capability is a comprehensive reflection of several conditions [42,43], which is reflected by factors such as management capabilities, economic and financial conditions, technology and information resources, infrastructure, and institutional environments [16,44–48]. In general, improving environmental conditions allows a species, population, or individual to better adapt to the environment. Due to the human field and social ecosystems, the standards of adaptation far exceed the ability to survive and reproduce, which includes the results of social and economic activities and the quality of life [39]. Smithers and Smit noted that the adaptability of the human system includes the capability of the social ecosystem to respond to environmental changes and to promote improved conditions related to the environment [49].

Kasperson et al. distinguished between adjustment and adaptation [50]. They believed that adjustments are a system's response to interference or pressure without fundamentally changing the system itself. This is a short-term and relatively small system adjustment as adaptation is the system's response to interference or pressure. The response to stress changes the system itself and can sometimes transform a system state to a new state [51]. The concepts involved in adaptability include the coping ability, management capacity, stability, robustness, and flexibility [28,49,52–54]. From the definition of ISDR (International Strategy for Disaster Reduction), the coping ability refers to all available forces and resources in a community or organization that can reduce the risk level or impact of a disaster [55]. Brooks et al. defined adaptive capability as natural, economic, institutional, or human resources that can be used for adaptation and included the availability of information, professional knowledge, social networks, and other resources [56]. Cultural values also play an important role in the construction of human adaptability [57,58].

The IPCC analyzed the relationship between the adaptability and capacity of a response for the social ecosystem and believed that the connotation of adaptability should be broader than that of the capacity of response. However, these all depend on the specific definition of adaptability and capacity of response in studies of coupled social ecosystems [58]. Adger [59], Smit and Wandel [39], and IPCC [19] all defined the system's coping ability or capacity of response as adaptability. Turner et al. distinguished the capacity of response from adaptability, considered that both are components of the system resilience, and regarded adaptation as a manifestation of the reconstruction of the system after a response [60]. Generally, the capacity of response is an inherent attribute of the system that describes its ability to respond to interference, mitigate potential damage, take advantage of opportunities, adjust to system changes, and respond to system transformation. The capacity of response is also an attribute that the system has priority over interference [58]. Vogel believed that the coping ability is a short-term behavior only for survival, while the adaptive ability is used for long-term or more continuous adjustments [61].

Boundaries between the medium- and long-term scales are blurred. The response refers to the actions taken by society or individuals when faced with the adverse consequences of climate change or natural disasters, which are considered short-term adjustments to extreme events [59]. Coping strategies usually occur naturally and cause varying degrees of vulnerability. Adaptation includes the stress response, such as changing sources of income, immigration, or other lifestyle changes, as well as long-term intervention by government agencies [62].

2.2.2. Adaptability, Vulnerability, and Resilience

Due to diversity and differences of views, the relationship between adaptability and resilience is unclear. According to Smit and Wandel [39], some scholars have equated adaptability with social resilience. Gunderson et al. regarded adaptability as the effectiveness of a system to changes in resilience [63]. Carpenter et al. regarded adaptability as a component of resilience, which reflects the response of system behaviors to disturbances [64]. Adger regarded adaptability as the collective ability of human activists to manage resilience, which includes reducing or eliminating undesirable factors, creating new expectation factors, and promoting the transformation of the current system to the desired state [59]. Folke et al. believed that vulnerability is the opposite or antithesis of resilience [65]. However, the vulnerability of a system with resilience is lower than that of a system without resilience. Resilience is related to the capacity of response in a vulnerable element, making it smaller than the negative range of vulnerability [58].

The most fundamental difference is that resilience is applied to the maintenance of system behaviors, and the opposite of vulnerability is the ability to resist interference and maintain the system structure. Therefore, for the elements of social ecosystems, resilience appears to be a true subset of the adaptive capacity. Adaptability includes not only the resilience of the system but also its ability to cope with impacts and take advantage of opportunities [66]. Adaptation is a measure taken by humans with a constant evaluation of vulnerability. In the formula $V_{ist} = f(E_{ist}, A_{ist})$ as modified by Smit and Pilifosova [42], V is the vulnerability, E is the exposure sensitivity, and A is the adaptive capacity. Here, i refers to the system, s refers to the climate stimulus, and t is time. The sensitivity refers to the degree to which the system suffers and responds to climate stimuli. Exposure is a factor of risk that is the total amount of hazard-affected bodies that are exposed to a hazard. Gallopin [58] used a systematic perspective to comprehensively analyze the relationship between the concepts of vulnerability, resilience, and adaptability (Figure 2), which has been adopted by some scholars [2,3,67,68].

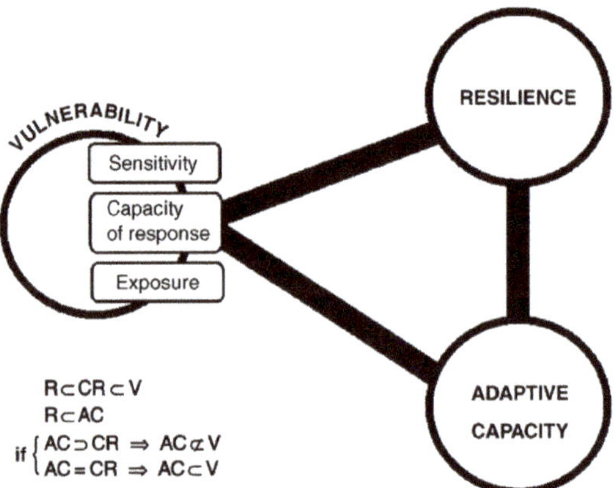

Figure 2. A diagrammatic summary of the conceptual relations between vulnerability, resilience, and adaptive capacity. (The R, V, AC, and CR stand for resilience, vulnerability, adaptive capacity, and capacity of response, respectively.)

2.2.3. Adaptation and Mitigation

Climate change risks can be managed through efforts to mitigate climate change forcers, adaptation of impacted systems, and remedial measures. Mitigation refers to efforts to reduce or prevent the emission of greenhouse gases or to enhance the absorption of gases already emitted, thus limiting the magnitude of future warming [33]. Mitigation avoids difficult to handle situations, while adaptation aims to manage the inevitable consequences [69]. There are uncertainties in adaptive strategies that require long-term perspectives, which may be unpopular for current governments. The current understanding of adaptation is to regard the adaptation period as placing future social public resources in danger. The adaptation strategies must be concurrent and complementary with mitigation efforts because, over the long term, emissions reduction choices will determine the severity of climate change, its impacts, and the degree of adaptation required in the future [70]. Unlike mitigation, adaptation is most practical at the local level.

3. Development Stages of Adaptation
3.1. Adaptation in Disasters and Other Fields

The concept of "cultural adaptation" was firstly used to describe the "cultural cores" to the physical environment [71,72]. To date, the adaptive example has been widely used in social sciences. Denevan defined cultural adaptation as the response to natural environmental changes and the associated humanities (such as population, economics, and organization) [71]. O'Brien et al. defined adaptation as an organizational or group enhancement environment or cultural grinding ability and believed that adaptation is the behavior selection result produced by cultural practice in changing environments [73]. In adapting to the subject, biological adaptation involves changes in individuals and populations, while social adaptation is the adjustment of individual and collective behaviors, and there are similarities between the two [49]. However, the ability of human systems to exhibit planned and managed adaptation makes it an important factor to contain natural environments and intrinsic stimulation double changes. This allows adaptation to be combined with environmental perception and risk assessment as important factors in adaptation strategies. In addition, as the human system is culturally adapted, human groups can create new and improved methods to process environmental issues into their culture. Thus, the pursuit of human systems is the adjustment of the target, not just the survival of the species, which includes an enhanced quality of life.

In the field of disasters, some scholars, such as Burton et al. [74], emphasized that adaptation should include risk cognition, adjustment, and disaster management. In addition, many scholars have focused mostly on adaptive adjustments and environmental disaster management [75]. Holling also used the concept of adaptation to study the recovery, balance, and adaptation management of natural environmental changes [76]. In the field of power theory and food safety, adaptation is seen as a resource acquisition and a response to people's behaviors, which forms a highlighted feature in this research area to reveal how individual or family adaptability is formed and how the social, political, and economic processes can be restrictive [77–79]. For adaptive research in political ecology, Kasperson believed that adaptation should be launched to research individual and family adaptability and considered how these can be shaped and restricted by social, political, and economic processes [50,80]. Currently, research on adaptation to disaster risk reduction is still in its initial stages, and there is no unified definition.

3.2. Adaptation in Climate Change

Adaptive research has continued to emerge with the constant concern of climate change. Early cases from Butzer [81] were based on predictable climate change and its expected impact on the world food supply while considering cultural adaptation (human wisdom from technological innovation and long-term planning). Since then, the adaptation analysis and research of climate change have gradually expanded [18,42,82]. Smit et al. proposed a schematic diagram for adapting to climate change and variations, which consists of several problems [18]. The Paris Agreement aims to strengthen the ability of countries to deal with the impacts of climate change through appropriate financial flows, a new technology framework, and an enhanced capacity-building framework. Saving lives and livelihoods requires urgent action to address the climate emergency and adaptation [83]. Through scientific guidance (understanding and prediction), the public participation (communication and education), scientific adaptation, adaptation management and methods, and decision-making (global convention and implementation of national strategies) are all interrelated.

4. Analysis Models and Methods of Adaptation

The research methods of adaptation are generally classified into two types: case analysis and mathematical models. Vulnerability research tends to focus on constructing and analyzing indicators. Research on resilience, especially in ecology, has developed several theories and mathematical models, while research on adaptation has focused on case analysis [84]. Table 2 summarizes the characteristics of all the methods of adaptation.

Table 2. An overview of analysis models and methods of adaptation.

Models and Methods	Characteristics
Scenario-Driven by Climate Change	Represented by the IPCC technical guidelines to evaluate climate change impacts and adaptation countermeasures.
Adaptation Decision Matrix	Suited for analyzing the cost–benefit of adaptation measures.
TEAM	A decision support system software. Suited for assessing the impact and adaptability of climate change in water resources, coastal areas, and agricultural sectors.
Multi-Criteria Evaluation Method	Various adaptation strategies can be compared and evaluated in an orderly and systematic manner.
Agent-Based Modeling Method	A useful policy tool to simulate the effects of different adaptation options toward reducing vulnerability.

4.1. Evaluation Standard

Although there have been some discussions on adaptation, such as the NFCCC (United Nations Framework Convention on Climate Change) and UNAPF (United Nations-Azerbaijan Partnership Framework), there is still no agreement on its goals; thus, the success of the adaptation behavior cannot be judged. Mercer [85] believed that the goals of adaptation should include the following aspects: maintaining risks associated with climate change at the current level, reducing risks to a lower level when existing risks are considered unacceptable, reduce the exposure of vulnerable populations, and others. A successful adaptation should consider the following points: cost–benefit, efficiency, distribution of costs and benefits, and legality of adaptation. Other aspects include sustainability, global and intergenerational fairness, and adaptation in harmony with cultural norms and socially recognized values [5,86].

The following situations may appear in some cases of adaptation to climate change. Success or failure is also affected by self-adaptation or adaptive behaviors. Individuals experiencing climate change can often distinguish whether they are well protected and better adapted, indicating there may also be different values that support various adaptation goals [87]. Temporary adaptation will only bring future challenges. Therefore, adaptation should be coordinated with the natural environment, including the consciousness formation of modern civilized society. Thus, it is necessary to improve the adaptability through capacity building at the local level [82].

The systematic analysis of potential adaptation options needs to consider their possible feasibility, cost, profit, effectiveness, execution speed, and the acceptability of relevant fund custodians [34,88]. The analysis primarily includes the following six aspects:

(1) The adaptive capacity. This can be measured by reducing the impact or exposure or by reducing disasters, preventing danger, and improving safety. However, in practice, the complex causal chain induces several problems when estimating the adaptive capacity [89].

(2) Effectiveness of adaptation. This is the extent to which the behavior reaches the goal. However, many issues need to be considered in the evaluation process. First, selecting a particular adaptation may be uncertain under the given circumstances. Second, the utility of an adaptation option introduced by the organizer may depend on other behaviors. Relying on the effectiveness of adaptation measures depends on individual behaviors and may be very difficult to evaluate. Third, the effectiveness of adaptation behaviors may depend on unknown future conditions. Fourth, an adaptation measure is effective as it reduces the impact of climate change or increases one location or period. Any adaptive behavior can potentially create unintended consequences on other natural and social systems [90].

(3) Adaptation efficiency. Adaptation to climate change needs to bear the cost but should generate significant benefits. The scale costs in the individual organization are the implementation measures, including transaction costs and inaccurately estimated costs, as well as benefits that reduce impacts or enhance opportunities. However, the analysis of efficiency adaptation at any scale is more economical than a simple and quantitative cost–benefit comparison [72].

(4) Cost of adaptation. Although there are many possible adaptation measures, it is necessary to understand the conditions that limit adaptability and the costs of improving adaptability. Fankhauser [52] believed that the impact cost is the sum of the costs of adaptation and residual losses. Any comprehensive evaluation of adaptation costs (including profits) not only considers economic indicators but also social welfare and equity. Adaptation costs can be divided into the direct cost of adaptation, cost of adapting to the state of adaptation, and cost of inadaptability [91].

(5) Fairness and legality of adaptation. Fair adaptation can be evaluated from the perspective of the main body whose income decides to adopt adaptation measures. There are many principles of income fairness, including the principle of deservedness and for fairness or need. Each principle has its own power, where the distribution of power in the income system could impact the legitimacy of decisions [22].

(6) A systematic view of adaptation. Adaptation assessment has become more inclusive over time, linking future climate change with current climate risks and other policy concerns [79]. Adaptation includes well-established practices from disaster risk management (e.g., early-warning systems), resource management (e.g., water rights allocation), spatial planning (e.g., flood zone protection), urban planning (e.g., building codes), public health (e.g., disease surveillance), and agricultural outreach (e.g., seasonal forecasts) [92]. The adaptation of farmers may extend beyond climate change and be multifaceted. For example, unstable land ownership affects the ability of farmers to develop agriculture sustainably. Therefore, economic and institutional factors impact adaptation. These factors gradually undermine the way farmers adapt and impact the accuracy and relevance of climate information they feel.

4.2. Evaluation Model and Method

The United Nations Framework Convention on Climate Change (UNFCCC) defines adaptability as being in two categories: spontaneous and planned [93]. The following describes general methods to evaluate adaptability for applications in many fields.

4.2.1. Scenario-Driven by Climate Change

To date, most research on the evaluation of climate change impacts and adaptation countermeasures has adopted so-called "scenario-driven" research methods. This approach is represented by the IPCC technical guidelines to evaluate climate change impacts and adaptation countermeasures. This is usually considered a standard research method or approach and consists of the following seven steps:

(1) Define the problem (clear research area, research content, select sensitive departments, etc.);

(2) Choose an evaluation method suitable for most problems;

(3) Select the test method and conduct sensitivity analysis;

(4) Select and apply climate change scenarios;

(5) Evaluate the impacts on biological, natural, and socio-economic systems;

(6) Evaluate spontaneous adjustment measures;

(7) Evaluate adaptation strategies.

The fourth step is the critical part of the entire evaluation process as it is driven by future climate change and socio-economic scenarios. Thus, an assessment of the impacts of climate change on humans and ecosystems can be performed. Once the ecosystem and socio-economic system are warned that the impact of climate change will be affected, these systems or departments spontaneously respond or adapt and reduce the losses caused by climate change through anticipated adaptation measures and countermeasures. This assessment approach represents a routine procedure that requires significant time, energy, and resources to select and apply climate change scenarios and impact assessments. In practice, there is often insufficient time and funds to conduct adaptation countermeasure assessment research.

In the climate change literature, most research has focused on the losses and impacts of climate change on specific aspects of human society and ecosystems. The main purpose of applying simulation models is to establish the future state of the ecosystem in relation to climatic conditions. For example, several different types of simulation models can be used to study the growth rate of crops or forests under various climate scenarios. Then, different adaptation strategies can be evaluated using the ecological simulation models. Changing the corresponding parameters of the simulation model reflects the adoption of certain adaptation countermeasures or measures under different climate change conditions. All applications have to take account of uncertainties in the information. Some of those are related to uncertainties in the climate models and future emissions, others are related to downscaling to the local scale, and still others are related to the lack of consistent data to verify the model at that local scale. The climate system is changing, so uncertainty about extremes is rising. For example, this could include using varied model parameters to indicate the adaptability of some new crops and tree species to future climate change and the development of production technology to adapt to future climate change [94–97].

4.2.2. Adaptation Decision Matrix

The adaptation decision matrix (ADM) is based on Excel or Lotus and is used to analyze the cost–benefit of adaptation measures. Researchers list policy goals in the upper part of the matrix and the adaptation strategies (including not taking any measures) in the lower part. Through expert diagnosis, research, and analysis, a value (from 1 to 5) is assigned to each adaptation policy to indicate the degree of satisfaction it can achieve for the specific goals under various adaptation strategies.

Researchers also have the authority to set different weights for each policy objective in the evaluation process and perform a weighted sum to calculate the cost when the benefit increases by one unit. For example, Mizina et al. [98] used an adaptive decision-making matrix and expert-scoring method (using arbitrary quantitative ratios and not monetary values) to analyze 12 adaptive factors that affect Kazakhstan's agriculture and screen out four important factors. Batima [99] evaluated the drought adaptation of pastures in Mongolia using a decision matrix to divide various measures into the three levels of high, medium, and low in terms of long-term effectiveness, short-term benefits, costs, and limitations. This method is useful when many of the benefits generated from the policy objectives are difficult to monetize or cannot be unified. However, conducting in-depth research requires detailed analysis results to provide researchers with basic information as the basis to evaluate and score. Otherwise, the scoring process relies too heavily on subjective judgment; however, if it is used as part of the questionnaire for statistical analysis, this error effect will be significantly reduced.

4.2.3. TEAM (Tools for Environmental Assessment and Management)

To evaluate the possible impacts and consequences of various adaptation countermeasures and planning to select suitable and satisfactory countermeasures, the United States Environmental Program has developed a decision support system software called TEAM (Tools for Environmental Assessment and Management) as a decision-making tool [100]. This system is based primarily on multi-criteria and multi-standard decision-making technologies and uses graphical means and man–machine dialogue to simplify and clarify the evaluation process. This evaluation method is suitable to assess the impact and adaptability of climate change in water resources, coastal areas, and agricultural sectors. The TEAM model is used as an analysis tool for the U.S. government to evaluate climate adaptation countermeasures based on the national unit's international climate change impact and adaptation countermeasure evaluation project [101].

The entire evaluation and research process based on the TEAM model includes five main steps, each of which provides a certain mechanism and function (Table 3). The first step is to determine the geographic location of the study area. The geographic location is used to analyze the conditions of various natural resources or ecosystems (such as water resources, agricultural systems, and coastal areas) and determine the characteristics of the vulnerability of these systems to climate variability or change. The second step is to select possible adaptation strategies and measures. The TEAM model recommends a series of adaptation countermeasures and measures to reduce the vulnerability of selected systems or departments for analysts or decision-makers. Analysts or decision-makers can select an appropriate number of adaptation countermeasures or add special countermeasures. The third step is to evaluate the adaptation strategies. The evaluation criteria must be carefully determined to comprehensively evaluate the economic, social, resource, and environmental effects of some climate change adaptation policies and programs. In the fourth step, analysts or decision-makers give scores to each standard or indicator based on the performance and benefits of each adaptation countermeasure in this standard. The TEAM model allows users to compare various quantitative data to determine scores. The fifth step is to display the evaluation results. The TEAM model provides users with several approaches, and the analysis results can be presented in a way that is more acceptable to both the general public and decision-makers, such as diagrams. The displayed evaluation results reflect the benefits of different adaptation strategies to various standards so that information on the associated advantages and disadvantages can be given.

Table 3. An overview of five steps of the TEAM model.

Number	Steps
Step 1	Determine the geographic location of the study area.
Step 2	Select possible adaptation strategies and measures.
Step 3	Evaluate the adaptation strategies.
Step 4	Give scores to each standard or indicator.
Step 5	Display the evaluation results.

4.2.4. Multi-Criteria Evaluation Method

When evaluating adaptation strategies for multi-standard and multi-group participation, the multi-standard evaluation method is a better analysis technique and can be used as an effective tool. In this way, various adaptation strategies can be compared and evaluated in an orderly and systematic manner. Multi-standard evaluation tools can determine satisfactory policies when given a series of possible adaptation policies to handle biological, natural, and socio-economic climatic vulnerabilities. Several methods and tools developed and established in the fields of decision science, multi-standard evaluation, and system analysis can also be used to evaluate adaptation measures. These can effectively link climate change impact assessments with regional sustainability and include goal planning (GP) [102], fuzzy pattern recognition (FPR) [103], neural network (NN) technology [104–106], and multi-level analysis process technology [107].

4.2.5. Agent-Based Modeling Method

Agent-based modeling is a useful policy tool to simulate the effects of different adaptation options toward reducing vulnerability, as it allows the representation of not only dynamic changes in climate and markets but also the dynamic adaptive process of different groups of communities on the impacts of these changes. Model simulations of adaptation options under various global change scenarios show that production support significantly reduces future vulnerability only if complemented with appropriate market support. Therefore, policies need to provide a complementary bundle of adaptation measures. These adaptation measures include developing knowledge on climate change impacts and adaptation; strengthening observations; promoting an approach appropriate for the different territories and utilizing legislative and regulatory instruments, implementing risk-reduction strategies in the insurance industry; etc. Lack of funds and information are the most important reasons to not apply available technical adaptation measures (Figure 3).

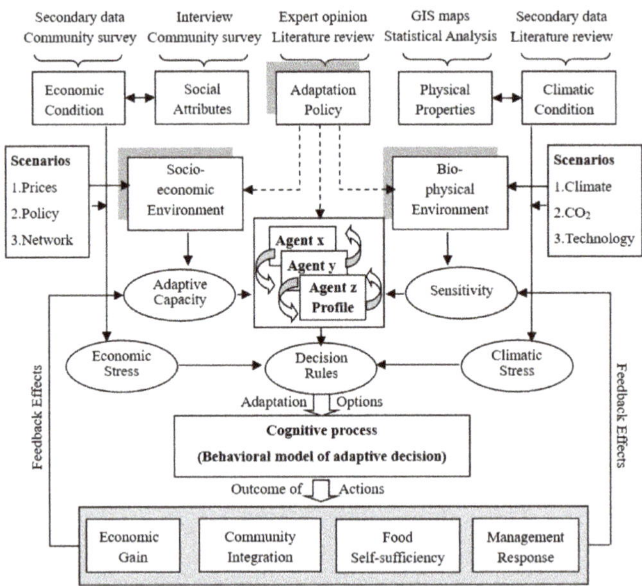

Figure 3. An agent-based inter-vulnerability framework to assess vulnerabilities and adaptations.

5. Regional Adaptation Evaluation

Adaptation has become one of the focuses of scientific debates in the field of climate change and disaster risk. This includes practical aspects and seeks ways to understand the methods and concepts that are conducive to human intervention [19].

5.1. Adaptation Research in the Field of Climate Change

During global research on climate change and its possible impact on human society, prevention was proposed in the 1970s, and activities that occurred as global-scale mitigations were discussed in the 1980s through now. Multi-scale adaptation research is generally accepted, and adaptation has become a major topic in climate change research [108]. In the United Nations Framework Convention on Climate Change (UNFCC), it is emphasized that mathematical–statistical models are used to estimate the impact of climate change at larger scales and under conditions of adaptation and non-adaptation. The purpose is to emphasize problems of risk under the existence of future climate change scenarios [109,110].

Physical climate information addresses how the climate system responds to the interplay between human influence, natural drivers, and internal variability. Knowledge of the climate response and the range of possible outcomes, including low-likelihood, high-impact outcomes, informs climate services—the assessment of climate-related risks and adaptation planning. Physical climate information at global, regional, and local scales is developed from multiple lines of evidence, including observational products, climate model outputs, and tailored diagnostics.

Based on current mainstream international research, climate change adaptation is focused primarily on the following four aspects.

(1) Adaptive assumptions can be adopted based on the impact conditions or parameter changes measured using the climate change scenario model to estimate and predict the effects of different adaptive methods [111]. However, this type of research has not performed actual investigations or verifications of adaptability [36].

(2) Another focus is primarily on the adaptation options and strategies of special systems under climate change conditions. In the UNFCCC [93] clause, it is emphasized

that each country should commit to applying appropriate climate change adaptability and structure while implementing effective response strategies to evaluate the advantages and utilities of adaptation and confirm the best strategy [112–114].

(3) A third focus is primarily on the relative adaptation of countries, regions, and communities and selection of certain standards, indicators, and variables for comparative evaluation and classification. This type of research is measured based on certain causal relationships and decisive factors. Through some indicators, scores, and grading processes, this evaluates the relative adaptation of a country, a region, and a community [115–117] and accumulates the adaptive capacity elements for each system to form the overall evaluation score [55].

(4) Related research focuses primarily on active adaptation practice strategies. To date, research on adaptive practice processes is not ubiquitous. There are not many targeted studies that directly considered the adaptive label or framework, especially at the regional level, while the community level is weaker [118]. However, adaptation at these levels is often the most practical and most complex. Differences in climate conditions, geographic locations, management systems, real estate, public facilities, resource availability, the implantation of local traditional cognition in the decision-making process, etc., all impact adaptation [119,120]. Nevertheless, many scholars in the fields of resource management, community development, regional planning, food safety, livelihood safety, and sustainable development are involved in studying the adaptive time and process.

Adaptation has temporal and spatial scales that range from household adaptation to government policy formulation. However, due to the impact of the spatial resolution, climate change prediction models are not suitable for farmers' planting management or other related activities and need to be downscaled [121]. Some scholars have also gradually realized the importance of the farm level in decisions to adapt to the process of adaptation, especially to understand climate extremes, and research has begun considering the role of humans when adapting to the impacts of climate change by investigating farmers' perceptions and risk management choices. Li and Chen [122] discussed the concept of vulnerability, sensitivity, adaptability, and their associated assessment methods. Vulnerability is affected not only by climate sensitivity but also by the structure, functioning, and succession of the system as well as its self-adjusting and recovering abilities. Regional empirical research on adaptation to global change as advocated by Ge et al. [70] includes research on adaptation to extreme events in the context of global change. Wang et al. [64] evaluated the adaptation of agriculture in response to global warming and drying scenarios in North China. This evaluation is focused primarily on the analysis of climate change with relatively little analysis of the socio-economic aspects. Yasuhara et al. [123] upgraded the methodology to estimate the effects on geo-disasters from combined events, e.g., global warming with increased typhoon and rainfall severity or the occurrence of large earthquakes. Olazabal et al. [124] tracked the progress of governments by analyzing the policies that provide insight into the goals and means of achieving adaptation targets. They identified 226 adaptation policies: 88 at the national level, 57 at the regional/state level, and 81 at the city/metropolitan level.

5.2. Adaptation Research in the Field of Natural Disaster Risk

In 1945, White proposed a "series of adjustments" in response to the intensification of flood disasters in the United States. For the first time, attention was given to expand disaster prevention and mitigation from hazards to human behavioral responses to disasters and noted that human behavior can be adjusted to reduce the impact and loss from disasters. This provided new ideas for subsequent comprehensive disaster reduction strategies [125] and established the current natural disaster field to explore the relationships between humankind and the environment by focusing on the impact of extreme events and human responses [17]. Scholars have done more research on the following three aspects of adaptability: disaster risk perception [50], farm-scale uncertainty risk management strategies [85,88], and individual decision making in agricultural systems [126–129].

To date, there are several evaluations of adaptation measures in the literature, but there are not many cases that evaluate adaptation by constructing an adaptability index system from the perspective of disaster systems. Dhyani and Thummarukuddy [130] proposed a method designed to assess the potential contributions of various adaptation options to improve a system's coping capacities by focusing attention directly on the underlying determinants of adaptive capacity. Then, they applied this method to expert judgments of six different adaptations that could reduce vulnerability in the Netherlands to increased flooding along the Rhine River.

The adaptation of agriculture is multi-scale and multi-perspective. Research on the adaptability of agriculture to climate change uses several research approaches that consider different scales (plants, sites, fields, farms, regions, countries, and even international) [126,131]. Agriculture is a complex system, and changes within the system are driven by the combined influence of economic, environmental, political, and social forces [132]. Studies have shown that decisions made at different levels of agricultural change are interrelated. Thus, adaptation is the result of individual decisions as determined by the internal forces of the farm family (such as the risk of income loss or environmental perception) and external influences on the power of agriculture systems (such as macroeconomic policies and institutional frameworks) [133].

As an adaptation model of agricultural activities, this is the product of multiple individual decisions (government, agricultural product business, and individual producers) [16,126], government policies, institutional arrangements, and macro social and economic conditions, which are continuously recognized in adaptation research [134]. The actual analysis of the adaptability of the Nile River and Rhine Delta is studied, and two analysis methods of adaptability are provided. One is to obtain the evaluation results through scores of various indicators and the other is to use the synthetic index for the evaluation. The Sahel drought adaptation strategy responds to the following five aspects: precipitation crisis (drought), food supply, livestock management, environmental degradation, and household handling capacity [135]. Therefore, a model of household economic diversity is proposed. This is the single-core structure of the planting industry to the dual-core structure composed of animal husbandry, which then leads to the three-core cycle model of migrant workers and complementary agriculture [94]. Shang [136] analyzed the vulnerability of typical farmers, and Wang et al. [129] analyzed the risk of agricultural drought. They obtained quantitative relationships between income and food production and measures taken to transfer drought risk. This analysis has two prerequisites. One is the factors that affect agricultural drought vulnerability and adaptation, and the other is that these factors can be quantitatively expressed. The factors that meet these two conditions should include the area of conversion of farmland to forests and water conservancy facilities. Wang et al. [121] proposed the vulnerability–adaptation (RA) model to diagnose the adaptation of rain-fed agriculture. The indicators used to evaluate the adaptability are the per capita arable land area, arable land flatness index, irrigation convenience index, irrigation water volume index, per capita food production, per capita proportion of large livestock, and non-agricultural income. Reid et al. [137] interviewed farmers in Boss, Ontario, Canada and focused on four types of farmers and recorded their responses to climate and weather conditions and risks. A broad advance and response management strategy was developed to manage climate risk. Slegers [118] conducted research in Tanzania through questionnaire surveys and interviews and found that farmers are aware of differences in soil types, soil location, land status, and land management practices in drought-vulnerable regions. In fact, farmers have accumulated significant experience in environmental adaptation, which is crucial to the adaptability of the entire agricultural system.

6. Discussions

Adaptation to disaster risk is a relatively complex concept. Its complexity is not only reflected in human perception, identification of risks, and the corresponding adaptive management/organizational behaviors but also in the process of adaptation [3,80,97]. The root of this complexity lies in the following. First, the variety of risk factors increases the risk complexity. Various hazard factors and chaining lead to a variety of exposure types under different risk factors, which are fundamentally strengthened. Second, the same hazard factor and its combination hit the same exposure in different regions, which results in different risk levels. The adaptation of the exposure under different risk levels can vary, which further aggravates the complexity. Third, at different spatial scales, the recovery process of the same exposure type will have various focuses that increase its complexity. Fourth, at different stages of historical development, there will be several qualitative changes in the exposures, which lead to a further increase in complexity. Fifth, human perception and the recognition of risks and adaptation differ, causing the process or speed of adaptation to the same disasters of the same intensity to vary, which also greatly increase the complexity.

The complexity of disaster adaptation leads to several influencing factors. The judgment of adaptation should be expressed with multiple indicators. A single indicator inevitably leads to misunderstandings and one-sided conclusions about disaster adaptation. At present, there are several research cases on disaster adaptation, and multiple indicators have been designed from different angles for comprehensive expressions [138–141]. The combination, division, and relationship determination of disaster adaptation factors gradually reveal the formation of disaster adaptation. At the same time, distinguishing the contribution rates of each influencing factor to disaster adaptation can provide the most powerful basis to strengthen disaster adaptability. To date, the determination of influencing factors for disaster adaptation is based more on the selection of the factors and indicators after understanding the process of disaster adaptation, which has greater subjectivity and uncertainty. Selecting the appropriate method and deciding how to apply reasonable methods to determine the influencing factors of adaptation and the associated contribution rate is the focus of future research.

The 2018 IPCC special report Global Warming of 1.5 °C projects that the climate system is heading off track into the territory of 2.9 to 3.4 °C warming [142]. If this happens, it would take future hydrometeorological hazard extremes well outside the known range of current experience and alter the loss and damage equations and fragility curves of almost all known human and natural systems, placing them at unknown levels of risk [142]. To reinforce SDG 13.1, It also means that it is no longer sufficient to address adaptation in isolation from development planning, and that sustainable socio-economic development, by definition, must include the mitigation of global warming [12].

Looking to the future, it is necessary to track adaptation to identify who is adapting to what, when, where, and why to understand the efficiency of assigned resources and adjust adaptation planning given that information. Adaptation and adaptive capacity should focus on four core issues (Figure 4). The evaluation criteria and adaptation evaluation scales are also shown in Figure 4.

"Adaptation to what" refers to climate change or variation, where adaptation can be a response to adverse effects or vulnerability, which can be a response to the current actual climate or climatic conditions for future predictions. "Who or what adapts" can be an individual, socio-economic department, managed or non-managed system, natural or ecological system, systematic practice, or operation and structure, where each system is distinguished by its adaptability and vulnerability. "How does adaptation occur" can be adaptable to the process or can result in consequences, which can be spontaneous or planned. The above three parts constitute "what is adaptation", which is a general concept and question that should be answered. The "how good is the adaptation" is an evaluation based on costs, interest, fairness, efficiency, emergency, executability, etc. As a response to countermeasures, the evaluation of choices and measures should be given.

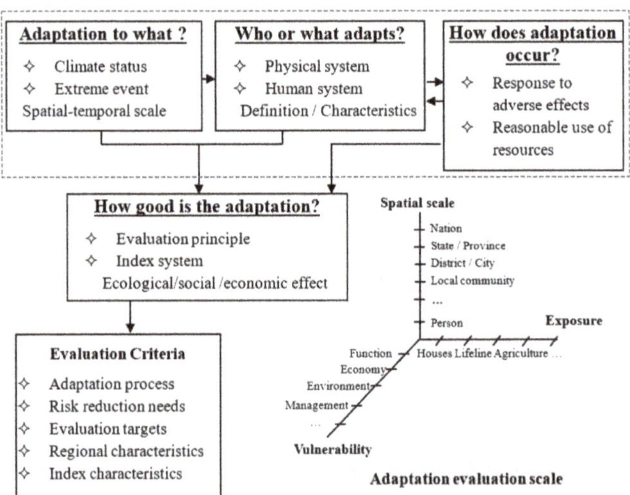

Figure 4. Adaptability constitutes and evaluation framework.

7. Conclusions

The adaptation research in the field of natural disasters has focused primarily on socio-economic responses, especially for farm-scale adaptation decision making and risk perception. These socio-economic responses to climate change have accumulated a significant literature basis for further research. However, there are relatively few studies on adaptation evaluation from the perspective of natural disaster systems. The evaluation cases of adaptation for disasters exist primarily as part of vulnerability evaluation. The indicators used at different scales can vary. As adaptability is a local feature, if the scale is too small, the data will be more restricted. If the scale is too large, the sensitivity of the adaptability index used to compare basic units needs to be considered. Learning from current adaptation practices and strengthening them through adaptive governance, lifestyle and behavioral change, and innovative financing mechanisms can help their mainstreaming within sustainable development practices.

Current thinking on disaster adaptation is the result of constructing theories and methods while refining and improving the disaster paradigm over the past few decades [143,144]. The introduction of adaptation in disasters has not been unanimously recognized and is full of controversy. Adaptation requires the integration of multiple disciplines and concepts. Therefore, to summarize the current discussions on disaster adaptation, distinguishing disaster vulnerability, resilience, and adaptation; constructing disaster adaptation assessment models; and exploring regional models for disaster adaptation assessments are important to construct reasonable disaster risk reduction frameworks.

Author Contributions: Conceptualization, F.C. and H.J.; methodology, H.J. and E.D.; formal analysis, F.C.; investigation, H.J. and E.D.; resources, F.C. and H.J.; writing—original draft preparation, H.J.; writing—review and editing, F.C. and E.D.; supervision, F.C. and H.J.; funding acquisition, F.C. and H.J. All authors have read and agreed to the published version of the manuscript.

Funding: This research was funded by the National Natural Science Foundation of China (41871345; 42171078), the Strategic Priority Research Program of the Chinese Academy of Sciences (Grant No. XDA19030101), and the International Partnership Program of the Chinese Academy of Sciences (131211KYSB20170046).

Institutional Review Board Statement: Not applicable.

Informed Consent Statement: Not applicable.

Acknowledgments: This research was funded by the National Natural Science Foundation of China (41871345; 42171078), the Strategic Priority Research Program of the Chinese Academy of Sciences (Grant No. XDA19030101), and the International Partnership Program of the Chinese Academy of Sciences (131211KYSB20170046). The authors would also like to thank the editors and the three anonymous reviewers for their constructive comments and advice.

Conflicts of Interest: The authors declare no conflict of interest.

References

1. Escaleras, M.; Anbarci, N.; Register, C.A. Public sector corruption and major earthquakes: A potentially deadly interaction. *Public Choice* **2007**, *132*, 209–230. [CrossRef]
2. Shi, P.J.; Li, N.; Ye, Q.; Dong, W.J.; Han, G.Y.; Fang, W.H. Research on integrated disaster risk governance in the context of global environmental change. *Int. J. Disaster Risk Sci.* **2010**, *1*, 17–23.
3. Zhou, H.J.; Wang, X.; Wang, J.A. A way to sustainability: Perspective of resilience and adaptation to disaster. *Sustainability* **2016**, *8*, 737. [CrossRef]
4. Guo, H.D. Big Earth data: A new frontier in Earth and information sciences. *Big Earth Data* **2017**, *1*, 4–20. [CrossRef]
5. Kirk, M.A.; Moore, J.E.; Stirman, S.W.; Birken, S.A. Towards a comprehensive model for understanding adaptations' impact: The model for adaptation design and impact (MADI). *Implement. Sci.* **2020**, *15*, 56. [CrossRef]
6. Winterhalder, B. Environmental analysis in human evolution and adaptation research. *Hum. Ecol.* **1980**, *8*, 135–170. [CrossRef]
7. Slobodkin, L.A.; Rappaport, A. An optimal strategy of evolution. *Q. Rev. Biol.* **1974**, *49*, 181–200. [CrossRef]
8. Pan, J.H.; Zheng, Y. Analytical framework and policy implications on adapting to climate change. China Population. *Resour. Environ.* **2010**, *20*, 1–5.
9. Acosta-Michlik, L.; Rounsevell, M. From Generic Indices to Adaptive Agents: Shift Foci in Assessing Vulnerability to the Combined Impacts of Climate Change and Globalization. *IHDP Newsl.* **2005**, *1*, 14–16.
10. UN (United Nations). Global Assessment Report on Disaster Risk Reduction 2019. Available online: http://gar.undrr.org/report-2019 (accessed on 20 December 2019).
11. UNISDR. *Sendai Framework for Disaster Risk Reduction 2015–2030*; UNISDR: New Delhi, India, 2015; pp. 24–26. Available online: http://www.cma.gov.cn/en2014/20150311/20151010/2015101002/201510/P020151012525690375817.pdf (accessed on 25 December 2015).
12. Nerini, F.; Sovacool, B.; Hughes, N.; Cozzi, L.; Cosgrave, E.; Howells, M.; Tavoni, M.; Tomei, J.; Zeriffi, H.; Milligan, B. Connecting climate action with other Sustainable Goals. *Nat. Sustain.* **2019**, *2*, 674–680. [CrossRef]
13. Yin, Y.Y. Adaptation Evaluation Tools and Analysis Methods for Climate Change. *J. Glaciol. Geocryol.* **2002**, *24*, 426–431.
14. Liu, C.Z. On some issues in studying climate change impact and adaptation. *Clim. Environ. Res.* **1999**, *4*, 129–134.
15. Osbahr, H.; Dorward, P.; Stern, R. Supporting agricultural innovation in Uganda to respond to climate risk: Linking climate change and variability with farmer perceptions. *Exp. Agric.* **2011**, *47*, 293–316. [CrossRef]
16. Watts, M.J.; Bohle, H.G. The space of vulnerability: The causal structure of hunger and famine. *Prog. Hum. Geogr.* **1993**, *17*, 43–67. [CrossRef]
17. Brooks, N. *Vulnerability, Risk and Adaptation: A Conceptual Framework*; Working Paper 38; Tyndall Centre for Climate Change Research, University of East Anglia: Norwich, UK, 2003.
18. Smit, B.; Burton, I.; Klein, R.; Wandel, J. An anatomy of adaptation to climate change and variability. *Clim. Chang.* **2000**, *45*, 223–251. [CrossRef]
19. IPCC. Technical Summary: Climate Change 2001: Impacts, Adaptation, and Vulnerability, A Report of Working Group II of the Intergovernmental Panel on Climate Change. Available online: http://www.grida.no/climate/ipcc_tar/wg2/pdf/wg2 TARtechsum.pdf (accessed on 22 December 2010).
20. Smit, B.; Skinner, M. Adaptation options in agriculture to climate change: A typology. *Mitig. Adapt. Strateg. Glob. Chang.* **2002**, *7*, 85–114. [CrossRef]
21. Burton, I. *Adapt and Thrive. Report of Environment Canada*; Downs View: Ontario, Canada, 1992; pp. 135–168.
22. Smith, J.B.; Lenhart, S.S. Climate Change Adaptation Policy Options. *Clim. Res.* **1993**, *6*, 193–201. [CrossRef]
23. Stakhiv, E. *Evaluation of IPCC Adaptation Strategies*; Institute for Water Resources, U.S. Army Corps of Engineers: Fort Belvoir, VA, USA, 1993; draft report.
24. Smit, B. *Adaptation to Climate Variability and Change*; Environment Canada: Guelph, Canada, 1993.
25. IPCC. *Fourth Assessment Report (AR2)*; Cambridge University Press: Cambridge, UK, 1996.
26. Smith, J.B.; Ragland, S.E.; Pitts, G.J. A process for evaluating anticipatory adaptation measures for climate change. *Water Air Soil Pollut.* **1996**, *92*, 229–238. [CrossRef]
27. Pielke, R.A.J. Rethinking the role of adaptation in climate policy. *Glob. Environ. Chang.* **1998**, *8*, 159–170. [CrossRef]
28. Adger, W.N.; Kelly, P.W.; Ninh, N.H. *Living with Environmental Change: Social Resilience, Adaptation and Vulnerability in Vietnam*; Routledge: London, UK, 2001.
29. IPCC. *Fourth Assessment Report (AR4)*; Cambridge University Press: Cambridge, UK, 2007.
30. Doria, M.D.; Boyd, E.; Tompkins, E.L.; Adger, W.N. Using expert elicitation to define successful adaptation to climate change. *Environ. Sci. Policy* **2009**, *12*, 810–819. [CrossRef]

31. UNISDR. Terminology on Disaster Risk Reduction. 2009. Available online: http://www.unisdr.org/eng/terminology/UNISDRterminology-2009-eng.pdf (accessed on 20 January 2010).
32. IPCC. *Managing the Risks of Extreme Events and Disasters to Advance Climate Change Adaptation*; Field, C.B., Barros, V., Stocker, T.F., Qin, D., Dokken, D.J., Ebi, K.L., Mastrandrea, M.D., Mach, K.J., Plattner, G.-K., Allen, S.K., et al., Eds.; Cambridge University Press: Cambridge, UK, 2012.
33. IPCC. Fifth Assessment Report. 2014. Available online: https://www.ipcc.ch/report/ar5/ (accessed on 22 September 2015).
34. Morgan, E.A.; Nalau, J.; Mackey, B. Assessing the alignment of national-level adaptation plans to the Paris Agreement. *Environ. Sci. Polic* **2019**, *93*, 208–220. [CrossRef]
35. He, C.Y.; Tung, C.; Lin, Y.J. Applying the DRCA Risk Template on the Flood-Prone Disaster Prevention Community Due to Climate Change. *Sustainability* **2021**, *13*, 891. [CrossRef]
36. Fankhauser, S. The Potential Costs of Climate Change Adaptation. In *Adapting to Climate Change: An International Perspective*; Smith, J.B., Bhatti, N., Menzhulin, G., Benioff, R., Budyko, M., Campos, M., Jallow, B., Rijsbcrman, F., Eds.; Springer: New York, NY, USA, 1996; pp. 80–96.
37. Bryant, C.R.; Smit, B.; Brklacich, M.; Johnston, T.R.; Smithers, J.; Chiotti, Q.; Singh, B. Adaptation in Canadian Agriculture to Climatic Variability and Change. *Clim. Chang.* **2000**, *45*, 181–201. [CrossRef]
38. Dobzhansky, T. Adaptness and fitness. In *Population Biology and Evolution*; Lewontin, R.C., Ed.; Syracuse Univ Press: Syracuse, NY, USA, 1968; pp. 109–121.
39. Smit, B.; Wandel, J. Adaptation, adaptive capacity and vulnerability. *Glob. Environ. Chang.* **2006**, *16*, 282–292. [CrossRef]
40. Huq, S.; Rahman, A.; Konate, N.; Sokona, Y.; Reid, H. Mainstreaming Adaptation to Climate changes in Least Developed Countries (LDCS), London International Institute for Environment and Development. *Bangladesh Ctry. Case Study* **2004**, *1*, 25–43.
41. Risbey, J.; Kandlikar, M.; Dowlatabadi, H.; Graetz, D. Scale, context and decision making in agricultural adaptation to climate variability and change. *Mitig. Adapt. Strateg. Glob. Chang.* **1999**, *4*, 137–164. [CrossRef]
42. Smit, B.; Pilifosova, O. From Adaptation to Adaptive Capacity and Vulnerability Reduction. In *Climate Change, Adaptive Capacity and Development*; Smith, J.B., Klein, R.J.T., Huq, S., Eds.; Imperial College Press: London, UK, 2003.
43. Yohe, G.; Strzepek, K.; Pau, T.; Yohe, C. Assessing vulnerability in the context of changing socioeconomic conditions: A study of Egypt. In *Climate Change, Adaptive Capacity and Development*; Smith, J.B., Kein, R.J., Huq, S., Eds.; Imperial College Press: London, UK, 2003.
44. Adger, W.N. Social vulnerability to climate change and extremes in coastal Vietnam. *World Dev.* **1999**, *27*, 249–269. [CrossRef]
45. Handmer, J.W.; Dovers, W.; Downing, T.E. Societal vulnerability to climate change and variability. *Mitig. Adapt. Strateg. Glob. Chang.* **1999**, *4*, 267–281. [CrossRef]
46. Wisner, B.; Blaikie, P.; Cannon, T.; Davis, I. *At Risk*; Routledge: London, UK, 2004.
47. Larsson, J.; Harrison, S.J. Spatial specificity and inheritance of adaptation in human visual cortex. *J. Neurophysiol.* **2015**, *114*, 1211–1226. [CrossRef]
48. Li, P.; Jing, T.; Wei, D.; Ying, L. Understanding the Resilience of Different Farming Strategies in Coping with Geo-Hazards: A Case Study in Chongqing, China. *Int. J. Environ. Res. Public Health* **2020**, *17*, 1226.
49. Smithers, J.; Smit, B. Human adaptation to climatic variability and change. *Glob. Environ. Chang.* **1997**, *7*, 129–146. [CrossRef]
50. Kasperson, R.E.; Dow, K.; Archer, E.; Cáceres, D.; Downing, T.; Elmqvist, T.; Eriksen, S.; Folke, C.; Han, G.; Iyengar, K.; et al. Vulnerable People and Places. In *Ecosystems and Human Well-being: Current State and Trends*; Hassan, R., Scholes, R., Ash, N., Eds.; Island Press: Washington, DC, USA, 2005; pp. 143–164.
51. Wang, J.A.; Su, Y.; Shang, Y.Y.; Hong, S.Q.; Wang, Z.Q.; Liu, Z. Vulnerability Identification and Assessment of Agriculture Drought Disaster in China. *Adv. Earth Sci.* **2006**, *21*, 161–169.
52. Jones, R. An environmental risk assessment/management framework for climate change impact assessments. *Nat. Hazards* **2001**, *23*, 197–230. [CrossRef]
53. Fraser, E.; Mabee, W.; Slaymaker, O. Mutual vulnerability, mutual dependence: The reflective notion between human society and the environment. *Glob. Environ. Chang.* **2003**, *13*, 137–144. [CrossRef]
54. Fussel, H.-M.; Klein, R.J.T. Climate change vulnerability assessment: An evolution of conceptual thinking. *Clim. Chang.* **2006**, *75*, 301–329. [CrossRef]
55. UN (United Nations). The Paris Agreement. 2015. Available online: https://unfccc.int/process-and-meetings/the-paris-agreement/the-paris-agreement (accessed on 25 December 2015).
56. Brooks, N.; Adger, W.N.; Kelly, P.M. The determinants of vulnerability and adaptive capacity at the national level and the implications for adaptation. *Glob. Environ. Chang.* **2005**, *15*, 151–163. [CrossRef]
57. Ge, Q.S.; Chen, P.Q.; Fang, X.Q.; Lin, H.; Ye, Q. Adaptation to Global Change: Challenge and Research Strategy. *Adv. Earth Sci.* **2004**, *19*, 516–524.
58. Gallopin, G.C. Linkages between vulnerability, resilience, and adaptive capacity. *Glob. Environ. Chang.* **2006**, *16*, 293–303. [CrossRef]
59. Adger, W.N. Vulnerability. *Glob. Environ. Chang.* **2006**, *16*, 268–281. [CrossRef]
60. Turner, B.L.; Kasperson, R.E.; Matson, P.A.; McCarthy, J.J.; Corell, R.W.; Christensen, L.; Eckley, N.; Kasperson, J.X.; Luers, A.; Martello, M.L.; et al. A framework for vulnerability analysis in sustainability science. *Proc. Natl. Acad. Sci. USA* **2003**, *100*, 8074–8079. [CrossRef]

61. Vogel, C. Vulnerability and global environmental change. *LUCC Newsl.* **1998**, *3*, 15–19.
62. Lioubimtseva, E. Human dimensions of post-soviet Central Asia. *Ann. Arid Zone* **2009**, *47*, 1–24.
63. Gunderson, L.H. Resilience in theory and practice. *Annu. Rev. Ecol. Syst.* **2000**, *31*, 425–439. [CrossRef]
64. Carpenter, S.; Walker, B.; Anderies, J.M.; Abel, N. From metaphor to measurement: Resilience of what to what? *Ecosystems* **2001**, *4*, 765–781. [CrossRef]
65. Folke, C.; Carpenter, S.; Elmqvist, T.; Gunderson, L.; Holling, C.S.; Waler, B. Resilience and sustainable development: Building adaptive captive capacity in a world of transformations. *AMBIO* **2002**, *31*, 437–440. [CrossRef]
66. Ford, J.D.; Berrang-Ford, L.; Biesbroek, R.; Araos, M.; Austin, S.E.; Lesnikowski, A. Adaptation tracking for a post-2015 climate agreement. *Nat. Clim. Chang.* **2015**, *5*, 967–969. [CrossRef]
67. Lei, Y.; Wang, J. A preliminary discussion on the opportunities and challenges of linking climate change adaptation with disaster risk reduction. *Nat. Hazards* **2014**, *71*, 1587–1597. [CrossRef]
68. Yu, J.L.; Sim, T.; Guo, C.L.; Han, Z.Q.; Lau, J.; Su, G.W. Household adaptation intentions to earthquake risks in rural China. *Int. J. Disaster Risk Reduct.* **2019**, *40*, 101253. [CrossRef]
69. Jaeger, C.; Shi, P.J. Core Science Initiative on Integrated Risk Governance. *IHDP Update* **2008**, *1*, 27–28.
70. Mastrandrea, M.D.; Heller, N.E.; Root, T.L.; Schneider, S.H. Bridging the gap: Linking climate-impacts research with adaptation planning and management. *Clim. Chang.* **2010**, *100*, 87–101. [CrossRef]
71. Denevan, W.M. Adaptation, variation and cultural geography. *Prof. Geogr.* **1983**, *35*, 399–406. [CrossRef]
72. Butzer, K.W. Cultural Ecology. In *Geography in America*; Gaile, G.L., Willmott, C.J., Eds.; Merrill Publishing Co.: Columbus, OH, USA, 1989.
73. O'Brien, M.; Holland, T.D. The role of adaptation in archeological explanation. *Am. Antiq.* **1992**, *57*, 36–69. [CrossRef]
74. Burton, I.; Kates, R.W.; White, G.F. *The Environment as Hazard*, 2nd ed.; Guilford Press: New York, NY, USA, 1993.
75. Burton, I.; Feenstra, J.F.; Smith, J.B.; Cohe, S.; Oude Essink, G.H.P.; Feenstra, J.F.; Hlohowskyj, I.; Hulme, M.; Iglesias, A.; Klein, R.J.T.; et al. *Handbook on Methods for Climate Change Impact Assessment and Adaptation Strategies*; United Nations Environment Programme and the Institute for Environmental studies, Free University of Amsterdam: Amsterdam, The Netherlands, 1998.
76. Holling, C.S. Resilience and stability of ecological systems. *Annu. Rev. Ecol. Syst.* **1973**, *4*, 1–23. [CrossRef]
77. Downing, T. Vulnerability to hunger in Africa: A climate change perspective. *Glob. Environ. Change* **1991**, *1*, 365–380. [CrossRef]
78. Adger, W.N.; Kelly, P.M. Social vulnerability to climate change and architecture of entitlements. *Mitig. Adapt. Strateg. Glob. Chang.* **1999**, *4*, 253–266. [CrossRef]
79. Adger, W.N. Institutional adaptation to environmental risk under the transition in Vietnam. *Ann. Assoc. Am. Geogr.* **2000**, *90*, 738–758. [CrossRef]
80. Kasperson, J.X.; Kasperson, R.E. *Climate Change, Vulnerability and Social Justice*; Stockholm Environment Institute: Stockholm, Sweeden, 2001.
81. Butzer, K.W. Adaptation to global environmental change. *Prof. Geogr.* **1980**, *32*, 269–278. [CrossRef]
82. Kelly, P.M.; Adger, W.N. Theory and practice in assessing vulnerability to climate change and facilitating adaptation. *Clim. Chang.* **2000**, *47*, 325–352. [CrossRef]
83. Owusu, V.; Yiridomoh, G.Y. Assessing the determinants of women farmers' targeted adaptation measures in response to climate extremes in rural Ghana. *Weather Clim. Extrem.* **2021**, *33*, 100353. [CrossRef]
84. Yu, X.; Li, C.; Zhao, W.X.; Chen, H. A novel case adaptation method based on differential evolution algorithm for disaster emergency. *Appl. Soft Comput. J.* **2020**, *92*, 106306. [CrossRef]
85. Mercer, J. Disaster risk reduction or climate change adaptation: Are we reinventing the wheel? *J. Int. Dev.* **2010**, *22*, 247–264. [CrossRef]
86. Dessai, S.; Adger, W.N.; Hulme, M.; Turnpenny, J.; Kohler, J.; Warren, R. *Defining and Experiencing Dangerous Climate Change*; University of East Anglia: Norwich, UK, 2003.
87. Hallegatte, S. Strategies to adapt to an uncertain climate change. *Glob. Environ. Chang.* **2009**, *19*, 240–247. [CrossRef]
88. Grothmanna, T.; Patt, A. Adaptive capacity and human cognition: The process of individual adaptation to climate change. *Glob. Environ. Chang.* **2005**, *15*, 199–213. [CrossRef]
89. Adger, W.N.; Nigel, W.; Tompkins, A.E.L. Successful adaptation to climate change across scales, *Glob. Environ. Chang.* **2005**, *15*, 77–86. [CrossRef]
90. Marten, G. Productivity, stability, sustainability, equitability and anatomy as properties for agroecosystem assessment. *Agric. Syst.* **1988**, *26*, 291–316. [CrossRef]
91. Rosenzwieg, C.; Parry, M.L. Potential impact of climate change on world food supply. *Nature* **1994**, *367*, 133–138. [CrossRef]
92. Fussel, H.M. Adaptation planning for climate change: Concepts, assessment approaches, and key lessons. *Sustain. Sci.* **2007**, *2*, 265–275. [CrossRef]
93. UNFCCC. Bali Action Plan (FCCC/CP/2007/6/Add.1 Decision 1/CP.13). 2007. Available online: http://unfccc.int/resource/docs/2007/cop13/eng/06a01.pdf#page=3 (accessed on 20 January 2021).
94. Yin, Y.Y.; Wang, G.X. *Assessment Methods and Applications of Global Climate Change*; Higher Education Press: Beijing, China, 2004.
95. Lei, Y.; Wang, J.; Yue, Y.; Zhou, H.; Yin, W. Rethinking the relationships of vulnerability, resilience, and adaptation from a disaster risk perspective. *Nat. Hazards* **2014**, *70*, 609–627. [CrossRef]

96. Jia, H.C.; Chen, F.; Zhang, J.; Du, E.Y. Vulnerability analysis to drought based on remote sensing indexes. *Int. J. Environ. Res. Public Health.* **2020**, *17*, 7660. [CrossRef]

97. Parida, Y.; Goel, P.A.; Chowdhury, J.R.; Sahoo, P.K.; Nayak, T. Do economic development and disaster adaptation measures reduce the impact of natural disasters? A district-level analysis, Odisha, India. *Environ. Dev. Sustain.* **2021**, *23*, 3487–3519. [CrossRef]

98. Mizina, S.V.; Smith, J.B.; Gossen, E.; Spiecker, K.F.; Witkowski, S.L. An evaluation of adaptation options for climate change impacts on agriculture in Kazakhstan. *Mitig. Adapt. Strateg. Glob. Chang.* **1999**, *4*, 25–41. [CrossRef]

99. Batima, P. *Climate Change Vulnerability and Adaptation in the Livestock Sector of Mongolia*; AIACC Reports; Institute of Meteorology and Hydrology: Ulaanbaatar, Mongolia, 2006.

100. Smith, A.E.; Chan, N.; Chu, H.Q.; Helman, C.J.; Kim, J.B. *Documentation of Adaptation Strategy Evaluator Systems*; Prepared for the U.S. Environmental Protection Agency under EPA Contract No.68- W2-0018; U.S. Environmental Protection Agency: Washington, DC, USA, 1995; Volume 1.

101. Scheraga, J.D.; Julius, S.H. Decision Support System for Evaluating Alternative Adaptation Strategies. Paper Presented at the U.S. Country Studies Vulnerability and Adaptation Workshop, St. Petersburg, Russia, 22–25 May 1995.

102. Yin, Y. Flood management and water resource sustainable development: The case of Great Lakes Basin. *Water Int.* **2000**, *26*, 197–205. [CrossRef]

103. Zadeh, L.A. Fuzzy sets. *Inf. Control* **1965**, *8*, 338–353. [CrossRef]

104. Onuma, H.; Shin, K.J.; Managi, S. Reduction of future disaster damages by learning from disaster experiences. *Nat Hazards* **2017**, *87*, 1435–1452. [CrossRef]

105. Yin, Y.; Xu, X. Applying neural net technology for multi objective land use planning. *J. Environ. Manag.* **1991**, *32*, 349–356. [CrossRef]

106. Chen, F.; Yu, B.; Li, B. A practical trial of landslide detection from single-temporal Landsat8 images using contour-based proposals and random forest: A case study of national Nepal. *Landslides* **2018**, *15*, 453–464. [CrossRef]

107. Saaty, L. *Decision Making for Leaders: The Analytical Hierarchy Process for Decisions in a Complex World*; McGraw–Hill: New York, NY, USA, 1982; pp. 56–92.

108. Ye, D.; Fu, C.; Dong, W. Progresses and future trends of global change sciences. *Adv. Earth Sci.* **2002**, *17*, 467–469.

109. Winters, P.; Murgai, R.; Sadoulet, E.; De Janvry, A.; Frisvold, G.B. Economic and welfare impacts of climate change on developing countries. *Eur. J. Agron.* **2000**, *13*, 179–189.

110. Parry, M. Scenarios for climate impact and adaptation assessment. *Glob. Environ. Chang.* **2002**, *12*, 149–153. [CrossRef]

111. Tol, R.S.J. The damage costs of climate change towards a dynamic representation. *Ecol. Econ.* **1996**, *19*, 67–90. [CrossRef]

112. Klein, R.J.T.; Nicholls, R.J.; Mimura, N. Coastal adaptation to climate change: Can the IPCC Technical guidelines be applied? *Mitig. Adapt. Strateg. Glob. Chang.* **1999**, *4*, 239–252. [CrossRef]

113. Dolan, A.H.; Smit, B.; Skinner, M.W.; Bradshaw, B.; Bryant, C.R. *Adaptation to Climate Change in Agriculture: Evaluation of Option*; Department of Geography: Guelph, ON, Canada, 2001.

114. Niang-Diop, I.; Bosch, H. Formulating an Adaptation Strategy. In *Adaptation Policy frameworks for Climate Change: Developing Strategies, Policies and Measures*; Lim, B., Spanger-Siegfried, E., Eds.; Cambridge University Press: Cambridge, UK, 2004.

115. Rayner, S.; Malone, E.L. Climate change, poverty, and intergenerational equity: The national level. *Int. J. Glob. Environ. Issues* **2001**, *1*, 175–202. [CrossRef]

116. O'Brien, K.; Leichenko, R.; Kelkar, U.; Venema, H.; Aandahl, G.; Tompkins, H.; Javed, A.; Bhadwal, S.; Barg, S.; Nygaard, L.; et al. Mapping vulnerability to multiple stressors: Climate change and globalization in India. *Glob. Environ. Chang.* **2004**, *14*, 303–313. [CrossRef]

117. Van der Veen, A.; Logtmeijer, C. Economic hotspots: Visualizing vulnerability to flooding. *Nat. Hazard* **2005**, *36*, 65–80. [CrossRef]

118. Slegers, M.F.W. If only it would rain Farmers' perceptions of rainfall and drought in semi-arid central Tanzania. *J. Arid Environ.* **2008**, *72*, 2106–2123. [CrossRef]

119. Olivier, D.; Greenstone, M. The economic impacts of climate change: Evidence from agricultural outputand random fluctuations in weather. *Am. Econ. Rev.* **2007**, *97*, 354–385.

120. Huh, T.; Park, Y.; Yang, J.Y. Multilateral Governance for Climate Change Adaptation in S. Korea: The Mechanisms of Formulating Adaptation Policies. *Sustainability* **2017**, *9*, 1364. [CrossRef]

121. Wang, Z.Q.; Yang, C.Y.; Wang, J.A.; Shang, Y.Y. Peasant household-based drought risk assessment and sustainable development of agriculture. *J. Nat. Disasters* **2005**, *14*, 94–99.

122. Li, K.R.; Chen, Y.F. Analysis of vulnerability of forest in China response to global climate change. *Acta Geogr. Sin.* **1996**, *51*, 40–49.

123. Yasuhara, K.; Komine, H.; Murakami, S.; Chen, G.; Mitani, Y.; Duc, D.M. Effects of climate change on geo-disasters in coastal zones and their adaptation. *Geotext. Geomembr.* **2012**, *30*, 24–34. [CrossRef]

124. Olazabal, M.; de Gopegui, M.R.; Tompkins, E.L.; Vennerj, K.; Smith, R. A cross-scale worldwide analysis of coastal adaptation planning. *Environ. Res. Lett.* **2019**, *14*, 124056. [CrossRef]

125. White, G.F. *Human Adjustments to Floods*; University of Chicago: Chicago, IL, USA, 1945.

126. Chiotti, Q.; Johnston, T.R.R.; Smit, B.; Ebel, B. Agricultural response to climate change: A preliminary investigation of farm-level adaptation in southern Alberta. In *Agricultural Restructuring and Sustainability: A Geographical Perspective*; Ilbery, B., Chiotti, Q., Rickard, T., Eds.; CAB International: Wallingford, UK, 1997; pp. 167–183.

127. Chen, F.; Zhang, M.M.; Guo, H.D.; Allen, S.; Kargel, J.; Haritashya, U.; Watson, C.S. Annual 30 m dataset for glacial lakes in High Mountain Asia from 2008 to 2017. *Earth System Science Data.* **2021**, *13*, 741–766. [CrossRef]
128. Yu, B.; Chen, F.; Xu, C. Landslide detection based on contour-based deep learning framework in case of national scale of Nepal in 2015. *Comput. Geosci.* **2020**, *135*, 104388. [CrossRef]
129. Wang, N.; Chen, F.; Yu, B.; Qin, Y.C. Segmentation of large-scale remotely sensed images on a Spark Platform: A strategy for handling massive image tiles with the MapReduce model. *ISPRS J. Photogramm. Remote Sens.* **2020**, *162*, 137–147. [CrossRef]
130. Dhyani, S.; Thummarukuddy, M. Ecological engineering for disaster risk reduction and climate change adaptation. *Env. Sci Pollut Res* **2016**, *23*, 20049–20052. [CrossRef]
131. Wang, K.; Zhang, A. Climate change, natural disasters and adaptation investments: Inter- and intra-port competition and cooperation. *Transp. Res. Part B* **2018**, *117*, 158–189. [CrossRef]
132. Sun, Y.H.; Zhou, H.J.; Zhang, L.Y.; Min, Q.W.; Yin, W.X. Adapting to droughts in Yuanyang Terrace of SW China: Insight from disaster risk reduction. *Mitig. Adapt. Strateg. Glob. Chang.* **2013**, *18*, 759–771. [CrossRef]
133. Chiotti, Q.P.; Johnston, T. Extending the boundaries of climate change research: A discussion on agriculture. *J. Rural Stud.* **1995**, *11*, 335–350. [CrossRef]
134. Kelman, I. Climate Change and the Sendai Framework for Disaster Risk Reduction. *Int. J. Disaster Risk Sci.* **2015**, *6*, 117–127. [CrossRef]
135. Yohe, G.; Richard, S.J. Indicators for social and economic coping capacity—Moving toward a working definition of adaptive capacity. *Glob. Environ. Chang.* **2002**, *12*, 25–40. [CrossRef]
136. Shang, Y.R. The analysis of drought, agricultural drought disaster and the farm houses' vulnerability: Taking the typical farmhouses of Xingtai County as an example. *J. Nat. Disasters* **2000**, *9*, 55–61.
137. Reid, S.; Smit, B.; Caldwell, W.; Belliveau, S. Vulnerability and adaptation to climate risks in Ontario agriculture. *Mitig. Adapt. Strat. Glob. Chang.* **2007**, *12*, 609–637. [CrossRef]
138. Renald, A.; Tjiptoherijanto, P.; Suganda, E.; Djakapermana, R.D. Toward resilient and sustainable city adaptation model for flood disaster prone city: Case study of Jakarta Capital Region. *Procedia Soc. Behav. Sci.* **2016**, *227*, 334–340. [CrossRef]
139. Radhakrishnan, M.; Islam, T.; Ashley, R.M.; Pathirana, A.; Quan, N.H.; Gersonius, B.; Zevenbergen, C. Context specific adaptation grammars for climate adaptation in urban areas. *Environ. Model. Softw.* **2018**, *102*, 73–83. [CrossRef]
140. Putra, G.A.Y.; Koestoer, R.H.; Lestari, I. Psycho-social performance towards understanding local adaptation of coastal flood in Cilincing Community, North Jakarta, Indonesia, *IOP Conf. Ser. Earth Environ. Sci.* **2019**, *243*, 012005.
141. Orimoloye, I.R.; Zhou, L.; Kalumba, A.M. Drought Disaster Risk adaptation through ecosystem services-based solutions: Way forward for South Africa. *Sustainability* **2021**, *13*, 4132. [CrossRef]
142. IPCC. SPECIAL REPORT: Global Warming of 1.5 °C. 2018. Available online: https://www.ipcc.ch/sr15/ (accessed on 20 September 2021).
143. Linnerooth-Bayer, J.; Hochrainer-Stigler, S. Financial instruments for disaster risk management and climate change adaptation. *Clim. Chang.* **2015**, *133*, 85–100. [CrossRef]
144. Rana, I.A. Disaster and climate change resilience: A bibliometric analysis. *Int. J. Disaster Risk Sci.* **2020**, *50*, 101839. [CrossRef]

International Journal of
*Environmental Research
and Public Health*

Article

Assessment on Temporal and Spatial Variation Analysis of Extreme Temperature Indices: A Case Study of the Yangtze River Basin

Guangxun Shi [1,2,3,4] and Peng Ye [5,6,*]

1   School of Geography, Nanjing Normal University, Nanjing 210023, China; 161301021@njnu.edu.cn
2   Key Laboratory of Virtual Geographic Environment, Ministry of Education, Nanjing Normal University, Nanjing 210023, China
3   Jiangsu Center for Collaborative Innovation in Geographical Information Resource Development and Application, Nanjing 210023, China
4   State Key Laboratory Cultivation Base of Geographical Environment Evolution (Jiangsu Province), Nanjing 210023, China
5   Urban Planning and Development Institute, Yangzhou University, Yangzhou 225127, China
6   College of Civil Science and Engineering, Yangzhou University, Yangzhou 225127, China
*   Correspondence: 007839@yzu.edu.cn; Tel.: +86-156-5195-8693

Citation: Shi, G.; Ye, P. Assessment on Temporal and Spatial Variation Analysis of Extreme Temperature Indices: A Case Study of the Yangtze River Basin. *Int. J. Environ. Res. Public Health* **2021**, *18*, 10936. https://doi.org/10.3390/ijerph182010936

Academic Editors: Mikio Ishiwatari and Daisuke Sasak

Received: 22 September 2021
Accepted: 15 October 2021
Published: 18 October 2021

Publisher's Note: MDPI stays neutral with regard to jurisdictional claims in published maps and institutional affiliations.

Abstract: Extreme temperature change is one of the most urgent challenges facing our society. In recent years, extreme temperature has exerted a considerable influence on society and the global ecosystem. The Yangtze River Basin is not only an important growth belt of China's social and economic development, but also the main commodity grain base in China. The purpose of this study is to study the extreme temperature indices in the Yangtze River Basin. In this study, the Mann–Kendall nonparametric test and R/S analysis method are used to analyze the spatial and temporal variation characteristics of major extreme temperature indices in the Yangtze River Basin from 1970 to 2014. The main conclusions are drawn as follows: (1) The occurrence of cold days (TX10), cold nights (TN10), ice days (ID), and frost days (FD) decrease at a rate of -0.66---2.5 d/10a, respectively, while the occurrence of warm days (TX90), warm nights (TN90), summer days (SU), and tropical nights (TR) show statistically significant increasing trends at a rate of 2.2–4.73 d/10a. (2) The trends of the coldest day (TXn), coldest night (TNn), warmest day (TXx), warmest night (TNx), and diurnal temperature range (DTR), range from -0.003 to 0.5 °C/10a. (3) Spatially, the main cold indices and warm indices increase and decrease the most in the upper and lower reaches of the Yangtze River Basin. (4) DTR and TN90 show no abrupt changes; the main cold indices changed abruptly in the 1980s and the main warm indices changed abruptly in the late 1990s and early 2000s. (5) The extreme temperature indices are affected by the atmospheric circulation and urban heat island effect in the Yangtze River Basin. Relative indices and absolute indices will continue to maintain the present trend in the future. In short, the main cold indices of extreme temperature indices show a decreasing trend, the main warm indices of extreme temperature indices show an increasing trend, and cold indices and warm indices will continue to maintain the present trend in the future in the Yangtze River Basin. Extreme temperature has an important impact on agriculture, social, and economic development. Therefore, extreme temperature prediction and monitoring must be strengthened to reduce losses caused by extreme temperature disasters and to promote the sustainable development in Yangtze River Basin.

Keywords: extreme temperature indices; spatial heterogeneity; abrupt; prediction; disaster risk; Yangtze River Basin

1. Introduction

According to the fifth assessment report of the Intergovernmental Panel on Climate Change, from 1880 to 2012, the global mean surface temperature increased significantly, with an increase of 0.85 °C. The annual mean temperature from 2003 to 2012 increased by 0.78 °C compared with that from 1850 to 1900 [1]. By the end of the century, global mean surface temperature may increase, and it is possible that the surface temperature will rise by a maximum of 2 °C [2]. The rise in temperature has become an indisputable fact. Global warming may increase the frequency and intensity of extreme weather and climate events [3]. Compared with the increase in mean temperature, regional extreme temperature changes have a more significant and direct impact on society and ecosystems.

At present, extreme temperature events have become commonplace on a global scale. Many scholars have analyzed the extreme temperatures in different regions of the world and they found that Australia's extremely cold days and frost days are on a downward trend, and extreme high temperature days are on the rise [4]. Some scholars have argued that mean annual temperature over southern Canada has increased by an average of 0.98 °C, with the largest warming during winter and early spring, since 1900 [5]. Other studies suggest that the number of warm days has increased, and the number of cold nights has decreased significantly, in Uruguay [6]. Some scholars believe that number of frost days in Europe shows a decreasing trend, and the number of summer days shows an increasing trend [7]. The number of cold days and cold nights in Morocco is decreasing, while the number of warm days and warm nights is increasing [8]. Similar work has found the percentage of warm nights/days in 70% of stations in South and Central Asia increased significantly, while the percentage of cold nights/days is decreased significantly [9]. Meanwhile, Chinese scholars have also carried out research on extreme temperatures: the rising tendency of the minimum temperature in Tianjin in winter and spring is the most obvious; the longest cold nights and frost days in winter and spring are significantly reduced, and the number of long cold nights is significantly reduced [10]. On a small regional scale, it was observed that the increased magnitude of extreme high temperatures in Beijing is significantly less than that of extreme low temperatures, and the increase in the diurnal indices is less than the night indices of extreme temperatures [11]. On a larger regional scale, it was observed that the indices of cold nights (days) in Northeast China, Loess Plateau, Xinjiang region, Pearl River Basin and Yellow River Basin showed a downward trend, and the indices of warm nights (days) showed an increasing trend, but their increased magnitudes and spatial distributions were varied [12–15]. In addition, some scholars believe that the extreme minimum temperature, annual mean temperature, and extreme maximum temperature in China's mainland have significantly increased [16,17]. All the above studies demonstrate that in the global warming environment, the occurrence of extreme high temperatures increases, while that of the cold indices decreases. However, there were varieties in the increasing magnitudes of extreme temperatures and their spatial patterns. Therefore, this research on regional extreme temperatures is of practical significance. The Yangtze River Basin is the largest basin in China, and it is an important economic belt, spanning the three major economic zones of southwest, central and eastern China [18]. Therefore, the study of extreme temperature in the Yangtze River Basin is very important.

However, in the past few decades, with the rapid economic development, population increase, urban expansion, and changes of land use/land cover, the climate conditions in the Yangtze River Basin have changed [19,20]. According to the previous studies, the average temperature in the Yangtze River Basin is increased significantly [21]. The extreme temperatures of the Yangtze River Basin in winter and summer show an upward tendency [22]. Some scholars analyzed the daily minimum temperature, daily average temperature, and daily maximum temperature in the Yangtze River Basin, and found indicators on an obvious increasing tendency in the late 1980s [23]. Other scholars have argued that the number of warm days and warm nights were increasing, while the number of cold days and cold nights were decreasing [24]. Moreover, other studies suggest extreme temperature indices of the Yangtze River Basin and concluded that the warming range of the cold indices was greater than that of the warm indices, and the warming range of the night indices was greater than that of the diurnal indices [25]. However, it is found that there is a strong correlation between altitude and the changing trend of extreme temperature. Above 350 m, the warming rate increases with the increase of altitude, while below 350 m, the warming rate decreases with the increase of altitude [26]. In terms of spatial pattern, the meteorological stations with obvious trends of extreme temperature indices were mostly in Sichuan Basin and the lower Yangtze River Basin, and almost all the extreme temperatures showed the maximum tendency in spring and winter [27]. In addition, many scholars have conducted research on sub basins or provincial administrative units in the Yangtze River Basin [28–30]. The above studies on the extreme temperature in the Yangtze River Basin mainly focus on (1) the average temperature, the minimum temperature, and the maximum temperature in the Yangtze River Basin; (2) the temporal and spatial variation of extreme temperature indices in the Yangtze River Basin, or the sub basins or provincial administrative units in the Yangtze River Basin. There is a lack of prediction of the future extreme temperature in the Yangtze River Basin.

Another reason to study extreme temperatures in the Yangtze Basin is that the frequent occurrence of heat waves has become an important public health problem [31]. The IPCC assessment report points out that since the mid-20th century, due to global warming and urban heat island effect, summer heat waves have occurred frequently in the world [1]. Studies show that heat waves are linked to heatstroke and heat-related illnesses, leading to an increase in deaths among residents. There are many studies on the health effects of high temperature heat waves in Europe and North America. During the heat wave, a large number of people die directly or indirectly. These deaths are mainly concentrated in the elderly and in persons with underlying diseases [32–34]. In China, studies on the health effects of heat waves have focused on large cities, such as Wuhan City, Nanjing City, and Shanghai City in the Yangtze River basin [35–37]. The Yangtze River basin is an important economic zone and urban zone in China. The study of extreme temperature can help to prevent and monitor the impact of heat wave on public health.

Therefore, this paper selected 131 meteorological stations in the Yangtze River Basin from 1970 to 2014 for comprehensive analyses. The main purposes of this study are (a) to know the time when there are abrupt changes of the main extreme temperature indices in the Yangtze River Basin, (b) to select abnormal extreme temperature indices by using Rescaled range analysis (R/S) and box-plot method to predict the main extreme temperatures, (c) to discuss the causes of extreme temperature in the Yangtze River Basin. (d) The study carries theoretical value for extreme temperatures research. It is also conducive to a better understanding of extreme temperatures and influencing factors in various regions of the Yangtze River Basin. In addition, the empirical analysis provides the basis for the government to formulate corresponding policies, to reduce losses caused by extreme temperature disasters, and to promote sustainable development in the Yangtze River Basin.

2. Study Area, Data, and Methods

2.1. Study Area

The Yangtze River Basin (90°33′~122°25′ E, 24°30′~35°45′ N) (Figure 1) refers to the vast area through which the main and tributaries of the Yangtze River pass. It spans 19 provincial administrative units in three economic zones of China (the eastern economic zone, central economic zone and western economic zone). It is the third largest basin in the world with a total area of 1.8 million square kilometers, which accounts for 18.8% of China's land area. The terrain of the basin is fluctuant, high in the west and low in the east, showing a three-step shape. Except for some high-altitude areas such as the western Sichuan Plateau and the headwaters of the Yangtze River, most of the basin belongs to the middle and north subtropical monsoon climate. The Yangtze River Basin is the main driving axis of China's economic development. The Yangtze River economic zone, together with the eastern coastal areas, has become the main axis of China's inverted "T" shape economy and plays an extremely important strategic role in China's social and economic development [38]. The Yangtze River Basin is also a densely populated area in China. Strengthening the research on extreme temperature is conducive to reducing the impact of extreme temperature on public health problems. Moreover, the Chengdu Plain in the Mintuo River Basin and the Hanjiang Plain in the Jianghan River Basin, etc., in the Yangtze River Basin, are important commodity grain bases, which play an important role in China's food security. In short, studying extreme temperature in the Yangtze River Basin is conducive to ensure the sustainable development of society and economy.

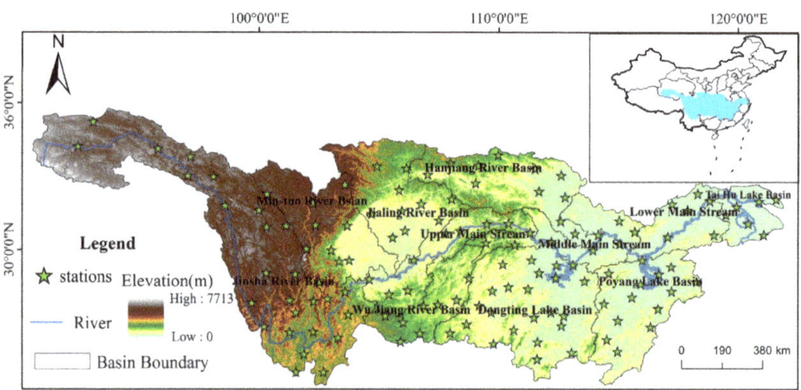

Figure 1. Location, basin boundaries, altitudinal variation range, and distribution of the meteorological stations in the Yangtze River Basin, China.

2.2. Data

There are 164 national meteorological stations in the Yangtze River Basin. The resource of daily observations of the stations utilized in this study were from the China Meteorological Data Service Center (CMDC). Considering the start and ending time of meteorological data records at the stations and that some stations have been demolished or relocated, this study selected real data from 131 stations in the basin from 1970 to 2014 so as to ensure the integrity and consistency of meteorological data (Figure 1). For a few stations with missing data less than 7 days within a month, the regression method, based on the adjacent stations, was used to interpolate [39]. The basin boundary data are extracted mainly based on the DEM data of 1 km resolution.

DMSP (Defense Meteorological Satellite Program)/OLS (Operational Linescan System) night light data were obtained from the NOAA National Geophysical Data Center. DMSP/OLS data are cloudless, non-radiation-calibrated night light images, including three kinds of annual average images: cloudless observation frequency, average light, and stable light. The spatial resolution of DMSP/OLS noctilucent data is 30" (arc second, about 1 km), and the pixel value is distributed in the interval (0, 63). It contains persistent light sources such as cities and towns, and excludes the influence of accidental noise such as moonlight, fire, and oil and gas combustion. DMSP/OLS data reflect the nighttime power consumption of public infrastructure, commerce, and residents. Therefore, it is closely related to the intensity of human economic activities and can reflect the development of urbanization, population, and industry. The DMSP/OLS data analyzed in this paper included F10 1992 and F18 2013 collected by two satellite (F10 and F18) images.

2.3. Methods

2.3.1. Extreme Temperature Indices

The definitions of extreme temperature indices are based on the detection and indicators of climate change defined by the World Meteorological Organization's Climatology Committee (WMO-CC), the World Climate Research Program (WCRP), Climate Variability and Predictability Program (CLIVAR), and Expert Team for Climate Change Detection Monitoring and Indices (ETCCDMI). This method has been widely used in extreme climate events research [40,41]. Thirteen extreme temperature indices were selected in this paper (Table 1).

Table 1. 13 Definitions of extreme temperature indices.

Category (1) [1]	Indices	Descriptive Name	Definitions	Units
Relative indices	TX10	Cold days	Number of days with T_{max} < 10th percentile	days
	TN10	Cold nights	Number of days with T_{min} < 10th percentile	days
	TX90	Warm days	Number of days with T_{max} > 90th percentile	days
	TN90	Warm nights	Number of days with T_{min} > 90th percentile	days
Absolute indices	ID	Ice days	Annual count when TX (daily minimum temperature) < 0 °C	days
	FD	Frost days	Annual count when TN (daily maximum temperature) < 0 °C	days
	SU	Summer days	Annual count when TX > 25 °C	days
	TR	Tropical nights	Annual count when TN > 20 °C	days
Extremal indices	TXn	Coldest day	Annual lowest TX	°C
	TNn	Coldest night	Annual lowest TN	°C
	TXx	Warmest day	Annual highest TX	°C
	TNx	Warmest night	Annual highest TN	°C
	DTR	Diurnal temperature range	Monthly mean difference between TX and TN	°C

[1] Category (2) Extremely cold temperature indices: TX10, TN10, ID, FD, TXn, and TNn. Extremely warm temperature indices: TX90, TN90, SU, TR, TXx, and TNx.

2.3.2. Mann–Kendall (M–K) Trend and Abrupt Changes Analysis

The nonparametric Mann–Kendall (M–K) test method is used here for trend analysis [42]. At present, the M–K test method is mainly used for analyses of abrupt changes in temperature when the significance level is $p < 0.05$. In the process of the M–K abrupt change test, if the positive and negative series have several obvious intersections in the confidence interval, by combining the moving t-test, the intersection can be determined to be a real abrupt change point [43]. The following is the introduction to the Mann–Kendall trend analysis method:

Provided that $x_1, x_2, \cdots, x_n$ are the data values in time series and n is the number of data points, in the Equation

$$S = \sum_{k=1}^{n-1} \sum_{j=k+1}^{n} \text{sgn}(x_j - x_k) \tag{1}$$

x_j, x_k are the measured values of j and k, and $k > j$. And,

$$\text{sgn}(x_j - x_k) = \begin{cases} 1 & , \quad x_j - x_k > 0 \\ 0 & , \quad x_j - x_k = 0 \\ -1 & , \quad x_j - x_k < 0 \end{cases} \tag{2}$$

Then,

$$Var(S) = \frac{n(n-1)(2n+5) - \sum_{k=1}^{n} k(k-1)(2k+5)}{18} \tag{3}$$

$$Z = \begin{cases} \frac{S+1}{\sqrt{Var(S)}}, & S < 0 \\ 0^{\circ\circ\circ}, & S = 0 \\ \frac{S-1}{\sqrt{Var(S)}}, & S > 0 \end{cases} \tag{4}$$

where Z is the standard normal test statistic and positive values of Z indicate increasing trends while negative Z values show decreasing trends. In this study, the specific significance level is $p < 0.05$, which means $Z > 1.96$ or $Z < -1.96$.

When Z does not equal zero, Sen's slope estimator is used to define the trends. The formula is:

$$f(t) = Qt + B \tag{5}$$

The value of Q indicates the steepness of the trend, B is a constant, and t is a data year.

$$Q = Median \frac{x_j - x_k}{j - k} \tag{6}$$

2.3.3. R/S Analysis

Rescaled range analysis (R/S) is a classification-structured analysis method for processing time series [44]. Studies have shown that natural phenomena such as precipitation, temperature, and sunspots all have a Hurst effect. H is the Hurst exponent. When $0.5 < H < 1$, it describes a dynamically persistent, or trend reinforcing series; the greater the H value is, the stronger and more persistent. When $H = 0.5$, it means that the time series is an independent random process, which indicates that the current trend will not affect the future trend. When $0 < H < 0.5$, it describes an anti-persistent, or a mean reverting system; the smaller the H value, the stronger the anti-persistent [45].

2.3.4. The Sliding *t*-Test

The sliding *t*-test detects mutations by examining whether there is a significant difference between the mean values of two groups of samples. For a time series x with n sample sizes, to set an artificial time as a reference point, the samples of two sequences x_1 and x_2 before and after the reference point, and are n_1 and n_2; the mean values of the two sequences are x_1 and x_2 and the variances are s_1^2 and S_2^2, respectively. Define statistics:

$$t = \frac{\overline{x_1} - \overline{x_2}}{s \cdot \sqrt{\frac{1}{n_1} + \frac{1}{n_2}}} \tag{7}$$

of which

$$s = \sqrt{\frac{n_1 s_1^2 + n_2 s_2^2}{n_1 + n_2 - 2}} \tag{8}$$

where the equation follows the *t*-distribution of freedom $V = n_1 + n_2 - 2$.

In order to avoid the drift of mutation points caused by arbitrary selection of subsequence length, the length of the subsequence can be changed repeatedly for experimental comparison when using this method to improve the accuracy of the calculation results.

3. Results

3.1. Temporal and Spatial Variations of Extreme Temperatures

3.1.1. Temporal Variation of Relative Indices

From the time scale representation, the relative indices of the Yangtze River Basin changed significantly from 1970 to 2014 (Figure 2). Cold days (TX10) and cold nights (TN10) were in a significant downward trend, and the trends were -2.2 d/10a ($p < 0.05$) and -3.6 d/10a ($p < 0.001$), respectively, tested by the significance level. The number of cold days reached its minimum in 1999, at about 21.7 d, and then increased in fluctuation. Warm days (TX90) and warm nights (TN90) obviously increased, and the trends were 4.73 d/10a ($p < 0.001$) and 3.81 d/10a ($p < 0.001$), respectively, among which the warm days (TX90) showed a significant upward trend after 2003. Besides, the number of warm nights reached a minimum in 1993, at about 21.8 d, and then rose in volatility.

3.1.2. Spatial Variation of Relative Indices

Spatially, cold days (Tx10) of more than 98% of the stations showed decreasing tendencies, of which 35.8% of the stations passed the significance level test ($p < 0.05$). The decreased magnitude of cold days in the Jinsha River basin was the most apparent, with the trends above -4 d/10a, followed by the middle reaches of the mainstream. More than 97% of stations had a decreasing trend in cold nights (TN10), with 74% of them passing the significance level test ($p < 0.05$). From the perspective of the whole river basin, there were particularly significant decreasing tendencies in the Jinsha River Basin and the middle reaches of the main stream area, and the Hanshui River Basin, with the trends exceeding -4.5 d/10a. However, the increasing magnitudes of extreme temperatures and their spatial patterns vary. The remaining 98.5% of the stations showed increasing tendencies, and 72.5% of the stations passed the significance level test ($p < 0.05$), which mainly concentrated in the Jinsha River Basin, the Taihu Lake Basin, and the southern stream of the Minjiang River and the Tuojiang River. In addition, warm nights (TN90) of 96.9% of the stations showed increasing tendencies, all of which passed the significance level test, and their change magnitudes were generally between 3 d/10a–6 d/10a. The change magnitudes of the regions in the basin were significant except for the slight increased magnitudes in the southern and upper mainstream of the Jialing River Basin (Figure 3).

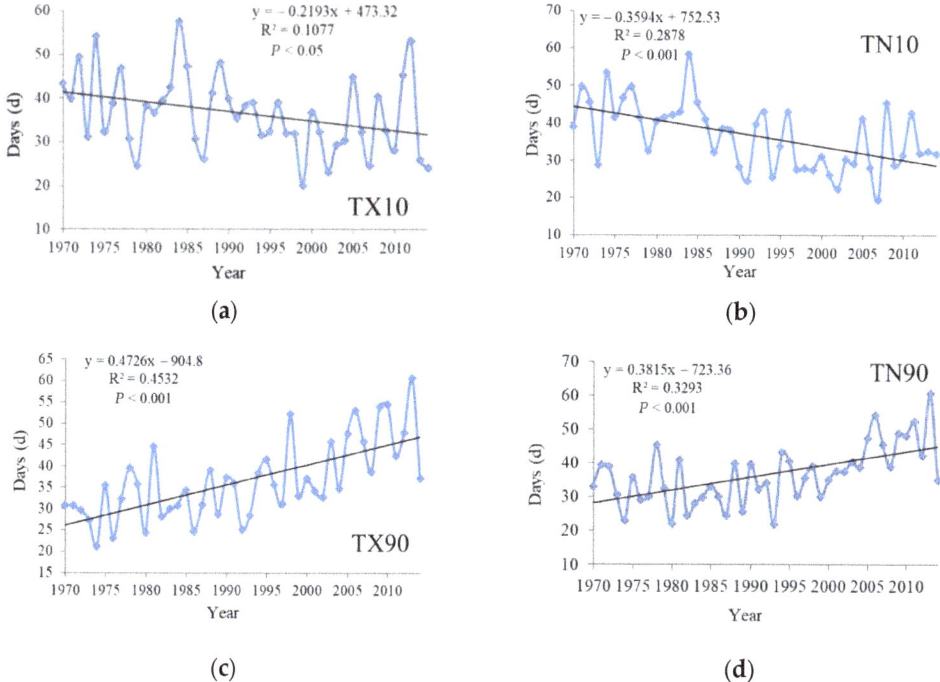

Figure 2. Temporal variation of extreme temperature relative indices in the Yangtze River Basin from 1970 to 2014. (**a**) is TX10, (**b**) is TN10, (**c**) is TX90 and (**d**) is TN90.

3.2. Temporal and Spatial Variations of Absolute Indices

3.2.1. Temporal Variation of Absolute Indices

The number of ice days (ID) showed a slight decreasing trend fluctuation, and the trend was −0.66/10a ($p < 0.001$). The number of frost days (FD) presented a fluctuating decreasing tendency with a trend of −2.5 d/10a ($p < 0.001$), the lowest value was 41.2 d in 1991, followed by a significant fluctuating decrease. Summer days (SU) showed a fluctuating upward trend with a trend of 2.2 d/10a ($p < 0.001$) reaching the minimum 142.5 d in 1982, followed by an increase in volatility. Tropical nights (TR) showed a fluctuating increasing trend with a trend of 2.8 d/10a ($p < 0.001$), reaching its lowest value 100.1 d in 1976 and then showing a significant increasing tendency in fluctuation (Figure 4).

3.2.2. Spatial Variation of Absolute Indices

From the regional scale representation, there were 19 stations with no changing trend in the number of ice days (ID), and there were no freezing days, which mainly distributed in the southern part of the Jinsha River Basin, the southern part of the Min-tuojiang River Basin, and most of the Jialing River Basin (Figure 5). A total of 76.3% of the stations showed a downward trend in ice days (ID), 32% of which passed the significance level test ($p < 0.05$), and the change magnitudes ranged from 0–0.5 d/10a. Stations in the Dongting Lake basin and its middle reaches of the mainstream showed an increasing trend, but it did not pass the significance level ($p < 0.05$). However, the stations that passed the test were mainly distributed in the Taihu Basin and the Poyang Lake Basin. The number of frost days (FD) decreased significantly in the whole basin and 96.9% of the stations showed a downward trend, among which 73.3% of the stations passed the significant level. The significant decreased magnitudes were mainly distributed in the Jinsha River Basin, Hanshui River Basin, Taihu Lake Basin, and the middle reaches, with the change magnitudes between −3−−7 d/10a. There were six stations in the basin with no change (the trend = 0) in summer days (SU), which mainly distributed in the high-altitude area of Qinghai Plateau. The summer days in the Taihu Lake Basin and most of the Poyang Lake Basin showed a decreasing trend with a trend of 0−−2.5 d/10a. Most of the rest of the region showed an increasing tendency, and they were distributed in the central part of the Yangtze River Basin, including Dongting Lake Basin, Hanshui River Basin, Jialing River Basin, Wujiang River Basin, Mintuojiang River Basin, and most of the southern part of the Jinsha River Basin.

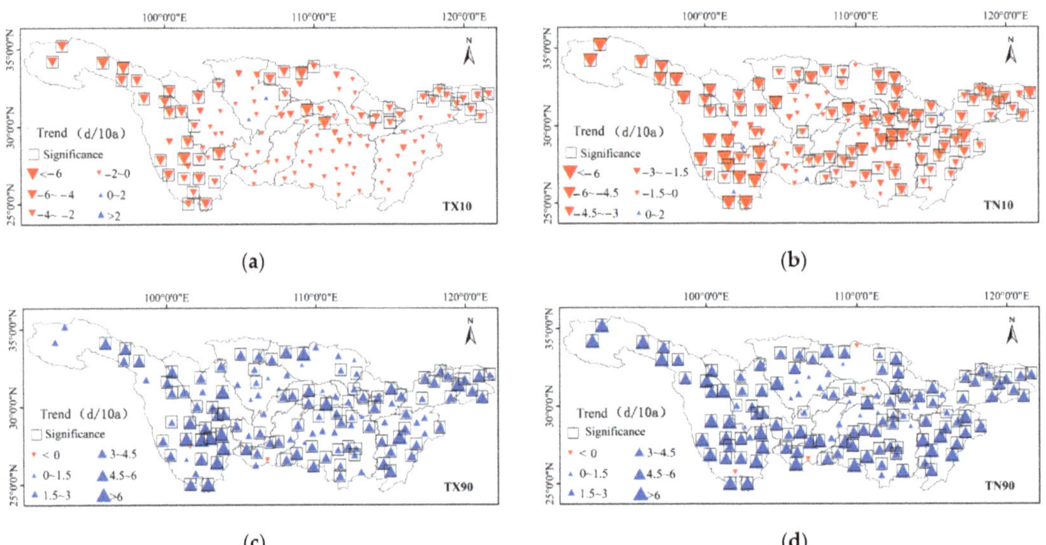

Figure 3. Spatial distribution of interannual variability of relative indices of extreme temperature in the Yangtze River Basin from 1970 to 2014. (**a**) is TX10, (**b**) is TN10, (**c**) is TX90 and (**d**) is TN90.

There were 21 stations showing no changes (the trend = 0) in terms of tropical nights (TR), and were mainly distributed in the Jinsha River basin. Tropical nights (TR) in other areas in the basin generally showed an increasing trend, which were mainly distributed in the middle and lower reaches of the Poyang Lake Basin, the Hanshui River Basin, the Middle and the Lower Mainstream Area, the Taihu Lake Basin, and the Wujiang River Basin, with a change magnitude of 2 d/10a–6 d/10a and 84.6% of the stations having passed the significance level test.

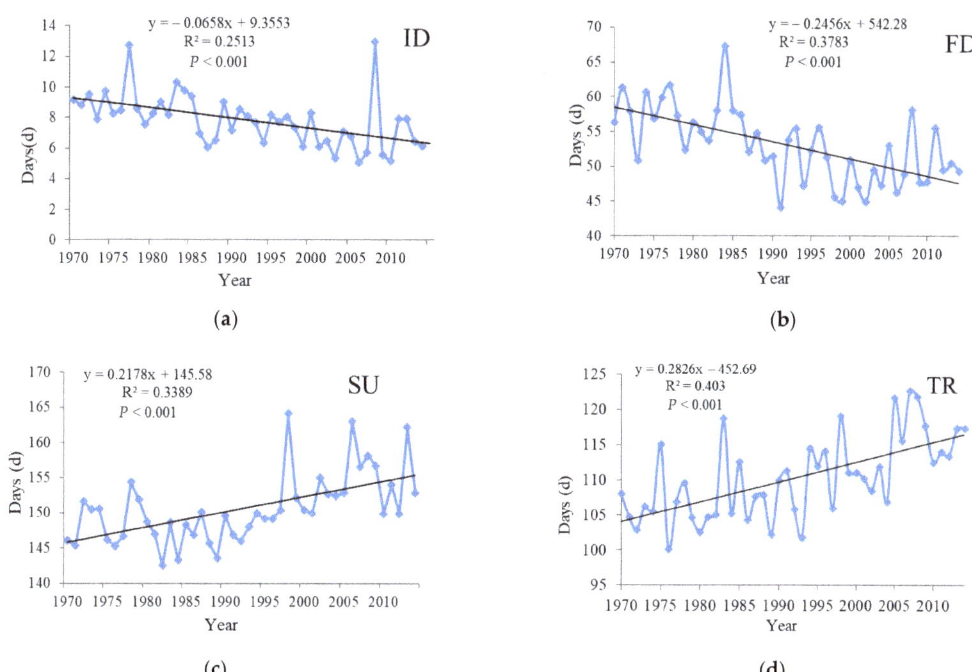

Figure 4. Temporal variation of the absolute indices of extreme temperature in the Yangtze River Basin from 1970 to 2014. (**a**) is ID, (**b**) is FD, (**c**) is SU and (**d**) is TR.

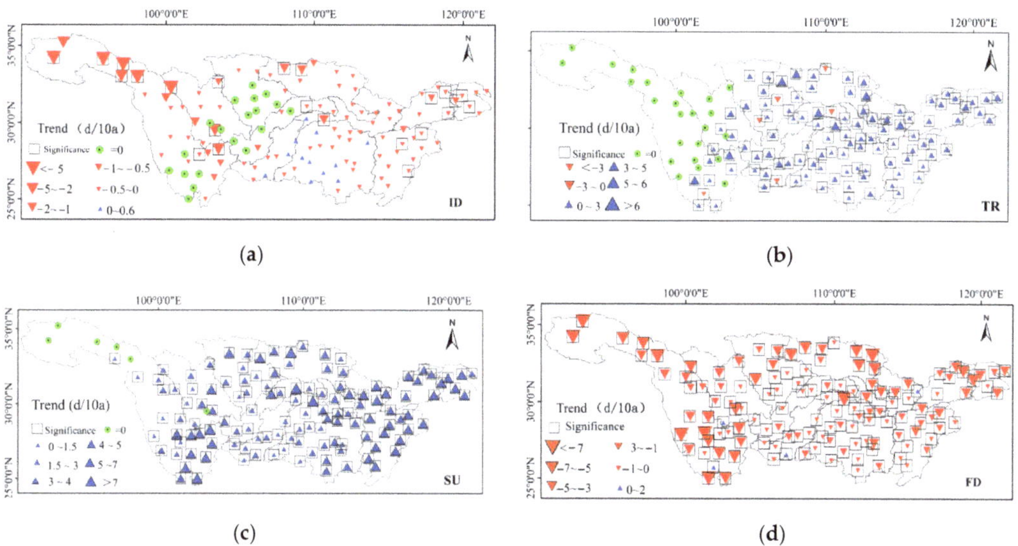

Figure 5. Spatial distribution of annual variation of absolute indices of extreme temperatures in the Yangtze River Basin from 1970 to 2014. (**a**) is ID, (**b**) is TR, (**c**) is SU and (**d**) is FD.

3.3. *Temporal and Spatial Variations of the Extremal Indices*

3.3.1. Temporal Variation of the Extremal Indices

The minimum value of daily maximum temperature (TXn) increased significantly with a trend of 0.39 °C/10a ($p < 0.001$), reached the lowest value of −3.5 °C in 1977 and increased in fluctuating tendencies after 1977 (Figure 6). The minimum value of daily minimum temperature (TNn) rose obviously with a trend of 0.5 °C/10a, and also showed the minimum in 1977, which was −9.1 °C, and after 1977 it increased in volatility. The maximum value of daily maximum temperature (TXx) evidently increased with a trend of 0.27 °C/10a ($p < 0.001$), which was extremely low in 1993 with a minimum of 34.7 °C, and increased in fluctuating tendencies after 1993. The maximum value of daily minimum temperature (TNx) showed a significant increasing trend, which was 0.24 °C/10a. In 1974, it reached its lowest value of 24.8 °C, and increased in volatility after that year. There were slight decreasing tendencies in diurnal temperature range (DTR) and the trend was −0.003 °C/10a ($p = 0.51$). In 1989, its value achieved the minimum of 6.48 °C and after that year it increased in volatility.

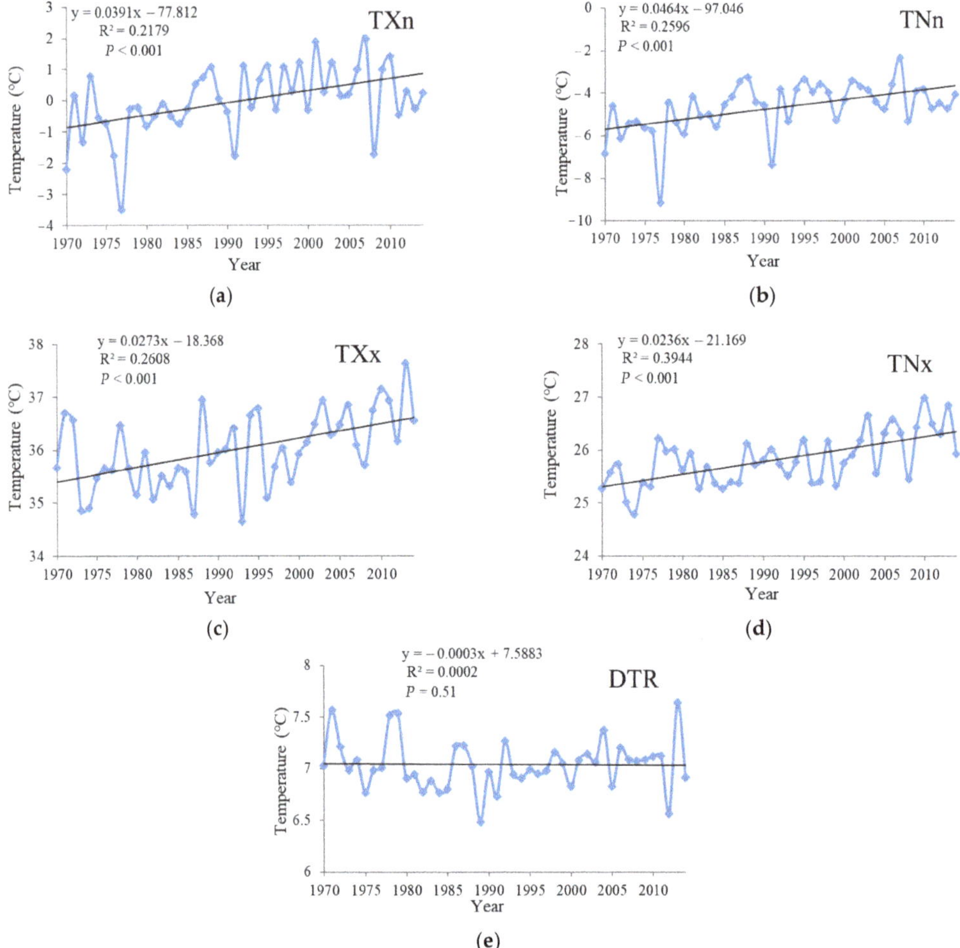

Figure 6. Temporal variation of extreme indices of the extreme temperatures in the Yangtze River Basin from 1970 to 2014. (**a**) is TXn, (**b**) is TNn, (**c**) is TXx, (**d**) is TNx and (**e**) is DTR.

3.3.2. Spatial Variation of Extremal Indices

The minimum value of daily maximum temperature (TXn) of 128 stations in the basin had increasing tendencies and 61% of the stations passed the significance level test (Figure 7). In particular, the increase was significant in most of the Dongting Lake Basin and the main stream of the middle reaches of Taihu Lake Basin, with a change magnitude of 0.4–0.6 °C/10a. The minimum value of the daily minimum temperature (TNn) of 95.4% of the stations showed an increasing trend, with 44% of the stations having passed the significance level test ($p < 0.05$). The areas with large increase of TNn were mainly distributed in the southern part of the Jinsha River Basin, the Hanshui River Basin, the middle reaches of the mainstream, the Poyang Lake Basin, and the Taihu Lake Basin, with a change magnitude of 0.2–0.6 °C/10a. Among them, the Dongting Lake Basin had a relatively slight change magnitude, which was between 0–0.2 d/10a. The maximum value of daily maximum temperature (TXx) was increased in all stations except for three stations, and 45% of the stations passed the significance level test ($p < 0.05$).

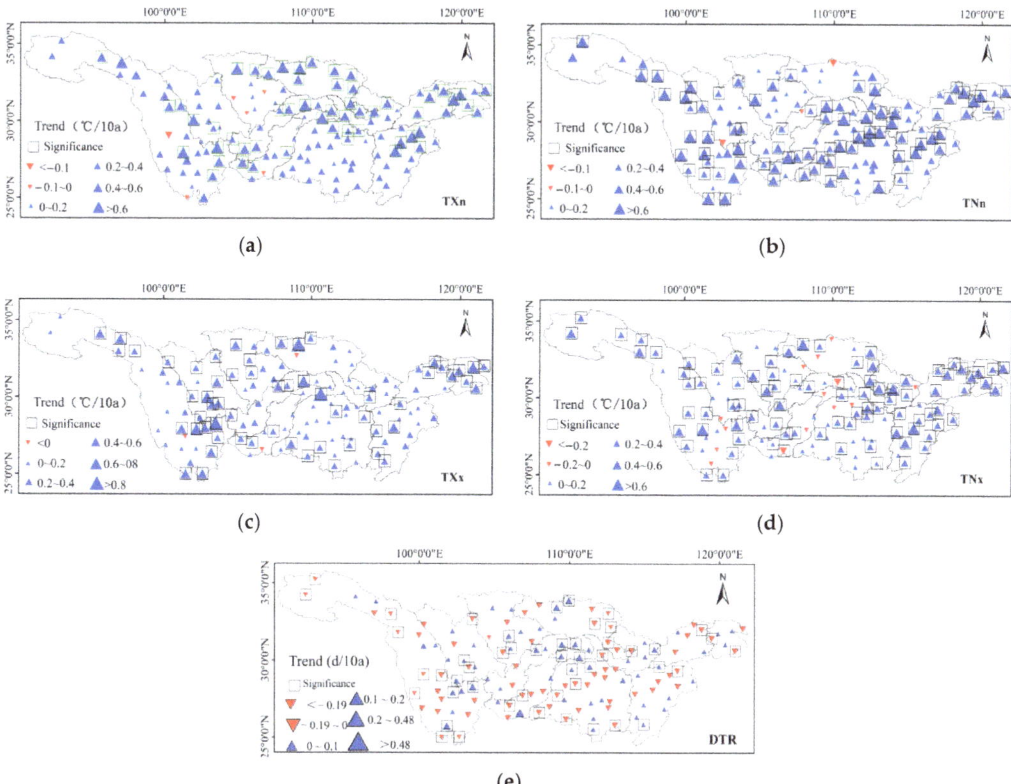

Figure 7. Spatial distribution of extreme indices of extreme temperatures in the Yangtze River Basin from 1970 to 2014. (**a**) is TXn, (**b**) is TNn, (**c**) is TXx, (**d**) is TNx and (**e**) is DTR.

The areas with a large increase of TXx were mainly concentrated in the Jinsha River Basin, Minjiang River Basin, and Taihu Lake Basin, with an increased magnitude of 0.6–0.8 °C/10a. The maximum value of the daily minimum temperature (TNx) of 90% of the sites in the basin showed increasing tendencies, of which 73.5% of the stations passed the significance level test ($p < 0.05$) with an increased magnitude of 0.2–0.6 °C/10a. The areas with a larger increase of TNx were mainly distributed in the middle stream, the downstream stream, the Poyang Lake basin, the Hanshui River Basin, and the Taihu

Lake Basin; diurnal temperature range (DTR) of 46.8% of the stations showed an increasing trend, and 33% of the stations were mainly distributed in the Jinsha River Basin, Wujiang River Basin, and Hanshui River Basin through the significance level test ($p < 0.05$), with a change magnitude of more than 0.15 °C/10a. DTR of 53.2% of the stations showed a downward tendency, of which 44.9% of the stations passed the significance level test ($p < 0.05$) and they were mainly concentrated in the Mintuojiang River Basin and the middle reaches of the mainstream, with a change magnitude of 0.15 °C–0.3 °C/10a.

3.4. Analysis of Abrupt Changes of Extreme Temperature Indices

3.4.1. Relative Indices

In this paper, the M–K test analysis is carried out on four relative indices in the Yangtze River Basin (Figure 8). There are two intersection points of the positive and negative serial curves of the cold days (TX10) within the confidence line, and a sliding *t*-test is performed in this paper, none of which show abrupt change points; positive and negative serial curves of other relative indices have only one intersection within the confidence line. Cold nights (TN10) has a crossing point in the confidence interval, and there was an abrupt change in 1987. There is an intersection point of warm days (TX90) in the confidence interval, and an abrupt change occurred in 2003. The warm nights (TN90) has an intersection out of the confidence interval and this point is not an abrupt change point.

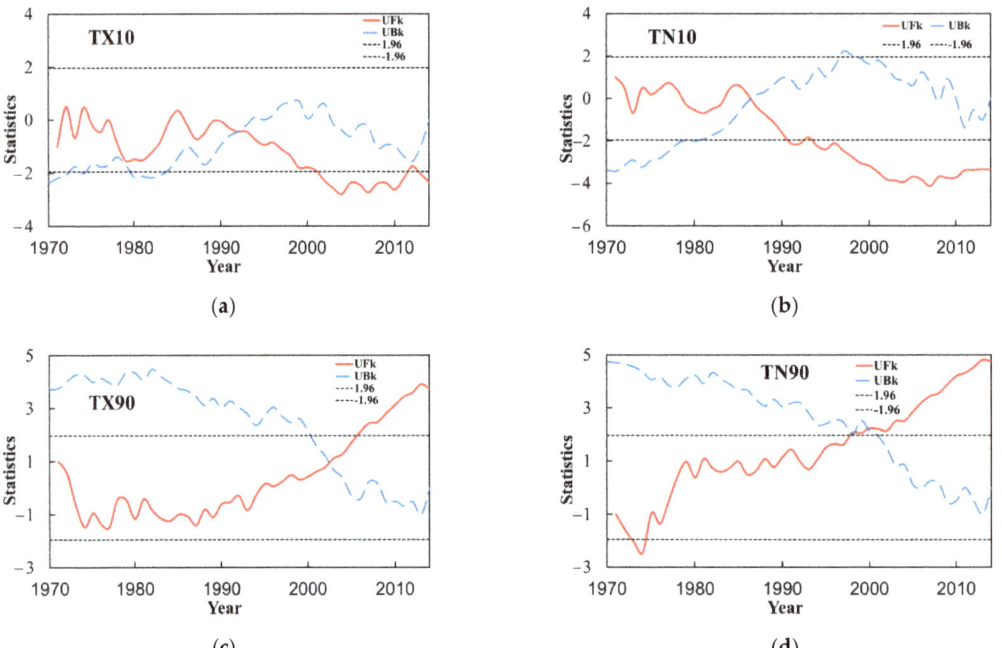

Figure 8. M–K abrupt change test of the relative temperature indices of the Yangtze River Basin from 1970 to 2014. (**a**) is TX10, (**b**) is TN10, (**c**) is TX90 and (**d**) is TN90.

3.4.2. Absolute Indices

In this paper, the M–K test analysis is carried out on the four absolute indices in the Yangtze River Basin (Figure 9). There is one intersection point in the positive and negative serial curves of ice days (ID) within the confidence line, and there was an abrupt change in 1992. The sliding *t*-test is performed in this paper, and the abrupt change point appeared in 1992. The positive and negative serial curves of other relative indices have only one intersection within the confidence line. Frost days (FD) has a crossing point in the confidence interval, and there was an abrupt change in 1988. In addition, there is an intersection of the summer days (SU) in the confidence interval and an abrupt change point occurred in 1998, tropical nights (TR) has an intersection, and there was an abrupt change in 1999.

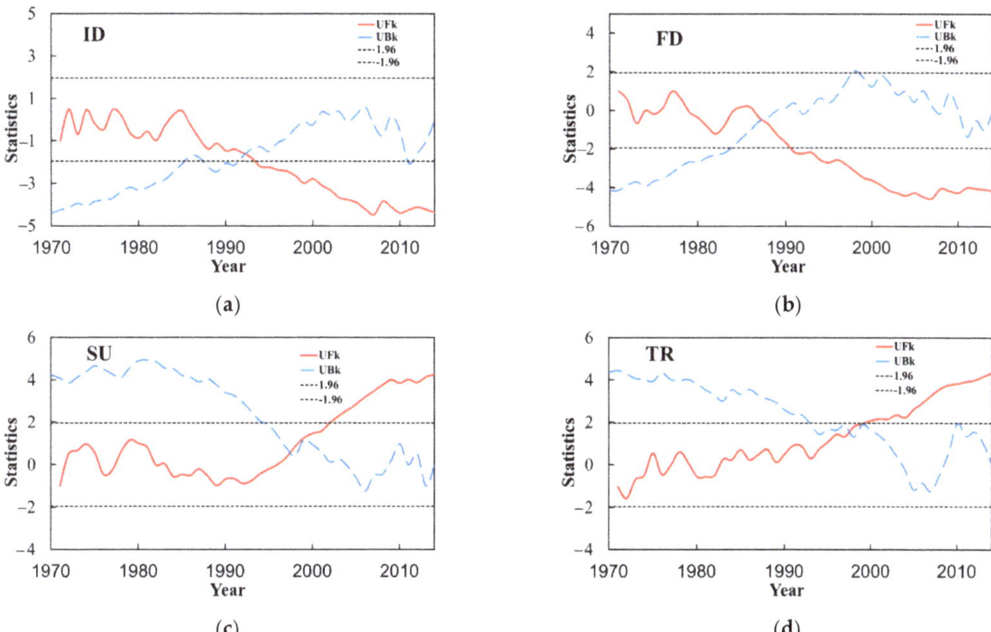

Figure 9. M–K abrupt change test of the absolute indices of extreme temperatures in the Yangtze River Basin from 1970 to 2014. (**a**) is ID, (**b**) is FD, (**c**) is SU and (**d**) is TR.

3.4.3. Extremal Indices

In this study, the M–K test analysis was carried out on the five extremal indices in the Yangtze River Basin (Figure 10). The positive and negative serial curves of the diurnal temperature range (DTR) have several intersection points within the confidence line. The sliding *t*-test was also performed and its intersections are not abrupt change points; the positive and negative serial curves of other relative indices have only one intersection point within the confidence line. The minimum value of daily maximum temperature (TXn) has a crossing point in the confidence interval, and there was an abrupt change in 1985. The minimum value of daily minimum temperature (TNn) has a crossing point in the confidence interval and the abrupt change occurred in 1981, the maximum value of daily maximum temperature (TXx) has an intersection, and there was an abrupt change in 2001, and the maximum value of daily minimum temperature (TNx) has an intersection in the confidence interval, and an abrupt change in 1998.

3.5. The Prediction of Extreme Temperature Indices

From 1970 to 2014, the stability of extreme temperature indices was significantly various (Figure 11). It can be seen from Figure 11 that SU and TR respectively have absolute indices for extreme temperature, and the distribution of each indicator data is relatively concentrated. The median and average values are greater than 0, the trend coefficient is positive, and the trend will continue to maintain the present trend.

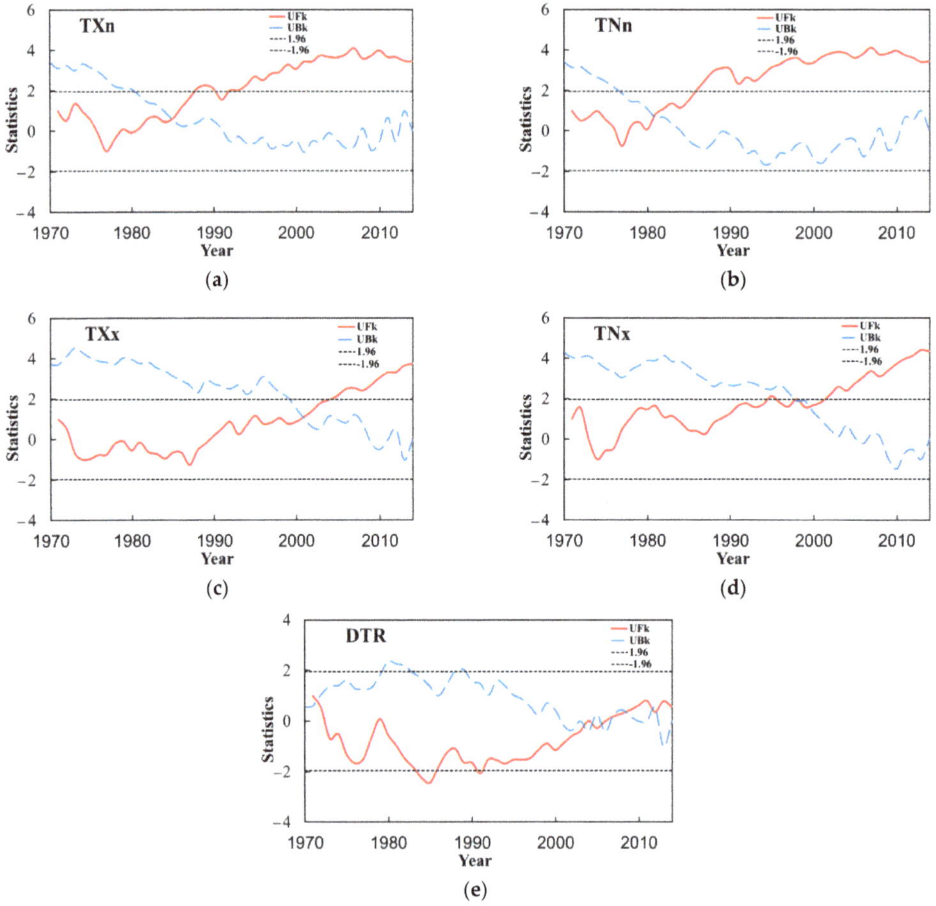

Figure 10. M–K abrupt change test of the extremal indices of extreme temperatures in the Yangtze River Basin from 1970 to 2014. (**a**) is TXn, (**b**) is TNn, (**c**) is TXx, (**d**) is TNx and (**e**) is DTR.

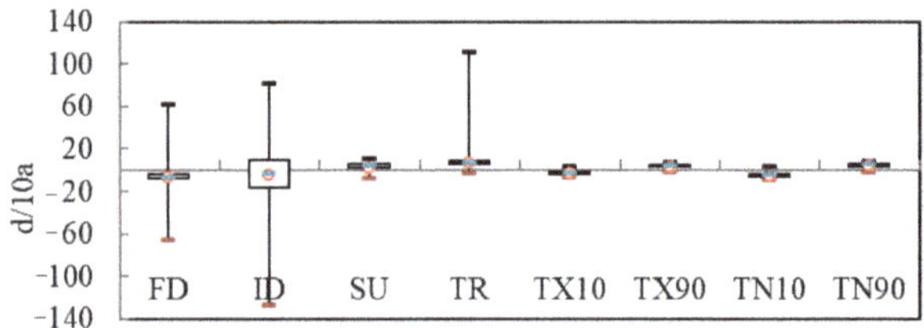

Figure 11. Box-plot of relative indices and absolute indices. At the top of the box-framed figure is the lower quartile value of the sequence, and at the bottom is the upper quartile value. Unit: FD, ID, SU, TR, TX10, RR1, TX90, TN10, TN90 (d/10a).

TNx, TXx, TNn, and TXn represent the extreme indices of extreme temperature, and the data distribution is relatively concentrated and relatively stable (Figure 12). The extreme indices of FD, TX10, and TN10 median and average are less than 0, the trend coefficient is negative and it shows an increasing trend, the data are rather scattered, and the downward tendencies and instability of the four indices are decreasing. The TX90 and TN90 indicate the extreme indices of relative indices, where the median and average values of TX90 and

TN90 are greater than 0, the trend coefficient is positive and it shows an increasing trend, the data are rather scattered, and the upward tendencies and instability of the four indices are increasing.

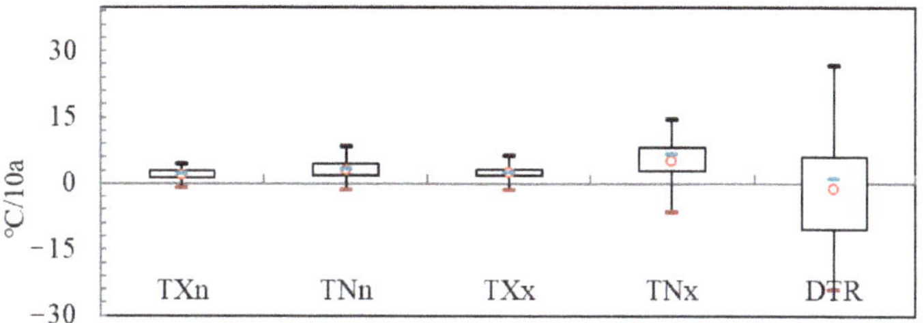

Figure 12. Box-plot of extremal indices. At the top of the box-framed figure is the lower quartile value of the sequence, and at the bottom is the upper quartile value. Unit: TXn, TNn, TXx, TNX, DTR (°C/10a).

Moreover, the R/S analysis (Figure 13) also shows that FD, TX10, and TN10 will continue to decrease in the future. TN90, TR, TX90, and SU will keep increasing for some years to come (Tables 2 and 3). As a result, the Yangtze River Basin continues to warm up, and the risk of extreme temperature events in the basin increases significantly.

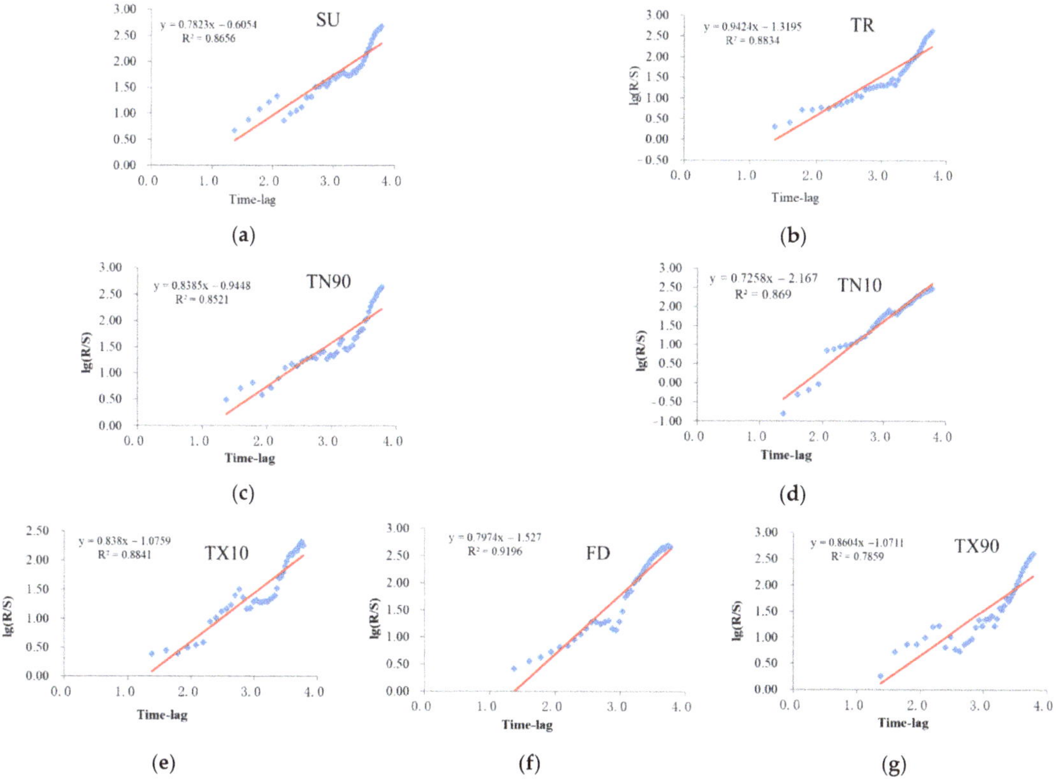

Figure 13. The extreme temperature indices of R/S analysis results in Yangtze River Basin from 1970 to 2014. (**a**) is SU, (**b**) is TR, (**c**) is TN90, (**d**) is TN10, (**e**) is TX10, (**f**) is FD and (**g**) is TX90.

Table 2. Results of R/S analysis of extreme temperature in indices.

Extreme Temperature Indices	H	R^2
FD	0.7974	0.9196
SU	0.7823	0.8656
TR	0.9424	0.8834
TX10	0.8380	0.8841
TX90	0.8604	0.7859
TN90	0.8385	0.8521
TN10	0.7258	0.8690

Table 3. The forecast of trends of extreme temperature indices.

Extreme Temperature Indices	Historical Change Tendency	H	Future Change Tendency
FD	decrease	0.7974	decrease
SU	increase	0.7823	increase
TR	increase	0.9424	increase
TX10	decrease	0.8380	decrease
TX90	increase	0.8604	increase
TN90	increase	0.8385	increase
TN10	decrease	0.7258	decrease

3.6. Possible Causes of Observed Changes in Temperature Extremes

The most areas in the Yangtze River Basin are located in the eastern monsoon region, so the temperatures in these areas are significantly affected by the atmospheric circulation. Studies have shown that Atlantic Multidecadal Oscillation (AMO) makes contributions to the warming of eastern Asia, strengthening the eastern Asian summer monsoon and weakening the eastern Asian winter monsoon, to a certain extent [46]. Through observation, analysis, and multi-mode simulation, it is found that the warm (positive) phase of AMO not only corresponds to the warm winter in most parts of China, but also warms eastern Asia in every season [47]. Niu et al., drew a similar conclusion by analyzing the correlation between the extreme temperature indices and AMO in the Yangtze River Basin [27]. Some scholars also consider that the number of high temperature days is positively correlated with the area and intensity of subtropical anticyclone. When the days with high temperature increase, the area of subtropical anticyclone increases, and the ridge point of subtropical anticyclone extends westward [48,49]. In addition, the activities of tropical cyclones or typhoons tend to weaken the Western Pacific subtropical anticyclone, and the number of high temperature days may be related to the number and influence the degree of typhoons [50].

On the other hand, human activities have greatly increased the risks of weather with extreme temperature indices [51,52]. Urbanization is the result of the continuous increase of population, the rapid development of economy, and the expansion of urban land. It has an impact on urban temperature, humidity, and precipitation. Among the effects, the heat island effect is the most prominent, which is one of the reasons for the rising trend of extreme temperature. The Yangtze River Basin is an important economic belt in China, with a high level of economic development and urbanization, including three major urban agglomerations in the Yangtze River Delta, the middle reaches of the Yangtze River, and the Mintuo River basin District. The Defense Meteorological Satellite Program (DMSP) has the operational line scanner (OLS), which provides a new data source for collecting data of the dynamic urban expansion on a large spatial scale [53]. The urban night lighting figure can reflect the dynamic expansion information, urbanization, its related land use change, and high-energy consumption. The following figures are the night lighting figures of the Yangtze River Basin in 1992 and 2013, respectively (Figure 14).

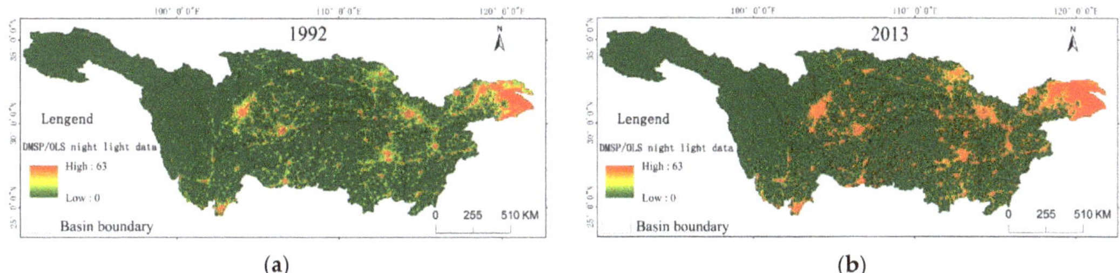

Figure 14. Distribution of DMSP/OLS light data in the Yangtze River Basin. (**a**) is the DMSP/OLS light data in 1992; (**b**) is the DMSP/OLS light data in 2013.

Among them, the DN (digital number) values of the night lighting intensity range from 0 to 63, the pixel DN value of the green area is 0, and the green area is the background area, indicating that there is no lighting; when the pixel DN value of the red area is greater than 0, it is the lighting area, and the color shades indicate the lighting intensity. In 1992, the city size in the Yangtze River Basin was relatively small and mostly concentrated in the core urban agglomerations, namely the urban agglomeration in the middle reaches of the Yangtze River, in the Mintuo River Basin, and in the Yangtze River Delta. In 2013, there were great changes. On the basis of the continuous expansion of the surrounding areas of the three major urban agglomerations, more remote areas had the observable lit pixels, and the urbanization level of the whole basin had been significantly improved. The main extreme temperature indices changed significantly in 2013. Except for TNn and TX10, the main extreme temperature indices, such as SU, TR, TNx, TXx, TXn, DTR, TX90, TN90, TN10, and TN90, showed an obvious upward or downward trend. FD and ID also maintained a downward trend (Table 4). This indicates that the urban heat island effect is one of the important reasons for the occurrence of extreme temperature in the Yangtze River Basin.

Table 4. The trend change of 13 extreme temperature indices was affected by urbanization in two different periods.

Indices	1970–1992	1993–2014
FD	−3.13 d/10a	−0.098 d/10a
ID	−0.729 d/10a	−0.27 d/10a
SU	−1 d/10a	2.74 d/10a
TR	1 d/10a	0.4 d/10a
TN_X	0.62 °C/10a	−0.16 °C/10a
TX_X	0.092 °C/10a	0.62 °C/10a
TXn	0.19 °C/10a	0.443 °C/10a
TNn	0.789 °C/10a	−0.092 °C/10a
DTR	−0.129 °C/10a	0.066 °C/10a
TX10	−0.989 d/10a	0.865 d/10a
TX90	−0.788 d/10a	8.479 d/10a
TN10	−4.51 d/10a	0.375 d/10a
TN90	1.49 d/10a	7.837 d/10a

4. Discussions

4.1. Comparison with Previous Studies

This paper mainly discusses the variation patterns of extreme temperature in the Yangtze River Basin from 1970 to 2014. In terms of temporal variation pattern, the main warm indices of meteorological stations in the Yangtze River Basin showed an upward trend, while the cold indices showed a downward trend, which is not only consistent with the previous research results in the same region, but is also in good agreement with that reported in Loess Plateau, China, Central Asia, Europe, and globally [47]. Spatially, the main warm indices also presented an increasing tendency, while the cold indices a decreasing tendency, which is consistent with the previous research results in the same region, and also in accordance with many studies in other regions [54]. The main cold indices changed abruptly in the 1980s and the main warm indices changed abruptly in the late 1990s and early 2000s, which is similar to the results of other scholars.

4.2. The Effects of Extreme Temperature Indices

Comparing with other regions, for the Plateau areas and the Taihu Lake Basin in the Yangtze River Basin, both the increasing trend shown by the main warming indices and the decreasing tendency presented by for the main cooling indices have remarkable changes. Extreme temperatures at high altitudes are more responsive to global warming [55,56]. In particular, the decreasing number of frost days and ice days is much more beneficial to the plateau grass turning green and livestock wintering. China's economy has been developing rapidly and the urbanization rate has been increasing since the reform development around 1980, which could be one reason why the extreme temperature indices began to be abrupt around the 1980s. Extreme temperatures affect human health. The frequent heat waves not only threaten lives and but also increase the risks of related diseases and lead to more and more people deceased every year [57]. In addition, they can affect agricultural development. The average annual yield loss of rice in the Yangtze River Basin increased significantly from 8.9% in the 1970s to 17.9% in the early 21st century due to the high temperatures [58,59]. Although the number of frost days and ice days in the river basin has decreased, the Yangtze River Basin, mainly located in the monsoon climate region with a large temperature variation, is prone to spring cold, and a large variation in the number of frost days and freezing days (Figure 4), which increases the affected area of crops [60,61]. In addition, the increase of extreme temperature is helpful to the winter of diseases and pests [62,63], which threaten agricultural production. Extreme temperatures will continue to occur in the future (Table 3). Therefore, extreme temperature prediction and monitoring must be strengthened. Promoting the sustainable development of agriculture, society, and economy in the Yangtze River Basin is important. This paper systematically and comprehensively expounds on the temporal and spatial variation trend of extreme temperature, the prediction of extreme temperature, and the causes and effects of extreme temperature in the Yangtze River Basin. The above methods have certain reference significance for the study of extreme temperature in economically developed and densely urban areas.

5. Conclusions

The Yangtze River Basin is an important economic and urban belt in China and the engine of China's social and economic development. The analysis of the extreme temperature is important for a high impact sustainable development of social economy in the Yangtze River Basin. The study of the characteristics of extreme temperature in this area could help governments and decision makers to make better informed decisions regarding urban construction planning and economic development planning. This paper assesses the temporal and spatial variation analysis of extreme temperature indices of the Yangtze River Basin. The main conclusions are as follows:

(1) The trend of cold days, cold nights, ice days, and frost days decreased by −2.2, −3.6, −0.66, and −2.5 d/10a, respectively, while the trend of TX90, TN90, SU, TXx, and TR shows trends of 4.73, 3.82, 2.2, 0.27, and 2.8 d/10a, respectively. The tendency rates of TXn, TNn, TNx, and DTR range is 0.39, 0.5, 0.24, and −0.003 °C/10a, respectively. Spatially, the main extremely warm indices of meteorological stations were increasing, while the extremely cold indices were decreasing in the Yangtze River Basin.

(2) Except for DTR and TN90, there were no abrupt changes; the other 11 extreme temperature indicators all had abrupt changes. TX10 changed abruptly in 1987 and TN10 changed abruptly in 2003; ID changed abruptly in 1982 and FD changed abruptly

in 1988; SU had an abrupt change point in 1988 and TR had an abrupt change point in 1985; the occurrences of abrupt changes of TXn and TNx were in 2001 and 1998, respectively. The main cold indices changed abruptly in the 1980s and the main warm indices changed abruptly in the late 1990s and early 2000s.

(3) The extreme temperature indices are affected by the atmospheric circulation and urban heat island effect in the Yangtze River Basin. Relative indices and absolute indices will continue to maintain the present trend in the future, which has a certain guiding significance for agricultural and social economic development.

In conclusion, the main cold indices of extreme temperature indices showed a decreasing trend, the main warm indices of extreme temperature indices showed an increasing trend, in the Yangtze River Basin, and cold indices and warm indices will continue to maintain the present trend in the future. Therefore, extreme temperature prediction and monitoring must be strengthened to reduce losses caused by extreme temperature disasters, and to promote the sustainable development in Yangtze River Basin.

Author Contributions: Conceptualization, G.S.; data curation, G.S.; formal analysis, G.S.; funding acquisition, P.Y.; investigation, G.S. and P.Y.; methodology, G.S.; project administration, P.Y.; validation, P.Y. and G.S.; visualization, G.S.; writing—original draft, G.S.; writing—review and editing, P.Y. and G.S. All authors have read and agreed to the published version of the manuscript.

Funding: This research was funded by the Postgraduate Research and Practice Innovation Program of the Jiangsu Province (grant no. KYCX18_1207).

Institutional Review Board Statement: Not applicable.

Informed Consent Statement: Not applicable.

Data Availability Statement: The data presented in this study are available on request from the corresponding author.

Acknowledgments: The authors thank Mingjun Ding of Jiangxi Normal University for his critical reviews and constructive comments.

Conflicts of Interest: The authors declare no conflict of interest.

References

1. IPCC. Working Group I Contribution to the IPCC Fifth Assessment Report, Climate Change 2013: The Physical Science Basis: Summary for Policymakers. Available online: http://www.cmcc.it/wp-content/uploads/2012/12/lista-autori-wgi-ar5.pdf (accessed on 25 July 2021).
2. IPCC. Managing the Risks of Extreme Events and Disasters to Advance Climate Change Adaptation: Special Report of the Intergovernmental Panel on Climate Change. Available online: https://www.ipcc.ch/site/assets/uploads/2018/03/SREX_Full_Report-1.pdf (accessed on 10 August 2021).
3. Alexander, L.V.; Zhang, X.; Peterson, T.C.; Caesar, J.; Gleason, B.; Klein Tank, A.M.G.; Haylock, M.; Collins, D.; Trewin, B.; Rahimzadeh, F.; et al. Global observed changes in daily climate extremes of temperature and precipitation. *J. Geophys. Res. Atmos.* **2006**, *111*, 1–22. [CrossRef]
4. Collins, D.A.; Della-Marta, P.M.; Plummer, N.; Trewin, B.C. Trends in annual frequencies of extreme temperature events in Australia. *Aust. Meteorol. Mag.* **2000**, *49*, 277–292.
5. Bonsal, B.R.; Zhang, X.; Vincent, L.A.; Hogg, W.D. Characteristics of daily and extreme temperatures over Canada. *J. Clim.* **2001**, *14*, 1959–1976. [CrossRef]
6. Rusticucci, M.; Renom, M. Variability and trends in indices of quality-controlled daily temperature extremes in Uruguay. *Int. J. Climatol.* **2010**, *28*, 1083–1095. [CrossRef]
7. Tank, A.K.; Können, G.P. Trends in indices of daily temperature and precipitation extremes in Europe, 1946–1999. *J. Clim.* **2003**, *16*, 3665–3680. [CrossRef]
8. Filahi, S.; Tanarhte, M.; Mouhir, L.; El Morhit, M.; Tramblay, Y. Trends in indices of daily temperature and precipitations extremes in Morocco. *Theor. Appl. Climatol.* **2016**, *124*, 959–972. [CrossRef]
9. Tank, A.M.; Peterson, T.C.; Quadir, D.A.; Dorji, S.; Zou, X.; Tang, H.; Santhosh, K.; Joshi, U.R.; Jaswal, A.K.; Kolli, R.K.; et al. Changes in daily temperature and precipitation extremes in central and south Asia. *J. Geophys. Res. Atmos.* **2006**, *111*, 709–720.
10. Guo, J.; Ren, G.Y.; Ren, Y. Changes of mean and extreme temperatures in Tianjin in recent 100 years. *Plateau Meteorol.* **2011**, *30*, 1399–1405.
11. Li, S.; Yang, S. Changes of extreme temperature events in Beijing during 1960–2014. *Chin. Geogr. Sci.* **2015**, *35*, 1640–1647.

12. Fischer, T.; Gemmer, M.; Lüliu, L.; Buda, S. Temperature and precipitation trends and dryness/wetness pattern in the Zhujiang River Basin, South China, 1961–2007. *Quat. Int.* **2011**, *244*, 138–148. [CrossRef]
13. Wang, B.; Zhang, M.; Wei, J.; Wang, S.; Li, S.; Ma, Q.; Li, X.; Pan, S. Changes in extreme events of temperature and precipitation over Xinjiang, northwest China, during 1960–2009. *Quat. Int.* **2013**, *298*, 141–151. [CrossRef]
14. Liang, K.; Bai, P.; Li, J.; Liu, C. Variability of temperature extremes in the Yellow River basin during 1961–2011. *Quat. Int.* **2014**, *336*, 52–64. [CrossRef]
15. Wang, Q.X.; Wang, M.B.; Fan, X.H.; Zhang, F.; Zhu, S.Z.; Zhao, T.L. Trends of temperature and precipitation extremes in the Loess Plateau Region of China, 1961–2010. *Theor. Appl. Climatol.* **2017**, *129*, 949–963. [CrossRef]
16. Ling, H.; Chen, A.F.; Zhu, Y.H.; Wang, H.L.; He, B. Trends of temperature extremes in summer and winter during 1971–2013 in China. *Atmos. Ocean. Sci. Lett.* **2015**, *8*, 220–225. [CrossRef]
17. Fang, S.; Qi, Y.; Han, G.; Li, Q.; Zhou, G. Changing trends and abrupt features of extreme temperature in mainland China from 1960 to 2010. *Atmosphere* **2016**, *7*, 22. [CrossRef]
18. Chen, X. The formation, evolvement and reorganization of spatial structure in Yangtze River Economic Zone. *Acta Geogr. Sin.* **2007**, *62*, 1266–1273.
19. Deng, X.; Huang, J.; Rozelle, S.; Uchida, E. Growth, population and industrialization, and urban land expansion of China. *J. Urban Econ.* **2008**, *63*, 96–115. [CrossRef]
20. Liu, Z.; He, C.; Zhang, Q.; Huang, Q.; Yang, Y. Extracting the dynamics of urban expansion in China using DMSP-OLS nighttime light data from 1992 to 2008. *Landsc. Urban Plan.* **2012**, *106*, 62–72. [CrossRef]
21. Sang, Y.F.; Wang, Z.; Liu, C. Spatial and temporal variability of daily temperature during 1961–2010 in the Yangtze River Basin, China. *Quat. Int.* **2013**, *304*, 33–42. [CrossRef]
22. Su, B.D.; Jiang, T.; Jin, W.B. Recent trends in observed temperature and precipitation extremes in the Yangtze River basin, China. *Theor. Appl. Climatol.* **2006**, *83*, 139–151. [CrossRef]
23. Sang, Y.F. Spatial and temporal variability of daily temperature in the Yangtze River Delta, China. *Atmos. Res.* **2012**, *112*, 12–24. [CrossRef]
24. Cui, L.; Wang, L.; Qu, S.; Singh, R.P.; Lai, Z.; Yao, R. Spatiotemporal extremes of temperature and precipitation during 1960–2015 in the Yangtze River Basin (China) and impacts on vegetation dynamics. *Theor. Appl. Climatol.* **2019**, *136*, 675–692. [CrossRef]
25. Wang, Q.; Zhang, M.; Wang, S.; Ma, Q.; Sun, M. Changes in temperature extremes in the Yangtze River Basin, 1962–2011. *J. Geogr. Sci.* **2014**, *24*, 59–75. [CrossRef]
26. Guan, Y.; Zhang, X.; Zheng, F.; Wang, B. Trends and variability of daily temperature extremes during 1960–2012 in the Yangtze River Basin, China. *Glob. Planet. Chang.* **2015**, *124*, 79–94. [CrossRef]
27. Niu, Z.; Wang, L.; Fang, L.; Li, J.; Yao, R. Analysis of spatiotemporal variability in temperature extremes in the Yellow and Yangtze River basins during 1961–2014 based on high-density gauge observations. *Int. J. Climatol.* **2020**, *40*, 1–21. [CrossRef]
28. Xia, F.; Liu, X.; Xu, J.; Wang, Z.; Huang, J.; Brookes, P.C. Trends in the daily and extreme temperatures in the Qiantang River basin, China. *Int. J. Climatol.* **2015**, *35*, 57–68. [CrossRef]
29. Chen, X.; Zhou, T. Relative contributions of external SST forcing and internal atmospheric variability to July–August heat waves over the Yangtze River valley. *Clim. Dynam.* **2018**, *51*, 4403–4419. [CrossRef]
30. Chen, R.; Wen, Z.; Lu, R.; Wang, C. Causes of the extreme hot midsummer in central and South China during 2017: Role of the western tropical Pacific warming. *Adv. Atmos. Sci.* **2019**, *36*, 465–478. [CrossRef]
31. Kovats, R.; Hajat, S. Heat Stress and Public Health: A Critical Review. *Annu. Rev. Public Health* **2008**, *29*, 41–55. [CrossRef] [PubMed]
32. Amengual, A.; Homar, V.; Romero, R.; Brooks, H.E.; Ramis, C.; Gordaliza, M.; Alonso, S. Projections of heat waves with high impact on human health in Europe. *Glob. Planet. Chang.* **2014**, *119*, 71–84. [CrossRef]
33. Sherbakov, T.; Malig, B.; Guirguis, K.; Gershunov, A.; Basu, R. Ambient temperature and added heat wave effects on hospitalizations in California from 1999 to 2009. *Environ. Res.* **2018**, *160*, 83–90. [CrossRef]
34. Martiello, M.A.; Giacchi, M.V. High temperatures and health outcomes: A review of the literature. *Scand. J. Public Health* **2010**, *38*, 826–837. [CrossRef]
35. Liu, G.; Zhang, L.; He, B.; Jin, X.; Zhang, Q.; Razafindrabe, B.; You, H. Temporal changes in extreme high temperature, heat waves and relevant disasters in Nanjing metropolitan region, China. *Nat. Hazards* **2015**, *76*, 1415–1430. [CrossRef]
36. Xu, X.; Ge, Q.; He, S.; Zhang, X.; Xu, X.; Liu, G. Impact of high temperature on the mortality in summer of Wuhan, China. *Environ. Earth Sci.* **2016**, *75*, 543.
37. Sun, X.; Sun, Q.; Yang, M.; Zhou, X.; Li, X.; Yu, A.; Geng, F.; Guo, Y. Effects of temperature and heat waves on emergency department visits and emergency ambulance dispatches in Pudong New Area, China: A time series analysis. *Environ. Health* **2014**, *13*, 76. [CrossRef] [PubMed]
38. Wang, H.; Tu, S.; Chen, Z. Interpolating Method for Missing Data of Daily Air Temperature and Its Error Analysis. *Meteorol. Mon.* **2008**, *34*, 83–91.
39. Yang, G.; Xu, X.; Li, P. Research on the construction of green ecological corridors in the Yangtze River Economic Belt. *Prog. Geogr.* **2015**, *34*, 1356–1367.
40. Moberg, A.; Jones, P.D. Trends in indices for extremes in daily temperature and precipitation in central and western Europe, 1901–1999. *Int. J. Climatol.* **2005**, *25*, 1149–1171. [CrossRef]

41. Toreti, A.; Desiato, F. Changes in temperature extremes over Italy in the last 44 years. *Int. J. Climatol.* **2008**, *28*, 733–745. [CrossRef]
42. Mann, H.B. Nonparametric tests against trend. *Econometrica* **1945**, *13*, 245–259. [CrossRef]
43. Wei, F. *Modern Diagnosis and Prediction of Climate Statistics*, 2nd ed.; Chinse Meteorological Press: Beijing, China, 2007.
44. Hurst, H.E. Long-term storage capacity of reservoirs. *Trans. Am. Soc. Civ. Eng.* **1951**, *116*, 770–799. [CrossRef]
45. Wang, G.; Pang, Z.; Boisvert, J.B.; Hao, Y.; Cao, Y.; Qu, J. Quantitative assessment of mineral resources by combining geostatistics and fractal methods in the Tongshan porphyry Cu deposit (China). *J. Geochem. Explor.* **2013**, *134*, 85–98. [CrossRef]
46. Li, S.; Bates, G.T. Influence of the Atlantic multidecadal oscillation on the winter climate of East China. *Adv. Atmos. Sci.* **2007**, *24*, 126–135. [CrossRef]
47. Wang, Y.; Li, S.; Luo, D. Seasonal response of Asian monsoonal climate to the Atlantic Multidecadal Oscillation. *J. Geophys. Res. Atmos.* **2009**, *114*, 1–15. [CrossRef]
48. Deng, H.; Zhao, F.; Zhao, X. Changes of extreme temperature events in Three Gorges area, China. *Environ. Earth Sci.* **2012**, *66*, 1783–1790. [CrossRef]
49. Chen, T.; Ao, T.; Zhang, X.; Li, X.; Yang, K. Climate change characteristics of extreme temperature in the Minjiang river basin. *Adv. Meteorol.* **2019**, *2019*, 1935719. [CrossRef]
50. Chan, J.C.L.; Gray, W.M. Tropical cyclone movement and surrounding flow relationships. *Mon. Weather Rev.* **1982**, *110*, 1354–1374. [CrossRef]
51. Zhong, S.; Qian, Y.; Zhao, C.; Leung, R.; Wang, H.; Yang, B.; Fan, J.; Yan, H.; Yang, X.; Liu, D. Urbanization-induced urban heat island and aerosol effects on climate extremes in the Yangtze River Delta region of China. *Atmos. Chem. Phys.* **2017**, *17*, 5439–5457. [CrossRef]
52. Yao, R.; Wang, L.; Huang, X.; Zhang, W.; Li, J.; Niu, Z. Interannual variations in surface urban heat island intensity and associated drivers in China. *J. Environ. Manag.* **2018**, *222*, 86–94. [CrossRef]
53. Yao, R.; Wang, L.; Gui, X.; Zheng, Y.; Zhang, H.; Huang, X. Urbanization effects on vegetation and surface urban heat islands in China's Yangtze River Basin. *Remote Sens.* **2017**, *9*, 540. [CrossRef]
54. Chen, W.; Huang, C.; Wang, L.; Li, D. Climate Extremes and Their Impacts on Interannual Vegetation Variabilities: A Case Study in Hubei Province of Central China. *Remote Sens.* **2018**, *10*, 477. [CrossRef]
55. Liu, X.; Chen, B. Climatic warming in the Tibetan Plateau during recent decades. *Int. J. Climatol.* **2000**, *20*, 1729–1742. [CrossRef]
56. IPCC. Contribution of Working Group I to the Fourth Assessment Report of the Intergovernmental Panel on Climate Change, Climate Change 2007: The Physical Science Basis. Available online: http://amper.ped.muni.cz/eave/old/SPM2feb07.pdf (accessed on 11 August 2021).
57. Huang, C.; Barnett, A.G.; Xu, Z.; Chu, C.; Wang, X.; Turner, L.R.; Tong, S. Managing the health effects of temperature in response to climate change: Challenges ahead. *Environ. Health Perspect.* **2013**, *121*, 415–419. [CrossRef] [PubMed]
58. Krishnan, P.; Ramakrishnan, B.; Reddy, K.R.; Reddy, V.R. High-temperature effects on rice growth, yield, and grain quality. *Adv. Agron.* **2011**, *111*, 87–206.
59. Xie, Z.Q.; Du, Y.; Gao, P.; Zeng, Y. Impact of high-temperature on single cropping rice over Yangtze-Huaihe River Valley and response measures. *Meteorol. Mon.* **2013**, *39*, 774–781.
60. Rapacz, M. Physiological effects of winter rape (*Brassica napus* var. oleifera) prehardening to frost. II. Growth, energy partitioning and water status during cold acclimation. *J. Agron. Crop Sci.* **1998**, *181*, 81–87. [CrossRef]
61. Allen, H.M.; Pumpa, J.K.; Batten, G.D. Effect of frost on the quality of samples of Janz wheat. *Aust. J. Exp. Agric.* **2001**, *41*, 641–647. [CrossRef]
62. Li, Y.; Wang, C.; Zhao, B.; Liu, W. Effects of climate change on agricultural meteorological disaster and crop insects diseases. *Trans. Chin. Soc. Agric. Eng.* **2010**, *26*, 263–271.
63. Jeffs, C.T.; Lewis, O.T. Effects of climate warming on host–parasitoid interactions. *Ecol. Entomol.* **2013**, *38*, 209–218. [CrossRef]

International Journal of
Environmental Research and Public Health

MDPI

Article

Comprehensive Risk Assessment of High Temperature Disaster to Kiwifruit in Shaanxi Province, China

Yining Ma [1,2,3], Suri Guga [1,2,3], Jie Xu [1,2,3], Jiquan Zhang [1,2,3], Zhijun Tong [1,2,3] and Xingpeng Liu [1,2,3,*]

[1] School of Environment, Northeast Normal University, Changchun 130024, China;
     mayn818@nenu.edu.cn (Y.M.); surgg146@nenu.edu.cn (S.G.); xuj463@nenu.edu.cn (J.X.);
     zhangjq022@nenu.edu.cn (J.Z.); gis@nenu.edu.cn (Z.T.)
[2] State Environmental Protection Key Laboratory of Wetland Ecology and Vegetation Restoration,
     Northeast Normal University, Changchun 130024, China
[3] Key Laboratory for Vegetation Ecology, Ministry of Education, Changchun 130024, China
*   Correspondence: liuxp912@nenu.edu.cn

Abstract: In recent years, the main kiwifruit producing region, central-south Shaanxi Province, has often suffered from the threat of extreme high temperatures. Assessing the risk of high-temperature disasters in the region is essential for the rational planning of agricultural production and the development of resilience measures. In this study, a database was established to assess the risk of a high-temperature disaster to kiwifruit. Then, four aspects, hazard, vulnerability, exposure and disaster prevention and mitigation capacity, were taken into account and 19 indexes were selected to make an assessment of the risk of a high-temperature disaster. At the same time, 16 indexes were selected for the assessment of the climatic suitability of kiwifruit in terms of light, heat, water, soil and topography, and were used as one of the indexes for exposure assessment. The analytic hierarchy process and the entropy weighting method were combined to solve the weights for each index. The results reveal that: (1) The Guanzhong Plain has a high climatic suitability for kiwifruit, accounting for 15.14% of the study area. (2) The central part of the study area and southern Shaanxi are at high risk, accounting for 22.7% of the study area. The major kiwifruit producing areas in Shaanxi Province (e.g., Baoji) are at a low risk level, which is conducive to the development of the kiwifruit industry. Our study is the first to provide a comprehensive assessment of the risk of a high-temperature disaster to the economic fruit kiwifruit, providing a reference for disaster resilience and mitigation.

Keywords: risk assessment; high-temperature disaster; kiwifruit; climatic suitability zoning; hazard; vulnerability; exposure; disaster prevention and mitigation capacity

Citation: Ma, Y.; Guga, S.; Xu, J.; Zhang, J.; Tong, Z.; Liu, X. Comprehensive Risk Assessment of High Temperature Disaster to Kiwifruit in Shaanxi Province, China. *Int. J. Environ. Res. Public Health* **2021**, *18*, 10437. https://doi.org/10.3390/ijerph181910437

Academic Editors: Daisuke Sasaki and Mikio Ishiwatari

Received: 23 August 2021
Accepted: 30 September 2021
Published: 4 October 2021

Publisher's Note: MDPI stays neutral with regard to jurisdictional claims in published maps and institutional affiliations.

1. Introduction

The first part of the Sixth Assessment Report (AR6) of the United Nations Intergovernmental Panel on Climate Change (IPCC) was released on 9 August 2021 [1]. The report states that the global surface temperature has increased by about 1.1 °C compared to 1850–1900, a level of warming not seen since 125,000 years ago. As a large agricultural country, China is experiencing extreme heat and weather caused by global warming [2], which has serious impacts on agro-ecosystems and national economic security, with losses increasing year by year [3–5]. The fruit tree industry, as an important part of agriculture, is often more vulnerable to extreme hot weather [6,7]. Therefore, reducing the impact of a high-temperature disaster on fruit tree production is of great importance in developing the agricultural economy, ensuring the supply of fruit and generating income and foreign exchange.

Shaanxi Province is the main production area of kiwifruit in China, with the highest planting area and yield year-round [8,9]. In 2019, the planting area of kiwifruit in Shaanxi Province was 87.67 km^2, the production input was 1980.9 yuan/mu, the economic income was 0.49 yuan/km^2, and the output was 1,072,400 tons, accounting for 6.05% of the province's fruit yield [10]. Frequent extreme high-temperature events have had a negative impact on the kiwifruit industry in Shaanxi Province [11]. June to August is the prime period for kiwifruit fruit growth. If suffering from high temperature at this time, it will accelerate transpiration of plants and evaporation of orchard soil water. When the fruit is suffering from the effects of high temperatures, the skin is sunken and easily becomes soft and rotten. Sunburned fruit are very susceptible to falling off and in severe cases this will lead to a significant reduction in yield. To reduce the adverse effects of high-temperature disasters on kiwifruit, we should actively think about the following issues: How do we reduce disaster risk? How can measures be taken to avoid disasters before they happen? How can we respond positively to disasters and minimize losses? With the continuous development of disaster science, risk assessment and risk management have become important research directions. An objective and reasonable risk assessment can help policy makers to prevent disasters and reduce disaster losses. It also plays an important role in agro-meteorological disaster insurance and crop production planning.

The risk assessment of agro-meteorological disasters in China started relatively late, and research on the risk assessment of economic fruit is lacking. On the one hand, compared with other major agro-meteorological disasters, such as drought [12,13], chilling injury [14,15] and waterlogging [16,17], high-temperature disasters have been relatively little studied. On the other hand, the current research is mainly concerned with field crops such as maize [18,19], wheat [20,21], rice [22,23], etc. More importantly, research methods are mostly based on vulnerability [24–26], the lack of risk assessments that integrate hazards, vulnerability, exposure and disaster prevention and mitigation capacity. Luo et al. consider four aspects in terms of hazards, vulnerability, exposure and disaster prevention and mitigation capacity. They proposed a grey cloud clustering model based on panel data to assess the agricultural drought disaster risk of Henan Province [27]. Liu et al. analyzed drought risk under different future scenarios and constructed a socio-economic risk model based on hazards, vulnerability and exposure. The study found that climate change will increase the risk of future droughts, with negative socio-economic impacts on countries [28]. In addition, an increase in the frequency and intensity of droughts can have a negative impact on global food security [29]. We can draw on the experience of previous studies to develop an indicator system and model for assessing the risk of high-temperature disasters in relation to fruit. The results of the study provide insight into the frequency, intensity and spatial and temporal patterns of high-temperature disasters. The aim is to effectively strengthen the preventive management of risks and to actively improve emergency mitigation measures.

As AR6 points out: "The future of the planet depends, in large part, on the choices that humanity makes today. Many of the most dire effects of climate change can still be avoided if aggressive action is taken now [1]". This is perhaps what risk assessment is all about: rather than actively remedying a disaster after it has occurred, it is better to effectively prevent it before it arrives. Avoiding and reducing the occurrence of disaster as much as possible is the way forward for disaster risk management, not just limited to agriculture. After all, "The future is in our hands [1]".

Our aims included the following: (1) We want to conduct an in-depth analysis of historical meteorological data in Shaanxi Province, based on the "Four Factors" theory. As far as possible, a more comprehensive range of factors was taken into account, leading to the selection of indexes for assessing the risk of high-temperature heat disasters. (2) We wanted to use climate suitability as one of the exposure indexes. A comprehensive consideration of light, heat, water, soil resources and topography has led to the construction of a more complete climate suitability zoning index system and the study of kiwifruit climate suitability zoning in the study area. (3) Building a risk assessment model for high-temperature disasters to kiwifruit. The weight of each index was determined using the combination weighting method. Conducting a high-temperature disaster risk assessment and mapping. The assessment results were also validated using historical disaster data.

(4) The results of the study can provide a scientific basis for disaster prevention and mitigation of kiwifruit and for achieving stable yields and increased income.

2. Study Area and Data Sources

Shaanxi Province is located in the northwest of China (31°42′–39°35′ N, 105°29′–111°15′ E) (Figure 1). Within the territory of rolling hills and rivers, bounded by the Beishan Mountains and the Qinling Mountains, the province is divided into three major landform areas: the northern Shanbei Plateau, the Guanzhong Plain, and the Qinba Mountains. The average annual temperature is 13.0 °C, the average precipitation is about 576.9 mm and the frost-free period is about 218 days. Shaanxi Province straddles the northern temperate and subtropical zones and has an overall continental monsoon climate. The heat and water resources from south to north gradually reduce, and due to its unique climate and geographical conditions, kiwifruit is grown in most areas except northern Shaanxi.

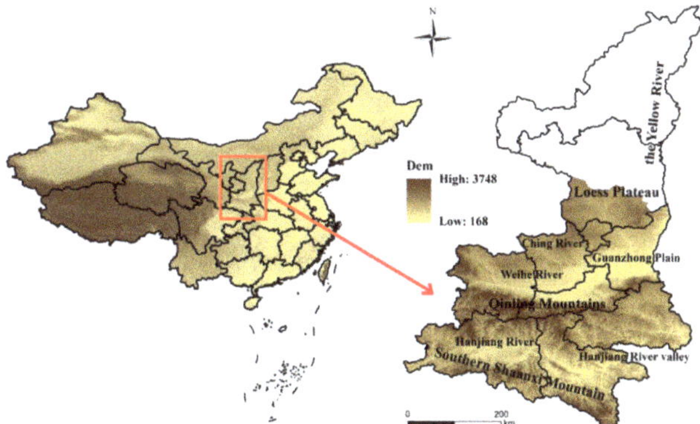

Figure 1. Location of the study area.

Meteorological data that included daily observations of the maximum temperature, minimum temperature, average temperature, precipitation, relative humidity and gale days from 37 meteorological stations in Shaanxi Province were collected from the National Meteorological Information Center (http://data.cma.cn/ (accessed on 25 July 2021)) for the period from 1960 to 2020. Data that were abnormal or missing longer time series were removed to ensure the integrity of the data for that time period. Historical disaster data were obtained from the statistical yearbooks of Shaanxi and the China Meteorological Disaster Dictionary—Shaanxi Volume. Data on agricultural production conditions, socio-economics and kiwifruit planting situation by county in Shaanxi Province are from the 1990–2019 Statistical Yearbook of Shaanxi. Soil erosion data are from the Geographical Information Monitoring Cloud Platform (http://www.dsac.cn/ (accessed on 25 July 2021)). Soil data are from the China Soil Database (http://vdb3.soil.csdb.cn/ (accessed on 18 July 2021)).

3. Research Methods

3.1. Framework for High-Temperature Disaster Risk Assessment

Figure 2 show the main research steps in the assessment of high-temperature disaster risk to kiwifruit in Shaanxi Province, including the following: (1) A comprehensive database for kiwifruit high-temperature disaster risk assessment was established by collecting relevant data such as meteorological data, historical disaster data and socio-economic data, etc. (2) Based on disaster risk assessment, the "Four Factors" theory, including selection of hazards, vulnerability, exposure and disaster prevention and mitigation capacity, was selected as the kiwifruit high-temperature disaster risk assessment method. The final 19 risk assessment indexes were selected, taking into account the environment of the study area and kiwifruit's growth and development needs. (3) An analysis of the spatial and temporal distribution characteristics of hazards, vulnerability, exposure and disaster prevention and mitigation capacity was conducted. Then, relevant assessment indexes' impact on the assessment of hazards, vulnerability, exposure and disaster prevention and mitigation capacity were analyzed. (4) Based on the above results, we conducted a comprehensive risk assessment and mapped high-temperature disaster to kiwifruit in Shaanxi Province and provide a scientific basis for decision making in response to high-temperature disaster.

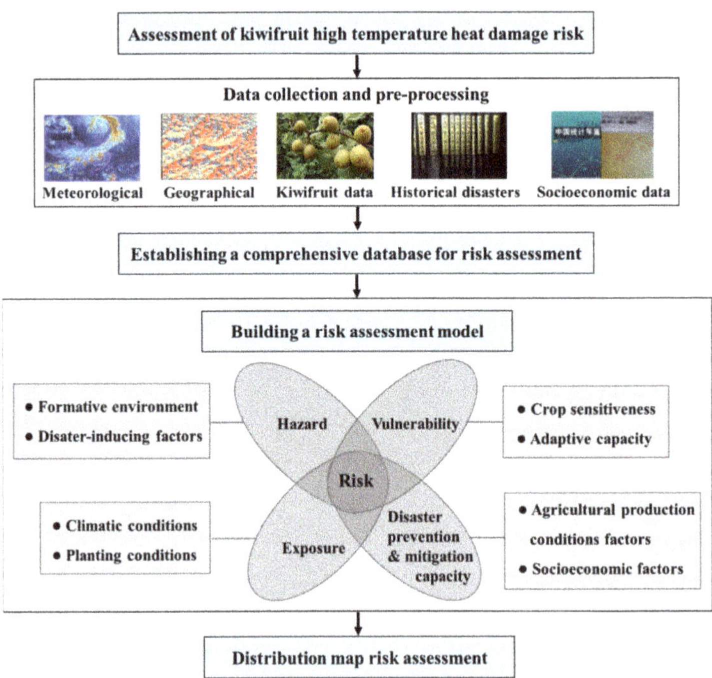

Figure 2. Kiwifruit high-temperature disaster risk assessment process for Shaanxi Province.

3.2. Selection Risk Assessment Indexes

3.2.1. Selection of Hazard Indexes

The selection of hazard indexes is based on both the disaster-inducing factors and the formative environment.

(1) The disaster-inducing factors use the maximum daily temperature and duration from June to August to classify these into three levels, as shown in Table 1.

Table 1. Classification of high-temperature disaster.

Period	Index	Disaster Level	Threshold
June–August	Daily maximum temperature (T_C/°C)	Light	$35 \leq T_C < 38$ (3–4 day)
		Moderate	$35 \leq T_C < 38$ (5–8 day)
		Severe	$35 \leq T_C < 38$ ($\geq$9 day) or $38 \leq T_C$ ($\geq$2 day)

(2) Soil erosion (Figure 3a) and gale days (Figure 3b) were selected for the formative environment hazard index. Soil erosion, which reduces the amount of water available, increases the loss of nutrients from the soil and reduces the organic matter content of the soil, is one of the most serious threats to the world's food production. Kiwifruit shoots are long and brittle, with large, thin leaves that are highly susceptible to high winds, causing

branches to dry out and break. The severity of soil erosion and the duration of gale days are directly proportional to the hazard of the formative environment.

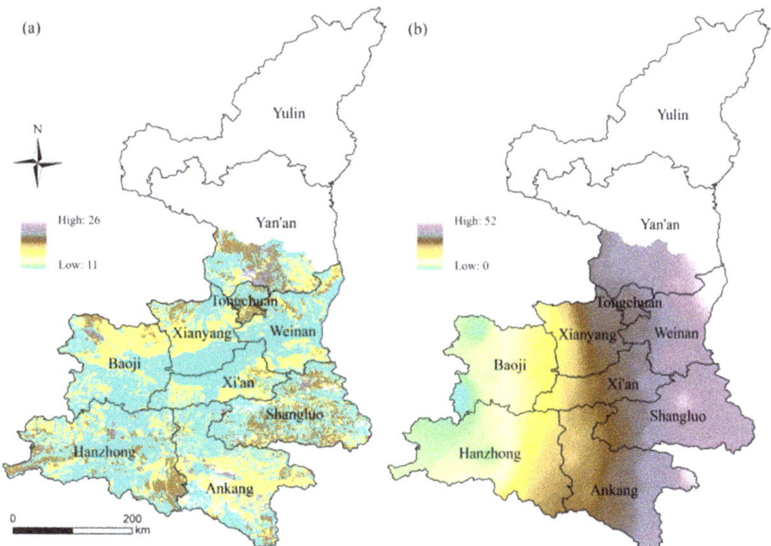

Figure 3. Spatial distribution of soil erosion (**a**) and gale days (day) (**b**).

3.2.2. Selection of Vulnerability Indexes

Vulnerability characterizes the degree of loss that may be caused by potential risks, based on both sensitivity and adaptability.

(1) Yield reduction rate (r) and yield reduction coefficient of variation (v)

Yield generally includes the trend yield, climate yield and random yield. The trend yield is determined by the level of social technology and the climate yield is influenced by climate factors. At the same time, variations in crop yields caused by changes in other factors are considered as random yield [30], calculated as follows:

$$Y = Y_t + Y_c + Y_e \qquad (1)$$

where Y is the actual unit yield (kg/hm^2), Y_t is the trend yield (kg/hm^2), Y_c is the climate yield (kg/hm^2), and Y_e is the random yield (kg/hm^2), which is generally negligible. In this study, trend yields were calculated using the 3a sliding average method. Then, we introduced the concept of relative meteorological yield (Y_w). This is a comparable relative value that is not influenced by differences in the level of agricultural technology in different historical periods. It can reflect more effectively the fluctuations in the actual yield affected by meteorological disaster [31].

$$Y_w = \frac{(Y - Y_t)}{Y_t} \qquad (2)$$

A year with a negative relative meteorological yield is defined as a yield reduction year, and the meteorological yield reduction rate is calculated as follows:

$$r = \frac{\sum x_i}{n} \qquad (3)$$

where $\sum x_i$ is the sum of the negative relative meteorological yield and n is the total number of samples. r is used to describe the location of the concentration of negative values in the relative meteorological yield, i.e., the concentration of the years of yield reduction, which characterizes the average level of yield reduction subject to natural risk for that subject. The higher the rate of meteorological yield reduction, the higher the degree of damage caused by the disaster, and vice versa, as shown below:

$$v = \sqrt{\frac{\sum(X_i - r)^2}{(n-1)}} / r \tag{4}$$

where v is the meteorological yield reduction coefficient of variation and X_i is the annual relative meteorological yield from year to year.

(2) Probability occurrence of yield reduction rate (p)

The probability occurrence of the yield reduction rate (p) is the cumulative probability that the relative meteorological yield will be less than a certain threshold value. An analysis of the variation of kiwifruit's actual unit yield by county in Shaanxi Province over the years shows that meteorological disasters often affect kiwifruit when relative meteorological yields reach 5%, resulting in large losses. In contrast, relative meteorological yields of <−10% are rare. Relative meteorological yields <5%, with a yield reduction rate >5% chance as a vulnerability assessment index, can reflect, to some extent, the strengths and weaknesses of the climatic conditions and the degree of occurrence of meteorological hazards in kiwifruit growing areas. In this study, SPSS was used to test the normality of the relative meteorological yield series for each county, with the majority of counties conforming to a normal distribution and samples that did not conform to a normal distribution being normalized. Therefore, using the sample mean (u) and the sample mean square error (σ) to establish a distribution function, it was calculated as follows:

$$F(x) = \int_{-\infty}^{x} \frac{1}{\sqrt{2\pi}\sigma} e^{\frac{1}{2\sigma}(x-u)^2} dx \tag{5}$$

where x is the relative meteorological yield. When x is less than the critical value of −5%, p is calculated as follows:

$$p(x < x_0) = \Phi\left(\frac{x_0 - u}{\sigma}\right) \tag{6}$$

(3) Meteorological sensitiveness index

The meteorological sensitiveness index is calculated from the climatic yield and climatic productivity. The calculation formula is as follows:

$$Km = Yw/Yv \tag{7}$$

where Km is the meteorological sensitiveness index, Yw is the actual productivity (kg·hm^2) of the year, and Yv is the climatic productivity (kg·hm^2); the Thornthwaite Memorial model [32,33] is used to calculate the climatic productivity of crops.

$$Yv = 30000\left(1 - e^{-0.000956(V-20)}\right) \tag{8}$$

$$V = \frac{1.05R}{\sqrt{1 + \left(\frac{1.05R}{L}\right)^2}} \tag{9}$$

$$L = 300 + 25t + 0.05t^3 \tag{10}$$

where 30,000 is the empirical coefficient, e = 2.718, V is the annual average evaporation (mm), R is the annual precipitation (mm), L is the annual average maximum evaporation (mm), and t is the annual average air temperature.

(4)   Selection of adaptive capacity indexes

Four indexes of adaptability were chosen (Figure 4): soil total nitrogen, phosphorus, potassium content and soil organic matter. The larger the index is, the stronger the soil adaptability is, and when the disaster occurs, it has stronger resistance and adaptability.

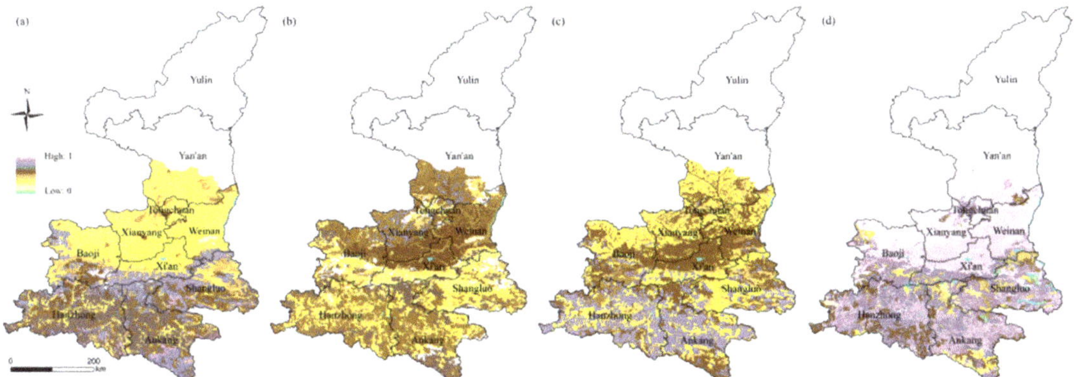

Figure 4. Spatial distribution of adaptive capacity indexes: (**a**) soil total nitrogen (g/kg), (**b**) soil total phosphorus, (**c**) soil total potassium content (g/kg) and (**d**) soil organic matter (g/kg).

3.2.3. Selection of Exposure Indexes

(1)   Planted area as a proportion of the province's planted area

$$X_A = \frac{Ar}{Aa} \tag{11}$$

where X_A is the proportion of kiwifruit planted area to kiwifruit planted area in Shaanxi Province. Ar is the area planted with kiwifruit in a county and Aa is the area planted with kiwifruit in the study area.

(2)   Climate suitability

Agro-climatic zoning is a regional spatial classification that clarifies the relationship between climate and agricultural production according to the specific climatic requirements of agriculture and is an important basis for making full use of climatic resources and optimizing the structure and layout of agricultural cultivation [34,35]. Climatic suitability was chosen as one of the indicators of exposure; the higher the suitability, the higher the exposure and the greater the risk will be. Shaanxi Province is the number one kiwifruit producing region in China. The main problem facing the use of climatic resources in the region is the lack of indicators for the zoning of kiwifruit cultivation. To solve this problem, we use the ANUSPLIN software [36–38]. For the refined interpolation of each climate zoning indexes, 1 km × 1 km DEM raster data were used as covariates. Through the analysis of the demand for kiwifruit growth and development in the study area [39–43], we selected five major categories of light, heat, water, soil and topography, with 16 climate suitability assessment indexes (Table 2).

Table 2. Standard of climatic suitability regionalization index.

	Suitable Ranks	Highly	Moderately	Hardly	Not
Heat Resource	Average annual temperature (°C)	[14, 16]	[13, 14)∪(16, 18]	[10, 13)∪(18, 20]	(−∞, 10)∪(20, +∞)
	Average temperature in January (°C)	[6, 8]	[4.5, 6)∪(8, 9]	[3.4, 5)∪(9, 10]	(−∞, 3)∪(10, +∞)
	Average temperature in March (°C)	[12, 14]	[11, 12)∪(14, 15]	[10, 11)∪(15, 16]	(−∞, 10)∪(16, +∞)
	Average temperature in July (°C)	[22, 24]	[20, 22)∪(24, 26]	[17, 20)∪(26, 28]	(−∞, 17)∪(28, +∞)
	≥10 °C accumulated temperature (°C)	[4500, 5200]	[4000, 4500)∪(5200, 5600]	[3500, 4000)∪(5600, 6000]	(−∞, 3500)∪(6000, +∞)
	Extreme minimum temperature (°C) (80% guaranteed)	[−3, +∞)	[−4, −3]	[−5, −4]	(−∞, −5)
	Average diurnal temperature amplitude in August (°C)	[8, 12]	[6, 8)	[5, 6)	(−∞, 5)
	Frost-free period (d)	[280, +∞)	[240, 280)	[200, 240)	(−∞, 200)
Light Resource	Annual sunshine hours (h)	[1200, 2000]	[900, 1200)∪(2000, 2200]	(−∞, 900)∪(2200, 2500]	(2500, +∞)
Water Resource	Annual average relative humidity (%)	[79, 82]	[75, 79)∪(82, 85]	[72, 75)	(−∞, 72)∪(85, +∞)
	Total annual precipitation (mm)	[1200, 1500]	[1100, 1200)∪(1500, 1600]	[900, 1100)∪(1600, 1700]	(−∞, 900)∪(1700, +∞)
Topography	Elevation (m)	[500, 1200]	[350, 500)∪(1200, 1500]	[200, 350)∪(1500, 2000]	(−∞, 200)∪(2000, +∞)
	Slope (°)	[10, 20]	[5, 10)∪(20, 25]	[3, 5)∪(25, 30]	(−∞, 3)∪(30, +∞)
	Slope direction	South/Southeast/Southwest	East/Northeast	West/Northwest	North
Soil Resource	Soil pH	[5.5, 6.5]	[5.0, 5.5)∪(6.5, 7.0]	[4.5, 5.0)∪(7.0, 7.5]	(−∞, 4.5)∪(7.5, +∞)
	Soil type	Sandy loamy	Light loamy / medium loamy	Heavy loamy	Clay/loamy clay

When zoning for climatic suitability, the not suitable areas are first eliminated according to the indexes to avoid compensatory effects between indexes. Subsequently, the climatic suitability regionalization indexes are scored with corresponding percentages according to the different zoning classifications, with the following formula:

For grid points located in a hardly suitable area:

$$G = g_{min} + (g_{max} - g_{min}) \times \frac{p - n_{min}}{N_{max} - n_{min}} \tag{12}$$

For grid points located in a moderately suitable area:

$$G = g_{min} + (g_{max} - g_{min}) \times \frac{p - N_{min}}{N_{max} - N_{min}} \tag{13}$$

For grid points located in a highly suitable area:

$$G = g_{min} + (g_{max} - g_{min}) \times \frac{p - N_{min}}{n_{max} - N_{min}} \tag{14}$$

where G is the percentage scoring result of the grid points; g_{max} and g_{min} are the maximum and minimum values of the range of scores corresponding to that zoning level, respectively; the range of scores corresponding to the different zoning classifications is shown in Table 2; N_{max} and N_{min} are the maximum and minimum values of the corresponding criteria, respectively; n_{max} and n_{min} are the maximum and minimum values, respectively, in the data set of grid points corresponding to the zoning classification; p is the actual value of the grid point. After scoring, the results were summed according to a certain weighting to obtain an overall score for the climatic zoning of kiwifruit in the study area. The range of scores corresponding to the different zoning classifications is shown in Table 3.

Table 3. The range of ranks corresponding to the different classifications.

Suitable Ranks	Highly	Moderately	Hardly
Threshold	10–15	5–10	0–5

The kiwifruit suitability assessment composite index ranges from 0 to 1. The higher the index, the higher the climatic suitability for kiwifruit planting. Climate suitability is classified into 4 classes according to the optimal partitioning method (Table 4).

Table 4. Classification of suitability ranks.

Suitable Ranks	Highly	Moderately	Hardly	Not
Threshold	0.65–1	0.36–0.65	0.12–0.36	<0.12

3.2.4. Selection of Disaster Prevention and Mitigation Capacity Indexes

Disaster prevention and mitigation capacity indicates the extent to which the study area can recover from a disaster in the long or short term. The higher the value, the less potential damage the study area may suffer and the lower the disaster risk.

3.3. High-Temperature Disaster Assessment Index System

The combination weighting method is used to determine the weight of each index, and the weighted comprehensive average method is used to construct risk assessment model (Table 5).

$$H = \sum_{i=1}^{n} W_{Hi} X_{Hi} \tag{15}$$

where H denotes the hazard, which is the degree of natural variability that causes the disaster. The higher the value, the more severe the loss caused by the disaster and the higher the risk of the disaster. X_{Hi} and W_{Hi} represent the hazard index and the corresponding weight, respectively.

$$S = \sum_{i=1}^{n} W_{Si} X_{Si} \tag{16}$$

$$A = \sum_{i=1}^{n} W_{Ai} X_{Ai} \tag{17}$$

$$V = S \times (1 - A) \tag{18}$$

where S and A denote crop sensitiveness and adaptive capacity, respectively, which are used to characterize vulnerability (V). X_{Si}, X_{Ai}, W_{Si} and W_{Ai} represent the assessment index and the corresponding

$$E = \sum_{i=1}^{n} W_{Ei} X_{Ei} \tag{19}$$

where X_{Ei} and W_{Ei} represent the exposure index and the corresponding weight, respectively.

$$C = \sum_{i=1}^{n} W_{Ci} X_{Ci} \tag{20}$$

where X_{Ci} and W_{Ci} represent the emergency response and recovery capability index and the corresponding weight, respectively.

Table 5. The weight of each risk assessment index.

Factor	Sub-Factor	Index	Weight
Hazard (H)	Disaster-inducing factors (0.832)	Light high-temperature disaster (XH1)	0.315
		Moderate high-temperature disaster (XH2)	0.324
		Severe high-temperature disaster (XH3)	0.361
	Formative Environment (0.168)	Gale days (XH4)	0.090
		Soil erosion (XH5)	0.078
Vulnerability (V)	Crop sensitiveness (S)	Yield reduction rate (XS1)	0.348
		Yield reduction coefficient of variation (XS2)	0.193
		Probability occurrence of yield reduction rate (XS3)	0.276
		Meteorological sensitivity index (XS4)	0.183
	Adaptive capacity (A)	Soil organic matter (XA5)	0.561
		Soil total nitrogen content (XA6)	0.142
		Soil total phosphorus content (XA7)	0.139
		Soil total potassium content (XA8)	0.158
Exposure (E)	Climatic conditions	Climate suitability (XE1)	0.481
	Planting conditions	Planted area as a proportion of the province's planted area (XE2)	0.519
Disaster prevention and mitigation capacity (C)	Agricultural production conditions factors	Total agricultural machinery power (XC1)	0.211
		Fertilizer consumption (XC2)	0.212
	Socioeconomic factors	Per capita disposable income of farmers (XC3)	0.361
		Rural electricity consumption per unit area (XC4)	0.216

3.4. High Temperature Disaster Risk Assessment Model

According to the "Four-Factors" theory of natural disaster risk formation, the four aspects of disaster are hazards, vulnerability, exposure and disaster prevention and mitigation capacity. We have established a kiwifruit high-temperature disaster risk assessment index to characterize the degree of hazard risk. The formula is as follows:

$$R = H^{W_H} \times V^{W_V} \times E^{W_E} \times (1 - C)^{W_C} \tag{21}$$

where R is the high-temperature disaster risk assessment index; H, E, V, C stand for for hazard, vulnerability, exposure and emergency response and recovery capability; W_H, W_V, W_E, W_C are the weight for hazard, vulnerability, exposure and disaster prevention and mitigation capacity, respectively. The combination weighting method was calculated as 0.384, 0.203, 0.221 and 0.192.

3.5. Standardized Treatment of Assessment Indexes

Due to the different dimensions of each assessment index, the assessment index must be standardized in weight calculation to eliminate the influence of different units and different measures among the indexes. In this paper, the range method chosen [44] for positive impact indicators is:

$$R = \frac{X_i - min(X_i)}{max(X_i) - min_j(X_i)} \tag{22}$$

and for negative impact indicators it is:

$$R = \frac{max(X_i) - X_i}{max(X_i) - min(X_i)} \tag{23}$$

where R is the normalized index value, X_i is the assessment indicator, $max(X_i)$ is the maximum value in the sequence, and $min(X_i)$ is the minimum value in the sequence.

3.6. Methodology

3.6.1. Combination Weighting Method

(1)   AHP method to determine subjective weights of indexes

The analytic hierarchy process (AHP) is a subjective weighting method that is suitable for quantitative analysis of qualitative problems under multi-criteria decision making and is commonly used in many fields [45].

(2)   Entropy weight method to determine objective weights of indexes

The entropy method is an objective weighting method [46]. In information theory, entropy measures the amount of valid information provided by the data. If the information entropy of an indicator is lower, the more information the indicator has the more weight it will have in the evaluation [47–49].

(3)   Combination weighting method

The combination weighting method combines the expert theoretical knowledge and rich experience (AHP method) with the full mining of data combination information of the objective weighting method (entropy weight method). To a certain extent, systematic and random errors can be reduced [50]. In order to scientifically assign weights to the combinations, reference is made to the principle of minimum discriminatory information. The objective function is defined as:

$$minJ(\omega) = \sum_{j=1}^{n}(\omega_j \ln \frac{\omega_j}{u_j} + \omega_j \ln \frac{\omega_j}{v_j}) \tag{24}$$

$$s.t. \sum_{j=1}^{n} \omega_j = 1, \ \omega_j \geq 0, \ j = 1, 2, \cdots, n$$

Solving this optimization model yields the combined weights as:

$$\omega_j = \frac{\sqrt{u_j v_j}}{\sum_{j=1}^{n} u_j v_j} \tag{25}$$

where ω_j is the combination weight and u_j and v_j are the subjective and objective weights of the indicator, respectively.

3.6.2. Mann–Kendall Method

The M-K test was originally proposed and developed by H.B. Mann [51] and M.G. Kendall [52], and it is an effective tool recommended by the World Meteorological Organization for extracting trends in series variability. The principle of the method and the calculation steps are detailed in [53].

4. Results and Discussion

4.1. Comprehensive Assessment of High-Temperature Disaster Hazards

4.1.1. Analysis of the Variation Characteristics of High-Temperature Events Frequency

Figure 5 shows the frequency of light (a), moderate (b) and severe (c) high-temperature events in the study area from 1980–2020, respectively. It can be seen that the change in frequency fluctuates, with averages of 0.18, 0.07 and 0.02, respectively. During 1980–1990, there were few high-temperature events, and no severe high-temperature events occurred for many years. Figure 5d shows the trend and M-K test of the disaster-inducing factors hazard index. The change shows a "U-shaped" increasing and decreasing trend, with the minimum value occurring in 1984 (0.0032) and the maximum value in 2017 (0.26). The UF and UB curves cross in 1961 and 2012, indicating a sudden change in the hazard index during those two years. Overall, the UF and UB statistics are basically >0, indicating an upward trend in variation, which also indicates that the adverse effects of high temperature on kiwifruit are also increasing year by year.

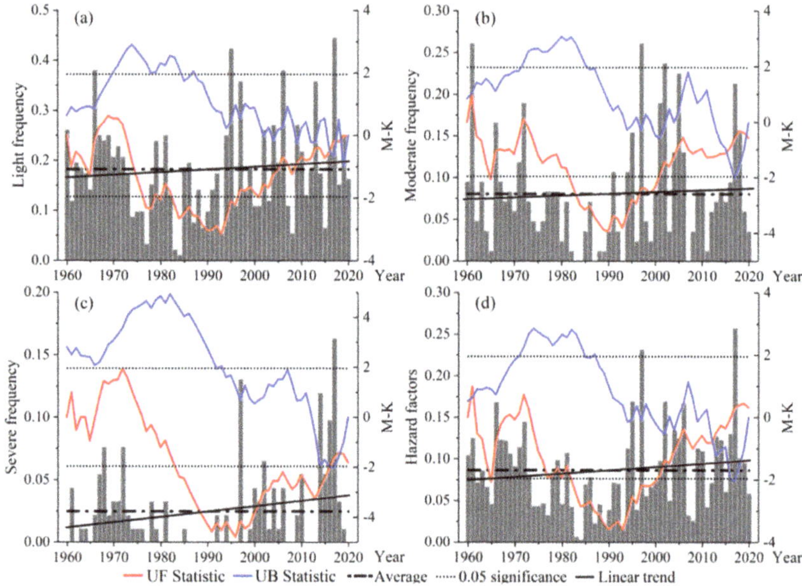

Figure 5. Variation and M-K test of the high-temperature events frequency and hazard index. (**a**) Light frequency. (**b**) Moderate frequency. (**c**) Severe frequency. (**d**) Hazard factors.

Figure 6 shows the spatial distribution of the frequency of light (a), moderate (b) and severe (c) high-temperature events in the study area. The high incidence of severe high-temperature events was relatively low and relatively concentrated in central Ankang, southwestern Weinan and southern Xianyang. Moderate high-temperature events are concentrated in central Ankang, northern and south-central Weinan, and southern Xianyang where it meets the northwestern part of the Xi'an. Light high-temperature events of varying degrees of intensity occurred in all areas except central Hanzhong and southern Baoji. Spatially, severe high-temperature events have a smaller impact area.

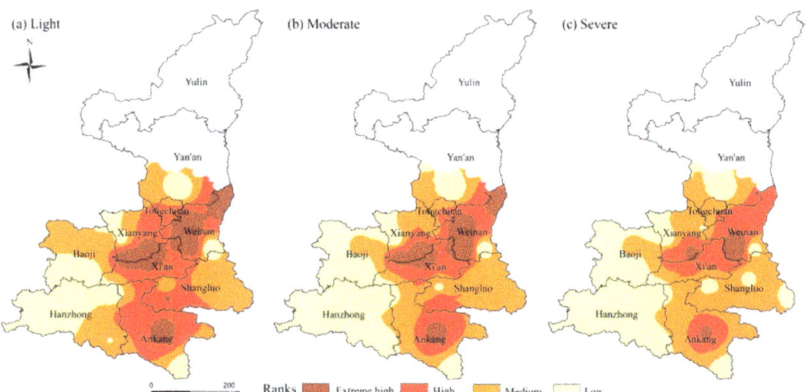

Figure 6. Spatial distribution of the frequency of high-temperature events.

4.1.2. Analysis of Spatial Patterns of High-Temperature Disaster Hazard

Figure 7 shows the spatial distribution of the hazard of high-temperature disaster in the study area. The spatial distribution is based on the Inverse Distance Weighted (IDW). The extreme high-hazard areas are concentrated in the southeast of the study area, influenced by the subtropical or warm temperate monsoon climate. Extreme high-hazard areas are concentrated in central Ankang, southern Xianyang, Xi'an and most of Weinan, accounting for 21.16% of the study area. High-hazard areas make up 37.72% of the study area, the largest area of any class. Covering the south-central Loess Plateau and the hinterland of the Guanzhong Plain. Most of these areas are major grain and fruit producing areas, and the extreme high-hazard areas are exposed to a higher frequency or intensity of high-temperature disasters, facing extreme high risk, and there are great potential economic losses. Areas of medium hazard are concentrated in the north of Baoji and parts of the border between Hanzhong and Ankang, accounting for 16.82% of the study area. The low-hazard areas are mainly in the south of Baoji and most of Hanzhong, and sporadically in Ankang, Weinan and Yan'an, accounting for 24.3% of the study area.

4.2. Comprehensive Assessment of High-Temperature Disaster Vulnerability

4.2.1. Changes in Kiwifruit Yield by County

Figure 8a–c show the yield reduction rate, the yield reduction coefficient of variation and the change in the probability occurrence of yield reduction rate, respectively. The counties with higher yield reduction rate were Luochuan (74.72%), Yijun (59.98%), Taibai (40.09%) and Baqiao (37.22%). The counties with higher yield reduction coefficients of variation were Liquan (120.64%), Baqiao (108.33%), Yijun (97.44%) and Zhouzhi (95.65%). The area with the highest probability occurrence of yield reduction rate was Baqiao, even reaching 85%. Baqiao, an emerging kiwifruit base county in Shaanxi Province, faces a higher risk of yield loss and is more affected by high-temperature disasters. Overall, kiwifruit yield was stable in all but a few counties. More than half of the areas have a probability occurrence of yield reduction rate of less than 50% and are less affected by meteorological disasters.

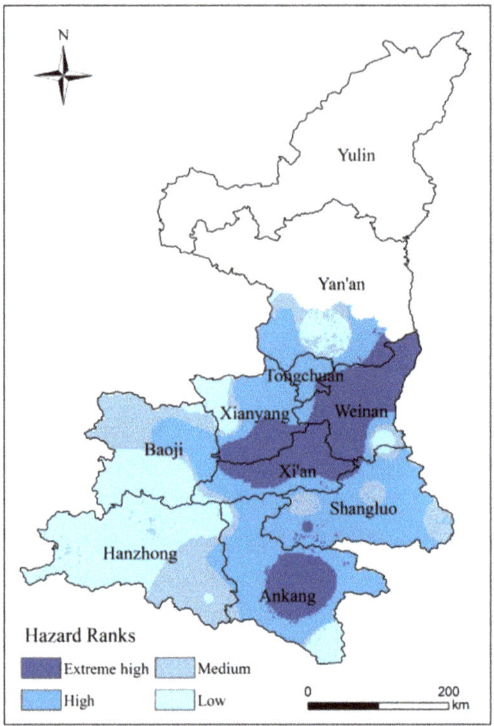

Figure 7. Distribution of hazard to kiwifruit in Shaanxi.

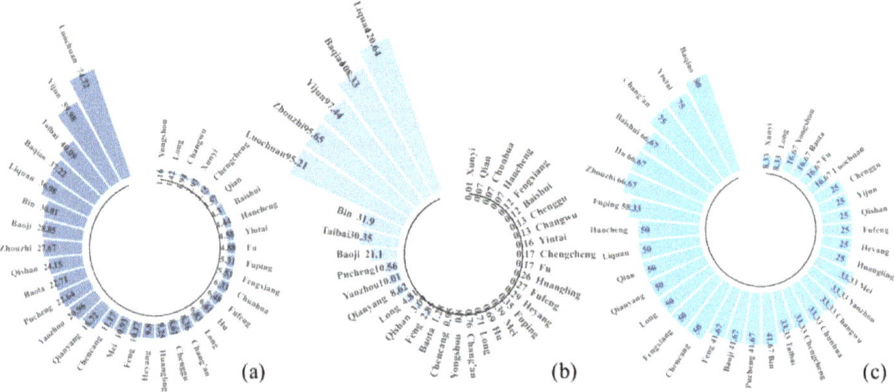

Figure 8. Changes in kiwifruit yield by county: (**a**) yield reduction rate (%), (**b**) yield reduction coefficient of variation and (**c**) the change in the probability occurrence of yield reduction rate (%).

4.2.2. Analysis of Spatial Patterns of High-Temperature Disaster Vulnerability

Areas of very extreme high, high, medium and low vulnerability represent 27.84%, 32.49%, 36.32% and 3.35% of the study area, respectively (Figure 9). The vulnerability of Hanzhong, the north of Baoji, the northwest of Xianyang and the east of Weinan is extremely high, and the ability to resist disaster in these areas is weak. Baoji, as the main producing area of kiwifruit in Shaanxi Province, has formed a centralized and continuous planting model of kiwifruit base county. If threatened by a high-temperature disaster, it can easily cause greater damage. Vulnerability in the southeast presents a gradually decreasing trend, and the eastern region in the study area is at low vulnerability. The Ankang and Shangluo kiwifruit planting areas have a lower degree of yield variability. Furthermore, these areas are at a higher altitude and close to the Qinling Mountains. The Qinling Mountains have a blocking effect on the warm and humid air currents from the south, making the nearby areas less affected by the high temperatures.

4.3. Comprehensive Assessment of High-Temperature Disaster Exposure
4.3.1. Analysis of the Climatic Suitability of Kiwifruit

Shaanxi kiwifruit planting areas are mainly distributed in the pre-mountain alluvium pro luvium fan area north of the Qinling Mountains, which has a warm temperate zone semi-humid and semi-dry climate. In recent years, with the quickening pace of rural industrial structure adjustment, the areas planted with kiwifruit and yield have continued to grow. A detailed climatic suitability zoning study is not only a reference for planting layout, but also a practical index of exposure (Figure 10). The highly suitable area is located in central Shaanxi, with a warm-temperate semi-humid and semi-dry climate, superior climatic resources, fertile soils and flat terrain. It accounts for 17.83% of the study area. The moderately suitable area extends from the highly suitable area outwards to below 1100 m above sea level in the Weibei Plateau and to higher elevations on both sides of the northern branch of the Qinling Mountains, and can be divided into two parts, north and south, accounting for 39.96% of the study area. In the north, temperatures are variable during spring, with more late frosts, less precipitation and more drought. There is abundant precipitation in the south, but they face higher temperatures. All areas are not very suitable for planting kiwifruit, except for Hanzhong and Ankang.

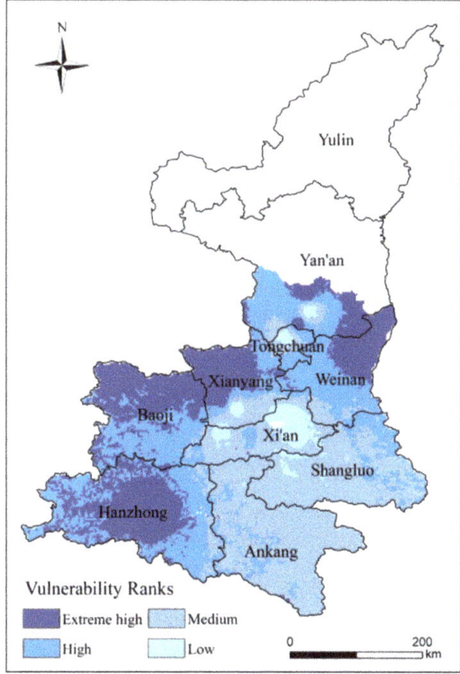

Figure 9. Distribution of vulnerability to kiwifruit in Shaanxi.

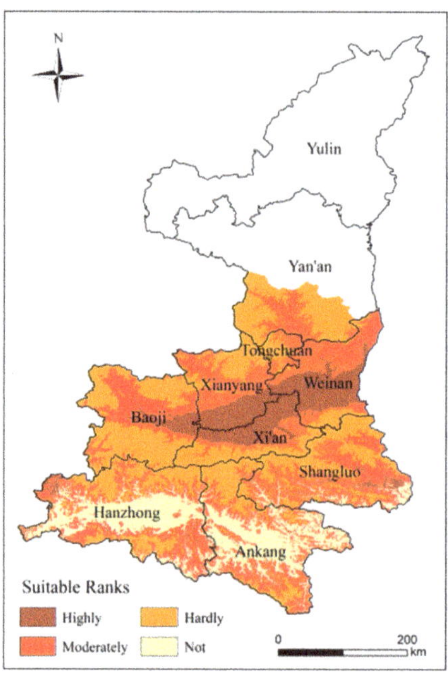

Figure 10. Climatic suitability zoning for kiwifruit in the study area.

4.3.2. Analysis of Spatial Patterns of High-Temperature Disaster Exposure

The exposed high-value area is located in the Guanzhong Plain region, including central Weinan, the part of Xianyang bordering Xi'an and Baoji, accounting for 14.54% of the study area (Figure 11). There is a large kiwifruit planting area in this area, and it has high climate suitability, so it is exposed. The central part of the study area has a climate suitable for the growth of kiwifruit and is suitable for extensive kiwifruit cultivation. It is also accompanied by a high potential risk. It is advisable to increase the cultivation of good kiwifruit varieties in the region and to strengthen pre-disaster prevention, response and post-disaster recovery in order to ensure the yield and quality of kiwifruit. Medium and low exposure areas represent 58.18% of the study area. Medium exposed areas are widely distributed and can affect kiwifruit yield if severe high-temperature events occur. Exposure in Hanzhong and Ankang is relatively low and potential losses from high-temperature disasters are likely to be low.

4.4. Comprehensive Assessment of High-Temperature Disaster Prevention and Mitigation Capacity

The total agricultural machinery power is high in the study area, with little difference between the north and south (Figure 12a). Weinan, Shangluo and Yan'an are at a high-value level, with a high level of agricultural modernization and therefore a high level of disaster resilience and relatively timely mitigation operations. In addition, the high total agricultural machinery power in the Shangluo region is related to the strong policy to develop agriculture in the region. The amount of fertilizer consumption can reflect the conditions of agricultural production in an area, and inputs of fertilizer to promote and improve crop growth can enhance crop resistance to disaster (Figure 12b). A comparison of the Figure 12a,b shows a certain consistency between fertilizer consumption and total agricultural machinery power. It shows a spatial distribution with more in the center and less in the north and south. Weinan, Shangluo, Yan'an and Baoji have relatively good agricultural production conditions. These areas are located in the plains and are suitable for cultivation and are relatively resilient to disaster. Per capita disposable income of farmers is another index of the comprehensive assessment of high-temperature disaster prevention and mitigation capacity (Figure 12c). Xianyang has a clear advantage in terms of disposing of funds and optimizing disaster prevention and mitigation measures. Rural electricity consumption per unit area gives an indication of the degree of modernization of the countryside and the affluence of farmers (Figure 12d). The high-value areas are Baoji, Yan'an and Xi'an. The low-value areas are Ankang and southeastern Hanzhong.

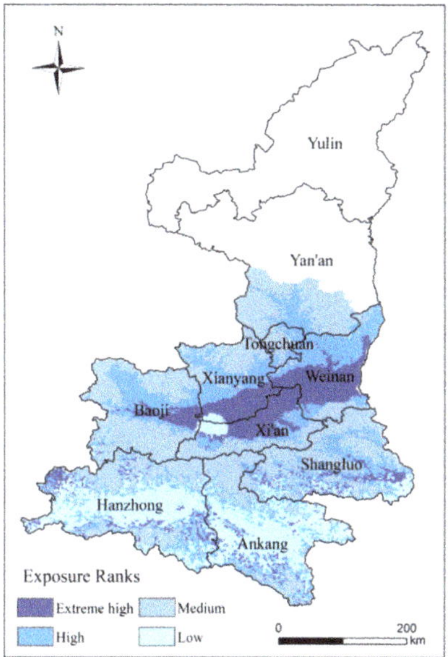

Figure 11. Distribution of exposure to kiwifruit in Shaanxi.

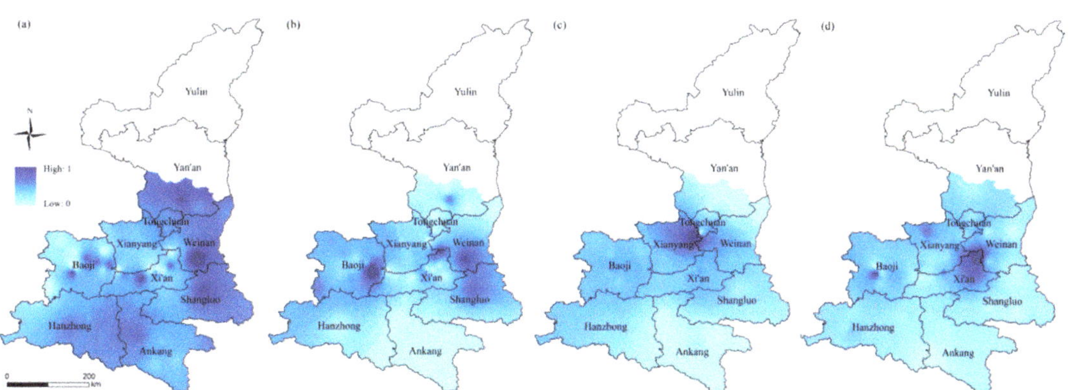

Figure 12. Spatial distribution of disaster prevention and mitigation capacity indexes: (**a**) total agricultural machinery power (KW), (**b**) fertilizer consumption (t), (**c**) per capita disposable income of farmers (yuan) and (**d**) rural electricity consumption per unit area (KW/h).

The study area has an overall medium level of disaster prevention and mitigation capacity, accounting for 47.43% (Figure 13). Ankang has a weak capacity for disaster prevention and mitigation, and its resistance and recovery from disaster need to be improved. The southern part of Weinan, the northeastern part of Xianyang and the central-eastern part of Baoji have extreme high disaster prevention and mitigation capacities, but this part of the area only accounts for 4.19% of the total area of the study area. The central part of the study area has a strong capacity for disaster prevention and mitigation, which is highly beneficial for fruit and food production, and have a better ability to cope with disaster.

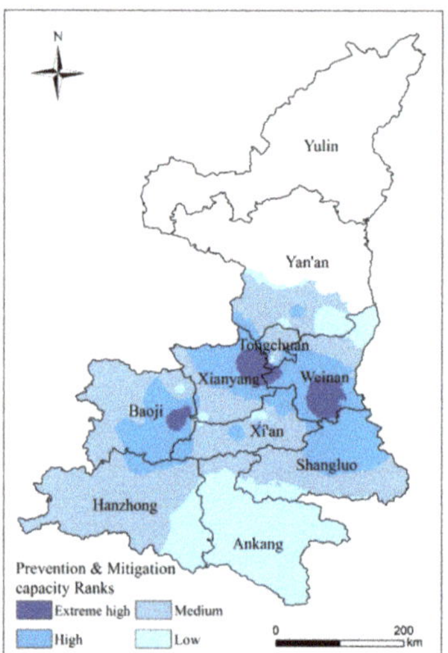

Figure 13. Distribution of high-temperature disaster prevention and mitigation capacity.

4.5. Risk Assessment of High-Temperature Disasters

4.5.1. Analysis of Spatial Patterns Risk of High-Temperature Disasters

We constructed risk assessment models using the results of hazard, vulnerability, exposure and disaster prevention and mitigation capacity. GIS was used to obtain results for high-temperature disaster risk assessment. The risk of high-temperature disasters is classified into four ranks from lowest to highest based on the optimal segmentation. Thus, the spatial distribution of high-temperature disaster risk was obtained (Figure 14).

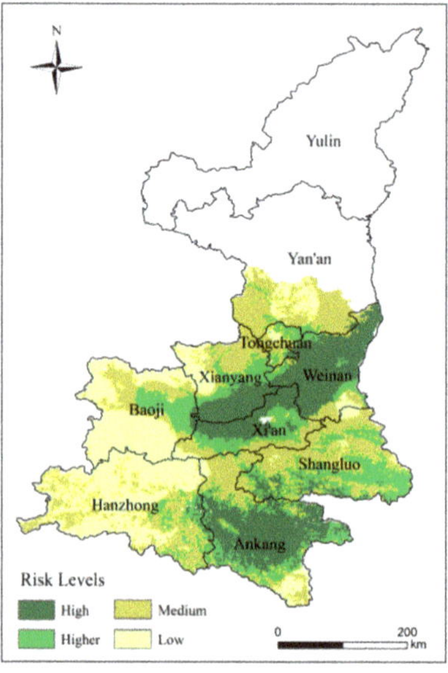

Figure 14. Distribution of high-temperature disaster to kiwifruit in Shaanxi.

The distribution of risk in kiwifruit planting areas shows clear regional differences. Extreme high-risk areas are concentrated in the central part of the study area and Ankang, accounting for 22.7% of the study area. These areas have a warm temperate monsoon climate with high temperature and rainy summer. The high-risk area spreads in a southerly and northerly direction, accounting for 19.22% of the study area. The medium risk area is located in the eastern part of the study area, accounting for 33.38%. The low level of risk areas covers Hanzhong and southern Baoji. This region is a major kiwifruit planting area and the low risk of high-temperature disasters is conducive to the development of kiwifruit farming.

4.5.2. Validation of Risk Evaluation Results

To verify the reliability of the assessment results, the data of high-temperature events in Shaanxi Province from 1960 to 2020 were queried. Six of the more serious heat events were collected and collated (Table 6). The results of the study are more in line with historical disaster data, with the high-risk areas located in parts of central and southeastern Shaanxi. Weinan experienced extreme heat events in 1971 and 2014.

Table 6. Historical disaster data for high temperature events in Shaanxi Province.

Sort	Time	Extreme Heat Event Records
1	1971.7.15–1971.7.28	From mid-July to mid-August 1971, there were high temperatures and little rain in Guanzhong and southern Shaanxi. The daily maximum temperature in the Weinan area was above 36 °C. The high temperature aggravated the drought and affected 133,000 hectares of farmland, accounting for 35 per cent of the total autumn field area. *
2	1972.8.7–1972.8.16	From late July to mid-August 1972, extreme heat events occurred in Hanzhong and southern Shaanxi. Cotton and maize were affected. *
3	1997.7.20–1997.7.27	From 1961 to 2006, the average number of high-temperature days in the northwest region was 2.4 days, with the most years being 1997, when there were 6.4 days. In Shaanxi Province, the annual number of high-temperature days is above 50, with daily maximum temperatures reaching 38–40 °C in Weinan and Xi'an. **
4	2001.7.11–2001.7.23	In mid to late July 2001, most areas in the middle and lower reaches of the Yangtze River and north of it in China experienced persistent extreme heat. The maximum temperature or the number of high-temperature days in many areas of southern Shaanxi Province exceeded the extreme values for the same period in history. The high temperatures and low rainfall were extremely detrimental to the growth of kiwifruit, citrus, grapes, apples and walnuts. ***
5	2014.7.27–2014.8.5	From 4 July to 10 August 2014, the longest consecutive run of high-temperature days in south-central Shaanxi was generally 10 days or more, reaching or exceeding historical extremes. Extreme high temperatures above 40 °C were experienced in local areas of Shaanxi. Intermittent hot weather occurred in southern Shaanxi, with Weinan reaching 39.1 °C. ****
6	2016.8.12–2016.8.21	July 20–August 26, a total of 30 provinces (autonomous regions and municipalities), 1653 counties (cities), had daily maximum temperatures of more than 35 °C high-temperature weather; the southeastern region of Shaanxi had amaximum temperature of 38–41 °C; Shaanxi Xunyang had one of up to 43.6 °C; 64 counties (cities) surpassed the local historical extreme; 11 provinces (autonomous regions and municipalities) in the south had an average number of high-temperature days of 19 days, the highest value since 1961. ****

* China Meteorological Disaster Dictionary—Shaanxi Volume. Wen Kegang. China Meteorological Press. ** Atlas of disastrous weather and climate in China (1961–2006). China meteorological administration. China Meteorological Press. *** Regional extreme events of drought, heavy precipitation, high and low temperatures in China. Ren Fumin. China Meteorological Press. **** China Meteorological Disaster Yearbook. China meteorological administration. China Meteorological Press.

4.6. Research Limitations and Prospects

The climate in Shaanxi Province varies greatly, gradually from north to south to the temperate zone, warm temperate zone and north subtropical zone. With the comprehensive effect of complex geographical environment and changeable climatic environment, not just one type of meteorological disaster occurs in the region. Many kinds of disaster may exist at the same time and interact with each other. We selected only the most serious high-temperature disaster to kiwifruit for our risk assessment and did not consider other disasters (freezing injury in early spring, extreme precipitation, low temperature and sparse sunlight, etc.). Following studies can screen for different disaster intensities by counting the occurrences of disasters in each county over the years. We also recommend conducting comprehensive dynamic risk assessments of different regions, different growth periods and multiple disasters. In addition, unlike field crops, kiwifruit, as a highly centrally managed economic fruit, is not just passively influenced by the climate. Orchard management, cropping systems and cultivation practices will all have an impact, and can even actively adapt to environmental changes through human regulation. In future research, therefore, further consideration needs to be taken of park management and other factors that affect the final risk assessment.

4.7. Recommendations

As an important economic fruit in Shaanxi Province, kiwifruit faces the threat of extreme high temperature mainly during the summer, which directly affects its yield and economic efficiency. Based on the above research, we propose three recommendations. Firstly, as an economic fruit, kiwifruit planting patterns are heavily influenced by policy. It is recommended to increase government investment, adjust the industrial structure and select and breed varieties with higher resistance to reduce the impact of high-temperature disasters. Secondly, due to the economic benefits of kiwifruit, which have come to the fore in recent years, there has been a blind introduction of planting. It is recommended that planting is carried out according to the results of the climatic suitability zone and adjust measures to local conditions. Finally, we recommend strengthening orchard management and infrastructure, while enhancing professional training for fruit farmers and improving cultivation techniques. Furthermore, risk should be considered to minimize losses.

5. Conclusions

We are the first to present a study to assessment the risk of high-temperature disasters affecting the kiwifruit. A system and model for assessing the risk of high-temperature disaster to kiwifruit was constructed from four aspects: hazard, vulnerability, exposure and disaster prevention and mitigation capacity. At the same time, a study on the climatic suitability of kiwifruit for zoning was carried out and used as one of the indexes of exposure. The results are as follows:

(1) From five aspects, heat, light, water, soil resources and topography, 16 indexes were selected that have an important influence on the growth and development of kiwifruit, and a climate suitability zoning of kiwifruit in Shaanxi Province was carried out. We found that the areas of high suitability were located in the Guanzhong Plain region, including Weinan, southern Xianyang, northern Xi'an and central-eastern Baoji. The highly and moderately suitable area accounted for 42.21% of the study area.

(2) Areas at high risk of high-temperature disaster are located in Ankang, Weinan, southern Xianyang and northern Xi'an, accounting for 22.7% of the study area. As a relatively concentrated area of agricultural production in the plains, the risk of high-temperature disasters poses a significant threat to the region's agricultural development. The south-western part of the study area has a low risk and is favorable to the kiwifruit industry.

(3) By comparing with historical disaster data, the areas where high-temperature disasters occurred are more consistent with our findings and the study has a high degree of confidence.

Author Contributions: Conceptualization, Y.M. and Z.T.; data curation, Y.M. and J.X.; funding acquisition, J.Z.; methodology, Y.M.; writing—original draft, Y.M. and Z.T.; writing—review and editing, S.G. and X.L. All authors have read and agreed to the published version of the manuscript.

Funding: This study is supported by the National Key Research and Development Program of China (2019YFD1002201); The National Natural Science Foundation of China (41877520, 42077443); The Science and Technology Development Planning of Jilin Province (20190303018SF); Key Research and Projects Development Planning of Jilin Province (20200403065SF); Industrial technology research and development project supported by Development and Reform Commission of Jilin Province (2021C044-5); The Science and Technology Planning of Changchun (19SS007).

Institutional Review Board Statement: Not applicable.

Informed Consent Statement: Not applicable.

Data Availability Statement: The data presented in this study are available upon request from the corresponding author.

Acknowledgments: We appreciate the editors and the reviewers for their constructive suggestions and insightful comments, which helped greatly to improve this manuscript.

Conflicts of Interest: The authors declare no conflict of interest.

References

1. Tollefson, J. IPCC climate report: Earth is warmer than it's been in 125,000 years. *Nature* **2021**, *596*, 171–172. [CrossRef] [PubMed]
2. Alexander, L.V.; Zhang, X.; Peterson, T.C.; Caesar, J.; Gleason, B.; Tank, A.K.; Haylock, M.; Collins, D.; Trewin, B.; Rahimzadeh, F.; et al. Global observed changes in daily climate extremes of temperature and precipitation. *J. Geophys. Res. Space Phys.* **2006**, *111*. [CrossRef]
3. Hatfield, J.L.; Boote, K.; Kimball, B.A.; Ziska, L.H.; Izaurralde, R.C.; Ort, D.; Thomson, A.M.; Wolfe, D. Climate impacts on agriculture: Implications for crop production. *Agron. J.* **2011**, *103*, 351–370. [CrossRef]
4. Luo, Q. Temperature thresholds and crop production: A review. *Clim. Chang.* **2011**, *109*, 583–598. [CrossRef]
5. Zhang, Y.; Qu, H.; Yang, X.; Wang, M.; Qin, N.; Zou, Y. Cropping system optimization for drought prevention and disaster reduction with a risk assessment model in Sichuan Province. *Glob. Ecol. Conserv.* **2020**, *23*, e01095. [CrossRef]
6. Hatfield, J.L.; Prueger, J.H. "Challenge for future agriculture.". In *Crop Wild Relatives and Climate Change*; Redden, R., Yadav, S.S., Maxted, N., Dulloo, M.E., Guarino, L., Smith, P., Eds.; Wiley Publishing: Hoboken, NJ, USA, 2015; pp. 24–43. [CrossRef]
7. Reyes, J.J.; Elias, E. Spatio-temporal variation of crop loss in the United States from 2001 to 2016. *Environ. Res. Lett.* **2019**, *14*, 074017. [CrossRef]
8. FAOSTAT. Kiwifruit Production of the World and China in 2014. Available online: http://faostat3.fao.org/browse/Q/QC/E (accessed on 1 August 2021).
9. Zheng, S.; Cui, N.; Gong, D.; Wang, Y.; Hu, X.; Feng, Y.; Zhang, Y. Relationship between stable carbon isotope discrimination and water use efficiency under deficit drip irrigation of kiwifruit in the humid areas of South China. *Agric. Water Manag.* **2020**, *240*, 106300. [CrossRef]
10. China Statistics Press. *The Statistical Yearbooks of Shaanxi*; China Statistics Press: Beijing, China, 2019.
11. He, B.; Wang, Q.J.; Wu, D.; Zhou, B.B. Spatio-temporal characteristics of agricultural drought in Shaanxi Province, China based on integrated disaster risk index. *Ying Yong Sheng Tai Xue Bao J. Appl. Ecol.* **2016**, *27*, 3299–3306.
12. Zhang, L.; Song, W.; Song, W. Assessment of agricultural drought risk in the lancang-mekong region, South East Asia. *Int. J. Environ. Res. Public Health* **2020**, *17*, 6153. [CrossRef] [PubMed]
13. Guo, H.; Chen, J.; Pan, C. Assessment on agricultural drought vulnerability and spatial heterogeneity study in China. *Int. J. Environ. Res. Public Health* **2021**, *18*, 4449. [CrossRef] [PubMed]
14. Cheng, Y.-X.; Huang, J.; Han, Z.-L.; Guo, J.-P.; Zhao, Y.-X.; Wang, X.-Z.; Guo, R.-F. Cold damage risk assessment of double cropping rice in hunan, China. *J. Integr. Agric.* **2013**, *12*, 352–363. [CrossRef]
15. Chatrabgoun, O.; Karimi, R.; Daneshkhah, A.; Abolfathi, S.; Nouri, H.; Esmaeilbeigi, M. Copula-based probabilistic assessment of intensity and duration of cold episodes: A case study of Malayer vineyard region. *Agric. For. Meteorol.* **2020**, *295*, 108150. [CrossRef]
16. Scorzini, A.R.; Di Bacco, M.; Manella, G. Regional flood risk analysis for agricultural crops: Insights from the implementation of AGRIDE-c in central Italy. *Int. J. Disaster Risk Reduct.* **2021**, *53*, 101999. [CrossRef]
17. Liu, Y.; You, M.; Zhu, J.; Wang, F.; Ran, R. Integrated risk assessment for agricultural drought and flood disasters based on entropy information diffusion theory in the middle and lower reaches of the Yangtze River, China. *Int. J. Disaster Risk Reduct.* **2019**, *38*, 101194. [CrossRef]
18. Zhang, F.; Chen, Y.; Zhang, J.; Guo, E.; Wang, R.; Li, D. Dynamic drought risk assessment for maize based on crop simulation model and multi-source drought indices. *J. Clean. Prod.* **2019**, *233*, 100–114. [CrossRef]
19. Zhang, J. Risk assessment of drought disaster in the maize-growing region of Songliao Plain, China. *Agric. Ecosyst. Environ.* **2004**, *102*, 133–153. [CrossRef]
20. Li, Z.; Zhang, Z.; Zhang, L. Improving regional wheat drought risk assessment for insurance application by integrating scenario-driven crop model, machine learning, and satellite data. *Agric. Syst.* **2021**, *191*, 103141. [CrossRef]
21. Zhang, L.; Chu, Q.-Q.; Jiang, Y.-L.; Chen, F.; Lei, Y.-D. Impacts of climate change on drought risk of winter wheat in the North China Plain. *J. Integr. Agric.* **2021**, *20*, 2601–2612. [CrossRef]
22. Wassmann, R.; Jagadish, S.V.K.; Heuer, S.; Ismail, A.; Redona, E.; Serraj, R.; Singh, R.K.; Howell, G.; Pathak, H.; Sumfleth, K. Climate change affecting rice production: The physiological and agronomic basis for possible adaptation strategies. *Adv. Agron.* **2009**, *101*, 59–122. [CrossRef]
23. Julia, C.; Dingkuhn, M. Predicting temperature induced sterility of rice spikelets requires simulation of crop-generated microclimate. *Eur. J. Agron.* **2013**, *49*, 50–60. [CrossRef]

24. Zhu, X.; Xu, K.; Liu, Y.; Guo, R.; Chen, L. Assessing the vulnerability and risk of maize to drought in China based on the AquaCrop model. *Agric. Syst.* **2021**, *189*, 103040. [CrossRef]
25. Farhangfar, S.; Bannayan, M.; Khazaei, H.R.; Baygi, M.M. Vulnerability assessment of wheat and maize production affected by drought and climate change. *Int. J. Disaster Risk Reduct.* **2015**, *13*, 37–51. [CrossRef]
26. Epule, T.E.; Ford, J.D.; Lwasa, S. Projections of maize yield vulnerability to droughts and adaptation options in Uganda. *Land Use Policy* **2017**, *65*, 154–163. [CrossRef]
27. Luo, D.; Ye, L.; Sun, D. Risk evaluation of agricultural drought disaster using a grey cloud clustering model in Henan province, China. *Int. J. Disaster Risk Reduct.* **2020**, *49*, 101759. [CrossRef]
28. Liu, Y.; Chen, J. Future global socioeconomic risk to droughts based on estimates of hazard, exposure, and vulnerability in a changing climate. *Sci. Total. Environ.* **2021**, *751*, 142159. [CrossRef] [PubMed]
29. Carrão, H.; Naumann, G.; Barbosa, P. Mapping global patterns of drought risk: An empirical framework based on sub-national estimates of hazard, exposure and vulnerability. *Glob. Environ. Chang.* **2016**, *39*, 108–124. [CrossRef]
30. Zhang, Y.; Wang, Y.; Niu, H. Effects of temperature, precipitation and carbon dioxide concentrations on the requirements for crop irrigation water in China under future climate scenarios. *Sci. Total. Environ.* **2019**, *656*, 373–387. [CrossRef] [PubMed]
31. Macholdt, J.; Hadasch, S.; Piepho, H.-P.; Reckling, M.; Taghizadeh-Toosi, A.; Christensen, B. Yield variability trends of winter wheat and spring barley grown during 1932–2019 in the Askov Long-term Experiment. *Field Crop. Res.* **2021**, *264*, 108083. [CrossRef]
32. Lieth, H.; Box, E. Evapotranspiration and primary productivity: C. W. Thornthwaite memorial model. *Publ. Climatol.* **1972**, *25*, 37–46.
33. Lieth, H. Modeling the Primary Productivity of the World. *Nature Res.* **1975**, *14*, 237–263. [CrossRef]
34. Yamada, E.S.M.; Sentelhas, P.C. Agro-climatic zoning of Jatropha curcas as a subside for crop planning and implementation in Brazil. *Int. J. Biometeorol.* **2014**, *58*, 1995–2010. [CrossRef]
35. Garcia-Barreda, S.; Sánchez, S.; Marco, P.; Serrano-Notivoli, R. Agro-climatic zoning of Spanish forests naturally producing black truffle. *Agric. For. Meteorol.* **2019**, *269–270*, 231–238. [CrossRef]
36. Hutchinson, M.F. Interpolating mean rainfall using thin plate smoothing splines. *Int. J. Geogr. Inf. Syst.* **1995**, *9*, 385–403. [CrossRef]
37. Hijmans, R.J.; Cameron, S.E.; Parra, J.L.; Jones, P.G.; Jarvis, A. Very high resolution interpolated climate surfaces for global land areas. *Int. J. Clim.* **2005**, *25*, 1965–1978. [CrossRef]
38. New, M.; Lister, D.; Hulme, M.; Makin, I. A high-resolution data set of surface climate over global land areas. *Clim. Res.* **2002**, *21*, 1–25. [CrossRef]
39. Gentile, R.M.; Malepfane, N.M.; Dijssel, C.V.D.; Arnold, N.; Liu, J.; Müller, K. Comparing deep soil organic carbon stocks under kiwifruit and pasture land uses in New Zealand. *Agric. Ecosyst. Environ.* **2021**, *306*, 107190. [CrossRef]
40. Huang, H. Biology, Genetic improvement, and cultivar development. *Kiwifruit* **2016**, *5*, 211–237. [CrossRef]
41. Ma, Y.; Lu, X.; Li, K.; Wang, C.; Guna, A.; Zhang, J. Prediction of potential geographical distribution patterns of *Actinidia arguta* under different climate scenarios. *Sustainability* **2021**, *13*, 3526. [CrossRef]
42. Wu, H.; Ma, T.; Kang, M.; Ai, F.; Zhang, J.; Dong, G.; Liu, J. A high-quality Actinidia chinensis (kiwifruit) genome. *Hortic. Res.* **2019**, *6*, 117. [CrossRef] [PubMed]
43. Zhu, H. *Kiwifruit*; China Forestry Publishing House: Beijing, China, 2009.
44. Liu, J.; Zhang, K. Comparative study on data standardization methods in comprehensive evaluation. *Digit. Technol. Appl.* **2018**, *36*, 84–85. [CrossRef]
45. Saaty, T.L. *The Analytical Hierrarchy Process*; McGraw Hill: New York, NY, USA, 1980.
46. Du, Y.; Zheng, Y.; Wu, G.; Tang, Y. Decision-making method of heavy-duty machine tool remanufacturing based on AHP-entropy weight and extension theory. *J. Clean. Prod.* **2020**, *252*, 119607. [CrossRef]
47. Mon, D.-L.; Cheng, C.-H.; Lin, J.-C. Evaluating weapon system using fuzzy analytic hierarchy process based on entropy weight. *Fuzzy Sets Syst.* **1994**, *62*, 127–134. [CrossRef]
48. Farhadinia, B. A Multiple criteria decision making model with entropy weight in an interval-transformed hesitant fuzzy environment. *Cogn. Comput.* **2017**, *9*, 513–525. [CrossRef]
49. Li, L.; Liu, F. Evaluation of airport aviation security based on combined weight. *J. Civ. Aviat. Univ.* **2014**, *32*, 55–58.
50. Guo, Y.; Wang, R.; Tong, Z.; Liu, X.; Zhang, J. Dynamic Evaluation and Regionalization of Maize Drought Vulnerability in the Midwest of Jilin Province. *Sustainability* **2019**, *11*, 4234. [CrossRef]
51. Mitra, S.; Srivastava, P. Spatiotemporal variability of meteorological droughts in southeastern USA. *Nat. Hazards* **2017**, *86*, 1007–1038. [CrossRef]
52. Mann, H.B. Nonparametric Tests Against Trend. *Econometrica* **1945**, *13*, 245–259. [CrossRef]
53. Guna, A.; Zhang, J.; Tong, S.; Bao, Y.; Han, A.; Li, K. Effect of Climate Change on Maize Yield in the Growing Season: A Case Study of the Songliao Plain Maize Belt. *Water* **2019**, *11*, 2108. [CrossRef]

International Journal of
***Environmental Research
and Public Health***

Article

Assessment on Agricultural Drought Vulnerability and Spatial Heterogeneity Study in China

Hongpeng Guo, Jia Chen and Chulin Pan *

College of Biological and Agricultural Engineering, Jilin University, 5988 Renmin Street, Changchun 130012, China; ghp@jlu.edu.cn (H.G.); chenjia19@mails.jlu.edu.cn (J.C.)
* Correspondence: pancl@jlu.edu.cn

Abstract: Reducing drought vulnerability is a basis to achieve sustainable development in agriculture. The study focuses on agricultural drought vulnerability in China by selecting 12 indicators from two aspects: drought sensitivity and resilience to drought. In this study, the degree of agricultural drought vulnerability in China has been evaluated by entropy weight method and weighted comprehensive scoring method. The influencing factors have also been analyzed by a contribution model. The results show that: (1) From 1978 to 2018, agricultural drought vulnerability showed a decreasing trend in China with more less vulnerable to mildly vulnerable cities, and less highly vulnerable cities. At the same time, there is a trend where highly vulnerable cities have been converted to mildly vulnerable cities, whereas mildly vulnerable cities have been converted to less vulnerable cities. (2) This paper analyzes the influencing factors of agricultural drought vulnerability by dividing China into six geographic regions. It reveals that the contribution rate of resilience index is over 50% in the central, southern, and eastern parts of China, where agricultural drought vulnerability is relatively low. However, the contribution rate of sensitivity is 75% in the Southwest and Northwest region, where the agricultural drought vulnerability is relatively high. Among influencing factors, the multiple-crop index, the proportion of the rural population and the forest coverage rate have higher contribution rate. This study carries reference significance for understanding the vulnerability of agricultural drought in China and it provides measures for drought prevention and mitigation.

Keywords: agricultural drought vulnerability; spatial heterogeneity; entropy weight method; contribution model; China

Citation: Guo, H.; Chen, J.; Pan, C. Assessment on Agricultural Drought Vulnerability and Spatial Heterogeneity Study in China. *Int. J. Environ. Res. Public Health* **2021**, *18*, 4449. https://doi.org/10.3390/ijerph18094449

Academic Editors: Mikio Ishiwatari and Daisuke Sasaki

Received: 22 March 2021
Accepted: 20 April 2021
Published: 22 April 2021

1. Introduction

Drought occurs frequently in China and there has been a long history of these occurrences. From 206 BC to 1949, 1056 droughts occurred in China [1]. From 1971 to 2016, the average annual disaster rates of droughts in Heilongjiang, Jilin, Liaoning, and Inner Mongolia Autonomous Region were 19.4%, 23.6%, 25.4% and 29.8%, respectively. The average annual disaster rates of droughts in Anhui, Hebei, Henan, Jiangsu, and Shandong provinces were 11.5%, 20.4%, 16.2%, 8.9%, and 18.3%, respectively [2]. The Ministry of Emergency Management of the People's Republic of China has notified that from July to November of 2019, droughts had affected a total of 1174 thousand hectares of crops in Jiangxi and Anhui provinces, resulting in a direct economic loss of 8.8 billion yuan [3]. From January to April of 2020, 2.433 million people had been affected in 81 counties of 16 cities (prefectures) in Yunnan Province. A total of 662 thousand people had requested for life assistance due to droughts, and 534 thousand hectares of crops were affected, leading to direct economic loss of 1.41 billion yuan [4].

Drought is considered as a slow-moving natural disaster that causes severe damage to water resources and to agriculture [5]. The characteristics of drought include, but are not limited to, high frequency, long duration, and large area being influenced [6]. Agricultural drought is a crucial part of drought and it refers to the situation where agricultural production is sensitive and vulnerable to drought stress [7]. Agriculture utilizes natural resources directly and it is also a national anchoring industry. Agriculture is less capable of resisting and dealing with disasters. The resistance and handling capacity of agriculture to disasters is low so the adverse impact on agricultural production is most severe when drought occurs. In the same way, droughts can be intensified by poor land management [8]. Therefore, the situation of agriculture and the extent of drought affect each other. According to the Assessment Report of the AR5 Climate Change 2014: Impacts, Adaptation, and Vulnerability: Vulnerability encompasses a variety of concepts and elements including sensitivity or susceptibility to harm and lack of capacity to cope and adapt [9]. Taking the initiative via human activity is an effective way to alleviate the loss caused by a drought disaster [5]. So, measuring agricultural drought vulnerability is a prerequisite for targeting interventions to improve and sustain the agricultural performance of both irrigated and rain-fed agriculture [10].

Climate change has an increasing impact on production and people's lives. In recent years, the topic of vulnerability to agricultural drought has gradually become the focus and research hotspot of scholars around the world.

Yi (2010) evaluated the agricultural droughts in Dalian, China. Ten evaluation indexes such as irrigation index, population density and proportion of paddy areas were selected [11]. Yuan (2016) proposed a comprehensive index of regional drought vulnerability that includes exposure, sensitivity, and adaptability [12]. The establishment of evaluation indicators cannot be applied to all since it is highly subjective to regional characteristics. However, different indexing systems provide more research possibilities in the field of drought vulnerability.

Yan (2012), Pang et al. (2013), Farhangfar et al. (2015), Liu et al. (2015), and others conducted quantitative evaluation on drought vulnerability of maize and wheat and obtained the severity and spatial changes of crops at different growth stages [13–16]. Kim et al. (2018) used multivariate statistical analysis method to assess the agricultural vulnerability to droughts in South Korea and the results showed that the Chungchongnam-Do area was most vulnerable [17]. Lestari et al. (2018) used Arc GIS spatial overlay analysis to evaluate the agricultural drought vulnerability of Semarang Port City in India. The results showed that high vulnerability in six villages, medium vulnerability in seven villages, and low vulnerability in three villages [18]. Based on super sufficiency DEA, Huang et al. (2019) evaluated the agricultural drought vulnerability of Hetao Irrigation Area in Inner Mongolia and the results showed that the drought vulnerability in the eastern part of Hetao Irrigation Area was much higher than that in the western part [19]. Frischen et al. (2019) combined the result from spatial analysis of expert consultation and determined the drought vulnerability of Zimbabwe's agricultural system. The results showed that the country's drought vulnerability and the degree of impact vary greatly. The northern and southern part of Matabeleland, a province in southwestern part, have higher vulnerability level [20]. Das et al. (2019) used Savitzky and Golay filtering methods to study the agricultural drought situation and vulnerability in India from 1982 to 2015. Results showed that the vulnerability of drought will continue to decrease over time [21]. On the basis of selecting the research areas and constructing the evaluation index system, scholars have adopted different methods to evaluate the agricultural vulnerability to droughts. For example: Data envelopment analysis [22,23], analytic hierarchy process [24–28], principal component analysis [29,30], entropy weight method [31–33], etc. STATA [34,35], ArcGIS [36–38] and other software have also been used to construct an evaluation model for quantitative analysis.

Rojas et al. (2011) and Zhang et al. (2016) used remote sensing technology to monitor and predict agricultural drought [39,40]. Guo et al. (2016) proposed a new method (vulnerability surfaces) for assessing vulnerability quantitatively and continuously by including the environmental variable as an additional perspective on exposure and assessed global drought risk of maize based on these surfaces [41]. Chen et al. (2017) and Zeng et al. (2019) conducted drought risk assessment on Yunnan Province and Gansu Province respectively [42,43]. All the above studies have provided scientific methods for drought risk assessment and they have since enriched the assessment system for agricultural drought vulnerability.

Basing on a wide range of research areas and research methods, there exists the differences in the natural geographical environment, economic and social conditions, which has led to different influencing factors and various degrees of agricultural drought vulnerability. For example: Zarafshani et al. (2012) argued that the vulnerability of wheat farmers in the western part of Iran is mainly affected by economical, socio-cultural, psychological, technological, and infrastructural factors [44]. Wu et al. (2017) believed that the water shortage rate and irrigation level in the growing season were the main factors affecting the vulnerability level of regional agricultural drought [45]. Kamali et al. (2019) believed that the fertilization level is an important factor affecting the vulnerability of crop to drought in sub-Saharan Africa. Generally, countries with a higher food production index and better infrastructure perform better in terms of withstanding drought [46].

To sum up, there are two methods namely qualitative research and quantitative research on agricultural drought vulnerability. Existing research on agricultural drought vulnerability in China mainly focused on certain regions for quantitative research [7,14,32,37,45,47–51]. There were only a few studies on the overall assessment of agricultural drought vulnerability and among those the research objects, conclusions and countermeasures are limited.

Therefore, this paper focuses on the agricultural drought vulnerability in China. Based on literature review and relative theories, the paper first constructs the vulnerability evaluation index system of agricultural drought. Then the paper uses entropy weight method, weighted comprehensive scoring method as well as k-means clustering algorithm to evaluate and categorize the vulnerability of agricultural drought in China. Finally, using the contribution model to analyze the influencing factors and the degrees of agricultural drought vulnerability in China, this paper proposes countermeasures to reduce agricultural drought vulnerability in China. In one aspect, the paper carries theoretical value for enriching vulnerability research. It is also conducive to a better understanding of drought conditions and influencing factors in various regions of China. In another aspect, the empirical analysis provides the basis for the government to formulate corresponding policies, to reduce losses caused by disasters, and to promote the sustainable development of agriculture in China.

2. Materials and Methods

2.1. Research Area Overview

The People's Republic of China is located in East Asia and to the west coast of the Pacific Ocean. Liberated on 1 October 1949, China's capital city is Beijing and the provincial administrative divisions are divided into twenty-three provinces, five autonomous regions, four municipalities, and two special administrative regions. China's land area is about 9.6 million square kilometers. China is the world's second largest economy, the world's largest industrial country, and the world's largest agricultural country. At the end of 2019, the total population of mainland China was more than 1.4 billion.

The terrain is high in the West and low in the East. Mountains, plateaus, and hills account for estimated 67% of the land area, basins, and plains account for around 33% of the total land area. The climate condition is complex and diverse.

Looking at the situation and distribution of China's agricultural natural resources as a whole, the light and heat conditions are superior. However, there is a great regional differences of dry and wet conditions. The total amount of river runoff is large; however, the coordination and distribution of soil and water is not even. The absolute amount of land resources is large; however, the land occupied per capita is small. Agriculture still serves as the basic industry of China's national economy.

2.2. Establishment of Indicator System and Data Sources

The establishment of evaluation index system is the prerequisite for evaluating agricultural drought vulnerability. Vulnerability is the root cause of drought disasters, which results from the interaction of natural environment and social economy system as well as the interactions of sensitivity and resilience in a certain space. Therefore, following the principles of science, comprehensiveness, pertinence, quantification, and availability of data [47], we select two first-level indicators, namely, sensitivity and resilience and 12 second-level indicators to conduct an evaluation on 31 provincial administrative units (except for Hong Kong, Macao, and Taiwan) in China to establish an indicator system (as shown in Table 1). The larger the indicator, the larger the vulnerability of agricultural drought. Hence, it is a positive indicator. On the contrary, it would be a negative indicator.

Sensitivity is the sum of all kinds of natural and social factors that would cause or aggravate drought and its impact on agricultural drought vulnerability is negative. That means the higher the sensitivity, the greater the vulnerability of agricultural drought. It includes agriculture in GDP proportion, multiple-crop index, rural population proportion, annual average temperature, annual sunshine duration, and annual precipitation.

Higher proportion of agriculture in GDP means that farmers rely heavily on agricultural income which is highly dependent on natural conditions. So the vulnerability of agricultural drought will increase. The higher the multiple-crop index, the more water the crop would need to grow. As a result, drought vulnerability will increase. The most severely impacted population at the time of drought is the agricultural population. Therefore, when the proportion of rural population increases, the degree of vulnerability will also increase. Moreover, higher the temperature and longer sunshine hours will lead to the increase of evaporation, and hence the agricultural drought vulnerability will increase together. Precipitation is the main factor affecting the growth of crops. The precipitation index can reflect the meteorological conditions of crops in this region and the impact of precipitation on vulnerability is negative.

Resilience refers to the ability of human society to prepare for, to respond to, and to recover from, disasters. It has a positive impact on agricultural drought vulnerability. That means the stronger the resilience, the lower the drought vulnerability. It includes forest coverage rate, net income per capita of rural residents, food production per capita, real GDP per capita, effective irrigation rate, and agricultural fertilizer per unit area.

The forest coverage rate reflects a country's (or region) actual level of forest resources and forestry possession. Net income per capita of rural residents reflects the group of people's economical ability to withstand and to resist drought. The higher the net income per capita of rural residents, the weaker the threats of agricultural drought. Food production per capita reflects the level of agricultural productivity. Real GDP per capita reflects the level of social and economic development. When the index is bigger, it means that the social and economic development level and the ability to withstand disasters is high. The effective irrigation rate reflects the degree of water conservancy and irrigation capacity. The increase of the amount of agricultural fertilizer per unit area is beneficial to enhance soil fertility, to improve soil structure and to increase the efficiency of land usage. The above indicators constitute the resilience of the agricultural system.

The agriculture in GDP proportion, the rural population proportion, the net income per capita of rural residents, the food production per capita, and the real GDP per capita affect the agricultural drought vulnerability from the economic and social perspectives. The multiple-crop index, the effective irrigation rate and the agricultural fertilizer per unit area affect the vulnerability of agricultural drought from the perspective of agricultural technology. The forest coverage rate, annual average temperature, annual sunshine duration, and precipitation affect the vulnerability of agriculture to drought from the perspective of natural conditions.

The indicator data in this paper comes from the website of the National Bureau of Statistics [52] and the China Meteorological Administration [53]. The annual precipitation, annual sunshine duration and annual average temperature are obtained from annual observations from 613 weather stations nationwide from China Meteorological Administration data network. In addition to the forest coverage rate, net income per capita of rural residents and real GDP (Gross Domestic Product) per capita can be directly obtained, other indicators need to be calculated. The descriptive statistical results of the complete sample are shown in Table 2.

Table 1. Index system and source of China's agricultural drought vulnerability assessment.

Indicators and Units	Calculation Formula	Source
Agriculture in GDP proportion (%)	Agricultural output value/GDP	[51,54]
Multiple-crop index (%)	Cultivated area of crops/Total cultivated area	[49]
Rural population proportion (%)	Rural population/Total population	[51,54]
Annual average temperature (°C)	Annual average value of each meteorological station	[32]
Annual sunshine duration (h)	Annual average value of each meteorological station	[55]
Annual precipitation (mm)	Annual average value of each meteorological station	[51,54,56]
The forest coverage rate (%)	Available directly	[56,57]
Net income per capita of rural residents (yuan/per)	Available directly	[22,58,59]
Food production per capita (kg/per)	Food production/Total population	[49]
Real GDP per capita (yuan/per)	Available directly	[32,51,59]
The effective irrigation rate (%)	Effective irrigation area/Total cultivated area	[31,56]
Agricultural fertilizer per unit area (ton/hm^2)	Amount of fertilizer used/Total cultivated area	[32]

Table 2. Descriptive statistical results of samples.

Variable	Obs	Mean	Std. Dev.	Min	Max
Agriculture in GDP proportion	279	20.51542	12.75119	0.3193709	59.28663
Multiple-crop index	279	1.371218	0.5034035	0.5117678	2.589842
Rural population proportion	279	60.47917	20.75162	10.39337	91.76649
Annual average temperature	279	13.00551	5.695829	0.5178571	25.18
Annual sunshine duration	279	2136.61	481.6672	703.8	3075.392
Annual precipitation	279	917.2576	495.0083	80.34242	2523
The forest coverage rate	279	23.74077	16.91269	0.3	66.8
Net income per capita of rural residents	279	4124.196	5357.99	100.93	30374.73
Food production per capita	279	377.0782	225.9848	15.84958	1989.61
Real GDP per capita	279	18,178.2	25,786.47	175	140,000
The effective irrigation rate	279	0.5104425	0.229718	0.0719334	1
Agricultural fertilizer per unit area	279	0.0388566	0.0252265	0.002069	0.1870795

2.3. Data Processing

From Table 1, each indicator has different dimensions; hence, direct comparison is not possible. Therefore, it is necessary to carry out the dimensionless standardization of each indicator. The positive and negative indicators have different influence directions on agricultural drought vulnerability so the treatment methods should be different.

Suppose there are k provinces, n years and m evaluation indicators; then $X_{\theta ij}$ represents the j indicator value of province i in year θ. The normalized value after treatment is expressed as $S_{\theta ij}$ ($0 < S_{\theta ij} < 1$). X_{min} is the minimum value of the j indicator and X_{max} is the maximum value of the j indicator.

Positive indicator:

$$S_{\theta ij} = \frac{X_{\theta ij} - X_{min}}{X_{max} - X_{min}} \tag{1}$$

Negative indicator:

$$S_{\theta ij} = \frac{X_{max} - X_{\theta ij}}{X_{max} - X_{min}} \tag{2}$$

2.4. Improved Entropy Weight Method

There are two methods to determine the weight: subjective weight method and objective weight method. This paper chooses the entropy weighting method (one of the objective weighting methods) for indicator weighting, which overcomes the subjective arbitrariness of the subjective weighting method and makes the weighting more scientific. The improved entropy weighting method has the following methods and steps [60,61]:

Build the matrix $Y_{\theta ij}$:

$$Y_{\theta ij} = \frac{S_{\theta ij}}{\sum_{\theta} \sum_{i} S_{\theta ij}} \tag{3}$$

Calculate indicator information entropy e_j:

$$e_j = -\frac{1}{\ln kn} \sum_{\theta} \sum_{i} Y_{ij} \ln(Y_{\theta ij}) \tag{4}$$

Find indicator difference coefficient (redundancy) g_j:

$$g_j = 1 - e_j \tag{5}$$

The weight of each indicator w_j:

$$W_j = \frac{g_j}{\sum_{j} g_j} \tag{6}$$

2.5. Vulnerability Assessment Model

This paper chooses the weighted comprehensive scoring method and uses $V_{\theta i}$ to represent the degree of vulnerability. The improved vulnerability assessment model of agricultural drought in China is as follows:

$$V_{\theta i} = \sum_{j} (W_j \times S_{\theta ij}) \tag{7}$$

2.6. K-Means Clustering Algorithm

According to the above steps, to calculate the degree of vulnerability of the target year of China's agricultural drought in various regions and put them in ascending order. After that, to use k-means clustering algorithm in Stata to grade the vulnerability of China's agricultural drought disaster [48,62].

Algorithms usually use Euclidean distance to calculate the distance between samples. The calculation formula is as follows:

$$d(x,y) = \sqrt{\sum_{i=1}^{n} (x_i - y_i)^2} \tag{8}$$

Suppose the class center of the k category is $center_k$, then the formula of $center_k$ is updated as follows:

$$center_k = \frac{1}{|c_k|} \sum_{x_i \in c_k} x_i \tag{9}$$

The clustering algorithm requires continuous iteration to re-classify and update $center_k$ value. Whenever the maximum number of iterations has been reached or the objective function is less than the threshold value, the iteration ends. The objective function is as follows:

$$J = \sum_{k=1}^{k} \sum_{x_i \in c_k} d(x_i, center_k) \tag{10}$$

2.7. Contribution Model

The main contributing factors of agricultural drought vulnerability in China are analyzed by contribution model. R_j is the weight of the j criterion level indicator; C_{ij} is the contribution degree of the j indicator factor to the vulnerability of the i evaluation object; U_r represents the contribution of the first level indicators to vulnerability; F_j is the weight of single indicator to total target; I_{ij} is the indicator membership degree (that is to say the proportion of Single factor indicator accounts for in vulnerability results. In the obstacle degree model, the indicator deviation degree is the difference between the individual index factor evaluation value and 100%. Therefore, the factor membership in the contribution degree model is the single indicator factor evaluation value ratio 100%) [32].

$$F_j = R_j \times W_j \tag{11}$$

$$I_{ij} = 1 - S_{\theta ij} \tag{12}$$

$$C_{ij} = \frac{F_j \times I_{ij}}{\sum_j (F_j \times I_{ij})} \tag{13}$$

$$U_r = \sum C_{ij} \tag{14}$$

3. Results and Discussion

3.1. Agricultural Drought Vulnerability in China

According to Formulas (1) and (2), after the data is being nondimensionalized and standardized, we use the calculation steps of the entropy weight method (Formulas (3)–(6)) to calculate the weight of each indicator, which is shown in Table 3.

It can be seen that, for the two first-class indicators, sensitivity index weight is 0.594 and resilience index weight is 0.406. Among them, multiple-crop index, annual average temperature, the forest coverage rate, the effective irrigation rate, and agriculture in GDP proportion have higher weight of over 0.1. Since the weight of agricultural in GDP proportion is 0.099, which is very close to 0.1, we also put significant important over this index.

According to Formula (7), the agricultural drought vulnerability degree of each region in 1978, 1983, 1988, 1993, 1998, 2003, 2008, 2013, and 2018 have been calculated and shown in Table 4.

Table 3. The weight of each indicator.

First-Level Indicator	Weight	Second-Level Indicatorand the Direction of Influence	Weight
A. Sensitivity	0.594	A1. Agriculture in GDP proportion (+)	0.099
		A2. Multiple-crop index (+)	0.145
		A3. Rural population proportion (+)	0.071
		A4. Annual average temperature (+)	0.106
		A5. Annual sunshine duration (+)	0.081
		A6. Annual precipitation (-)	0.091
B. Resilience	0.406	B1. The forest coverage rate (-)	0.102
		B2. Net income per capita of rural residents (-)	0.049
		B3. Food production per capita (-)	0.049
		B4. Real GDP per capita (-)	0.045
		B5. The effective irrigation rate (-)	0.101
		B6. Agricultural fertilizer per unit area (-)	0.061

Table 4. Vulnerability of agricultural drought in different provinces.

Province	Years									Level	Sort
	1978	1983	1988	1993	1998	2003	2008	2013	2018		
Shanghai	0.492	0.476	0.491	0.438	0.419	0.378	0.426	0.469	0.303	0.432	1
Beijing	0.505	0.517	0.492	0.513	0.427	0.417	0.399	0.408	0.541	0.469	2
Zhejiang	0.579	0.553	0.407	0.528	0.502	0.443	0.419	0.417	0.375	0.469	3
Guangdong	0.499	0.476	0.501	0.474	0.545	0.500	0.446	0.464	0.345	0.472	4
Fujian	0.549	0.520	0.524	0.521	0.485	0.434	0.467	0.444	0.349	0.477	5
Tianjin	0.514	0.536	0.508	0.522	0.493	0.434	0.460	0.457	0.483	0.490	6
Jilin	0.503	0.491	0.493	0.508	0.515	0.487	0.510	0.492	0.535	0.504	7
Heilongjiang	0.485	0.500	0.493	0.501	0.499	0.479	0.526	0.493	0.562	0.504	8
Liaoning	0.520	0.520	0.534	0.535	0.509	0.494	0.521	0.509	0.582	0.525	9
Jiangsu	0.593	0.612	0.609	0.558	0.526	0.442	0.524	0.530	0.392	0.532	10
Hunan	0.526	0.527	0.554	0.539	0.576	0.508	0.592	0.585	0.398	0.534	11
Jiangxi	0.598	0.571	0.598	0.591	0.547	0.496	0.577	0.538	0.422	0.549	12
Shaanxi	0.590	0.581	0.572	0.570	0.560	0.529	0.586	0.573	0.545	0.567	13
Hubei	0.607	0.587	0.630	0.610	0.585	0.512	0.585	0.574	0.463	0.573	14
Sichuan	0.562	0.581	0.568	0.572	0.650	0.572	0.635	0.616	0.516	0.586	15
Guangxi	0.613	0.636	0.645	0.638	0.587	0.573	0.574	0.581	0.507	0.595	16
Shandong	0.663	0.662	0.667	0.622	0.587	0.528	0.567	0.578	0.495	0.597	17
Inner Mongolia	0.635	0.638	0.634	0.615	0.613	0.568	0.558	0.548	0.579	0.599	18
Shanxi	0.623	0.628	0.615	0.621	0.607	0.551	0.582	0.580	0.598	0.601	19
Hebei	0.639	0.662	0.640	0.632	0.592	0.556	0.589	0.593	0.543	0.605	20
Anhui	0.699	0.646	0.681	0.630	0.612	0.533	0.603	0.585	0.462	0.606	21
Yunnan	0.608	0.614	0.616	0.631	0.614	0.588	0.607	0.642	0.577	0.611	22
Xinjiang	0.634	0.633	0.612	0.587	0.581	0.599	0.634	0.633	0.601	0.613	23
Chongqing	0.674	0.666	0.675	0.681	0.616	0.618	0.583	0.588	0.484	0.620	24
Qinghai	0.583	0.621	0.614	0.618	0.636	0.598	0.619	0.648	0.649	0.621	25
Henan	0.690	0.681	0.710	0.669	0.653	0.543	0.676	0.622	0.483	0.636	26
Hainan	0.657	0.710	0.683	0.649	0.705	0.606	0.584	0.576	0.586	0.640	27
Tibet	0.615	0.674	0.681	0.678	0.647	0.606	0.597	0.610	0.694	0.645	28
Guizhou	0.634	0.659	0.699	0.659	0.734	0.593	0.619	0.641	0.583	0.647	29
Ningxia	0.656	0.686	0.679	0.669	0.648	0.615	0.644	0.622	0.643	0.651	30
Gansu	0.629	0.660	0.645	0.667	0.687	0.643	0.674	0.681	0.735	0.669	31

It can be noticed that from 1978 to 2018, the vulnerability of agricultural drought in China has decreased year by year. Agricultural drought vulnerability in Gansu, Ningxia, Guizhou, and Tibet is relatively high with an average value of more than 0.648. Agricultural drought vulnerability in Shanghai, Beijing, and Zhejiang is low with the average value less than 0.47.

3.1.1. Classification of Agricultural Drought Vulnerability

In order to accurately classify China's agricultural drought vulnerability, according to Formulas (8)–(10) by using k-means clustering algorithm in Stata, the China Agricultural Drought Vulnerability Index (ADVI) is divided into four ranges between 0 to 1 and they are shown in Table 5.

Table 5. Classification of agricultural drought vulnerability.

Grade	Range
Low vulnerability	(0, 0.463)
Mild vulnerability	(0.463, 0.552)
Middle vulnerability	(0.552, 0.628)
High vulnerability	(0.628, 1)

3.1.2. Spatial Distribution and Evolution

According to the classification of agricultural drought vulnerability in Table 5, in order to express the results more clearly, this study uses ArcGIS to display the research results. The assessment results of agricultural drought vulnerability in China are shown in Figure 1a–i.

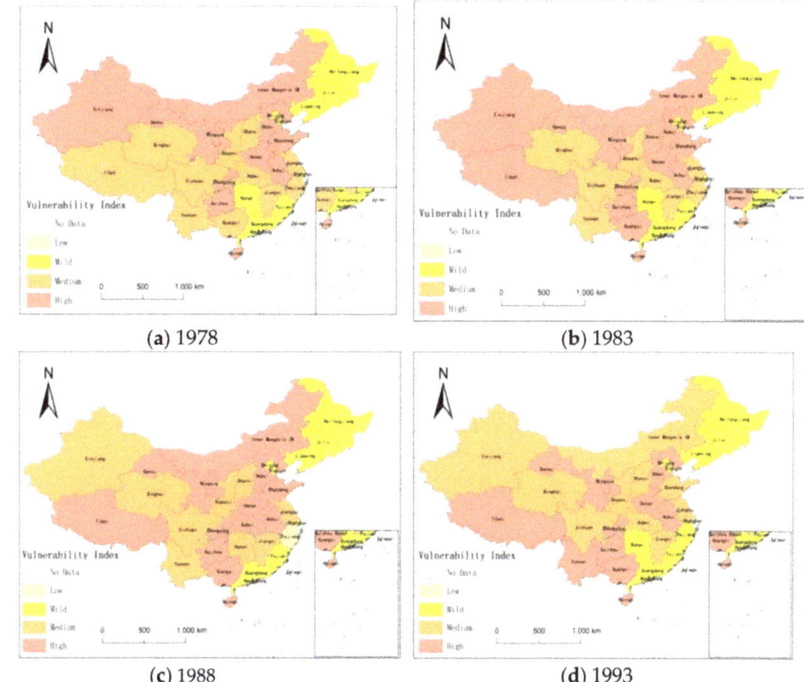

(a) 1978

(b) 1983

(c) 1988

(d) 1993

Figure 1. *Cont.*

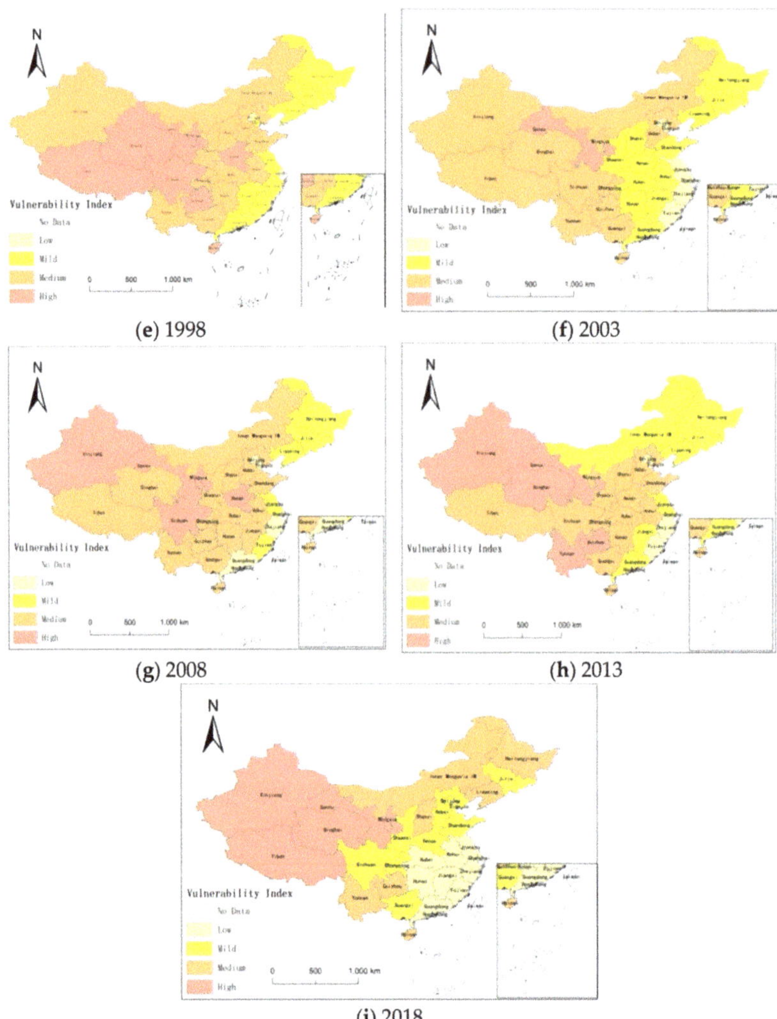

Figure 1. Spatiotemporal evolution of agricultural drought vulnerability in China (**a–i**).

It can be seen that the vulnerability of agricultural drought in China has changed significantly over time: from 1978 to 2018, the number of provinces and cities in low and mild vulnerability state has been increasing.

From 1978 to 2018, the spatial distribution pattern of agricultural drought vulnerability in China was obvious:

(1) The overall agricultural drought vulnerability in China is 0.569, which is at a moderate fragile level. This is in line with the characteristics of frequent droughts and serious losses in China [63–65].

(2) Highly vulnerability level: (0.628 < ADVI < 1) Over time, the number of cities in highly vulnerable areas has decreased, which mainly included Xizang, Guizhou, Ningxia, Gansu, etc. Among them, Gansu, Ningxia have higher vulnerability to drought, which is consistent with the research results of other scholars [59,66]. Firstly, most of these areas have complex terrain conditions and less precipitation. Drought

is their main natural feature. Secondly, the region is less developed compared with other regions and real GDP per capita is low while agriculture in GDP proportion and rural population proportion is high. It means that farmers are highly dependent on agricultural and natural conditions. With high sensitivity and weak resilience when drought occurs, the number of highly vulnerable provinces and cities are inevitably high.

(3) Middle vulnerability level: (0.552 < ADVI < 0.628) The number of provinces and cities in this region is stable and it accounts for nearly half of the total number of provinces and cities in China and most of them are concentrated in Central China. It included Inner Mongolia, Sichuan, Hebei, Anhui, etc. Most of them are important grain production bases in China and major agricultural provinces. Agriculture in GDP proportion, multiple-crop index and rural population proportion are high. It reflects that the region has a strong dependence on agriculture with high land utilization rate and heavy water demand.

(4) Low and mild vulnerability level: (0 < ADVI < 0.552) Although there has seen a small fluctuation in the number of slightly vulnerable provinces and cities, the overall trend shows a stable and marginal increase. This is consistent with the research results of some scholars [22,67]. The provinces and cities in this region such as Shanghai, Zhejiang, Beijing, and Tianjin have a high level of economic development. Their high real GDP per capita gives them better response ability and post disaster recovery ability when disasters occur. At the same time, those provinces and cities tend to have a small agricultural planting area multiple-crop index, agriculture in GDP proportion and rural population proportion are also low. When we turn to those provinces and cities in the Northeast China like Heilongjiang, Jilin, and Liaoning, their land is sparsely populated and the food production per capita is high. They also have high latitude, low average temperature, and less evaporation. The annual sunshine duration is long and the crops normally harvest once a year. With lower multiple-crop index, the water demand is lesser and the sensitivity of disaster is weak.

3.2. Analysis on the Influencing Factors of Agricultural Drought Vulnerability in China

3.2.1. Factor Contribution Analysis of First Level Index

According to the research results of agricultural drought vulnerability assessment in China, it can be noticed that the distribution of provinces and cities in different vulnerability levels has certain regional characteristics. Therefore, this paper studies the influencing factors of vulnerability in different regions. It adopts China's six geographic regions: North China, Northeast China, Northwest China, East China, Central and Southern China, and Southwest China. According to the Formulas (11)–(14), the contribution of sensitivity and resilience is shown in Figures 2 and 3.

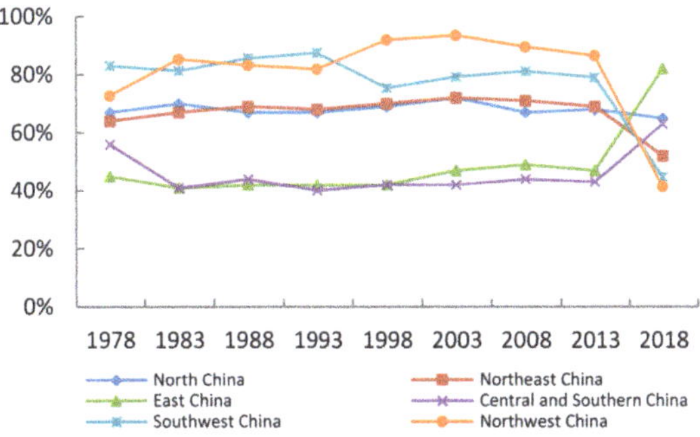

Figure 2. Changes in the contribution of sensitivity indicators.

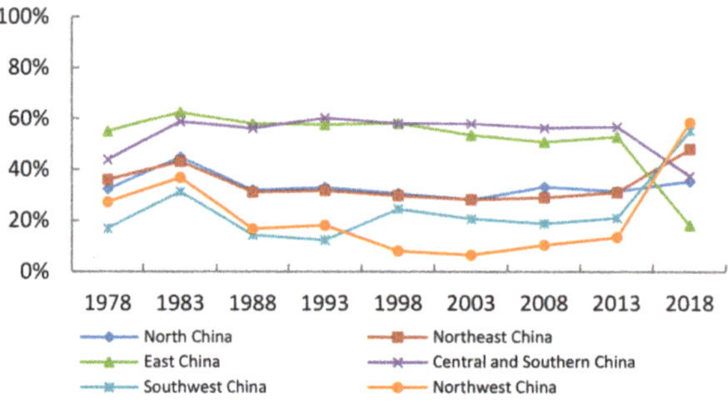

Figure 3. Changes in the contribution of resilience indicators.

As shown in Figure 2, looking at the trends as a whole, during the period of 1978–2013, the contribution of sensitivity indexes in different regions was relatively stable. From 2013 to 2018, except for North China, the contribution ratio of sensitivity indicators in other regions changed dramatically.

The contribution of sensitivity indicators in Northeast China, Northwest China, and Southwest China have declined. Possible reasons are as follows: agriculture in GDP proportion, multiple-crop index, and annual sunshine duration has decreased drastically while annual precipitation has increased significantly. Farmers in these regions have become less dependent on agricultural income. Lower land use has reduced water demand, and hence there is less evaporation.

The contribution of sensitivity indexes in East China and Central South have increased significantly. Possible reasons are as follows: agriculture in GDP proportion and multiple-crop index have increased while land use in the region has improved and water demand has increased.

As shown in Figure 3, looking at the trends as a whole, during the period of 1978–2013, the contribution of resilience indexes in different regions was relatively stable. From 2013 to 2018, except for North China, the contribution ratio of resilience indicators in other regions changed dramatically.

The contribution of resilience indicators in Northeast China, Northwest China, and Southwest China have increased. I think the possible reasons are as follows: the contribution of forest coverage rate, food production per capita, and agricultural fertilizer per unit area have increased while net income per capita of rural residents has declined. The region is less dependent on agriculture, needs to improve its ability to conserve water, and is less able to cope with disasters and recover from disasters.

The contribution of resilience indicators in East China, Central China, and South China have decreased evidently. Possible reasons are as follows: the forest coverage rate, the effective irrigation rate and agricultural fertilizer per unit area have increased. The region's ability to conserve water is better, so the recovery capacity after the disaster is improved.

On the whole, the contribution of resilience in East China, Central, and Southern China is more than 50%, which is greater than the contribution of sensitivity. Cities in the central and southern region: the Yellow River valley passes through Henan, and the Yangtze River valley passes through Hubei and Hunan. Guangxi, Guangdong, and Hainan are adjacent to the sea. Therefore, the relative abundance of water resources and the contribution of sensitivity indicators are less. East China includes Shanghai, Jiangsu, Zhejiang, Anhui, Fujian, Jiangxi, and Shandong. These several provinces and cities are adjacent to the sea, or within the territory of the river flow through, and the level of economic development is at the forefront of China. Therefore, the contribution degree of resilience index is relatively high.

Northwest China, Northeast China, North China, and Southwest China have higher sensitivity contributions than that of resilience. The sensitivity contribution of the Southwest China and Northwest China is as high as 75%. Desert is widespread in the northwest and annual precipitation about 200–400 mm. Deep inland and blocking the arrival of moist air. Northeast China is an important grain production base with a large area of cultivated land. Compared with the coastal areas, its economic development level is not high.

3.2.2. Factor Contribution Analysis of Secondary Index

Select the top four indicators of contribution among the 12 indicators and the calculated results are shown in Table 6.

Table 6. The top four indicators and contribution of agricultural drought vulnerability in different regions.

Regions Years	North China		Northeast China		East China		Central and Southern China		Southwest China		Northwest China	
1978	A3	A1	A2	A3	A5	B4	A5	A6	A5	A3	A2	A4
	19.14	17.33	17.69	15.33	16.78	13.51	25.73	19.51	26.02	17.52	30.17	24.05
	A2	B4	A4	A1	B5	A6	B6	B5	A2	A6	A1	B6
	15.34	13.98	15.03	12.26	13.39	11.64	12.95	11.22	16.48	12.99	18.47	10.72
1983	A1	A6	A2	A3	A5	B5	A5	B6	A5	A6	A2	A6
	16.04	15.46	16.01	14.30	21.32	15.00	32.92	17.32	32.83	12.67	26.31	22.12
	A2	A3	A4	B1	B4	A1	B1	B5	A3	A2	A4	A1
	14.50	13.29	13.60	10.88	14.01	10.93	14.13	13.64	12.60	11.74	20.80	12.16
1988	A1	A3	A2	A3	A5	B2	A5	B6	A5	A3	A2	A6
	15.47	15.17	16.48	14.72	15.04	13.80	29.65	16.36	33.60	19.91	21.14	19.18
	A2	B4	A4	A1	B5	B4	B1	A3	A2	A4	A4	A3
	13.92	12.48	14.00	13.28	13.03	12.94	14.00	12.46	11.73	11.01	17.64	14.32
1993	A1	A2	A2	A4	A5	B4	A5	B1	A5	A3	A2	A6
	15.51	15.33	16.44	13.97	20.21	14.89	27.93	15.28	31.28	26.59	21.74	19.60
	A6	B4	A3	A1	A1	B2	B6	B5 10.7	A6	A4	A4	A1
	13.82	12.51	12.92	11.21	13.02	12.97	15.26		10.58	10.54	18.43	10.98
1998	A1	A2	A2	A3	A5	A1	A5	B1	A5	A4	A2	A6
	14.71	14.60	17.43	15.57	19.33	13.14	29.65	17.82	40.25	14.11	24.87	21.12
	A3	A6	A4	B1	B2	B1	B6	B5	A6	A2	A4	A3
	12.84	12.15	14.81	11.85	11.48	10.88	15.27	10.36	11.14	9.93	19.22	18.46
2003	A3	A2	A2	A3	A5	B4	A5	B1	A5	B1	A2	A6
	15.25	13.85	16.24	15.17	16.10	12.34	31.60	21.12	36.68	13.65	28.07	22.37
	A1	A6	A4	B1	A1	B2	B6	A3	A6	A2	A4	A3
	13.80	12.06	15.03	12.02	12.30	12.34	15.45	10.32	13.37	16.61	20.78	11.94
2008	A3	A1	A2	A4	A5	A3	A5	B1	A5	A2	A2	A6
	15.20	13.76	18.76	15.94	16.13	13.64	31.02	21.16	32.99	21.60	27.46	23.87
	A2	A6	A3	A6	A1	B2	B6	A3	B1	A6	A4	A1
	11.95	11.87	14.19	11.90	11.56	11.23	16.46	11.95	14.10	11.82	22.45	9.96
2013	A3	A1	A2	A4	A3	A5	A5	B1	A5	A2	A2	A6
	15.30	13.85	19.32	16.41	15.52	14.56	31.32	20.26	28.93	19.64	26.18	25.94
	A6	A2	A3	B1	A1	B2	B6	A3	B1	A6	A4	A3
	13.85	11.94	12.89	11.16	13.52	11.92	15.76	11.84	15.33	14.52	20.60	7.40
2018	A3	A1	A4	A6	A2	A5	A5	A2	A5	B4	A6	A4
	18.97	17.17	19.15	14.24	21.93	19.19	25.82	25.37	21.36	13.26	15.32	13.55
	A6	A2	B4	A3	A3	A1	B4	A3	B2	A6	B1	B4
	15.76	12.10	12.87	12.78	18.00	17.68	12.02	10.09	10.30	10.14	13.06	11.94

On the whole, referring to the calculation results above, it can be noticed that the contribution factors of agricultural drought vulnerability in China mainly focus on sensitivity. Among them, A2 (multiple-crop index) and A3 (rural population proportion) are more important. It shows that these two indicators have a greater impact on the vulnerability of agricultural drought. We should sustainably reduce the land utilization rate, reduce the water demand, strengthen the vocational skills training of rural residents, supervise and protect the legitimate rights and interests of migrant workers, and promote the transfer of rural population to cities. Hence, the proportion of rural population can be effectively reduced, and the vulnerability of agricultural drought can also be mitigated.

Among the indexes of resilience, B1 (the forest coverage rate) and B4 (real GDP per capita) are more important. According to the data from the Ninth National Forest Resources Inventory, China's forest coverage rate is still lower than the world average level. Strengthening afforestation is highly effective for soil and water conservation, hence reducing water evaporation and improving the forest coverage rate. The adverse impact from drought can also be reduced significantly. Similarly, the higher the real GDP per capita, the easier the recovery would be after droughts. The GDP per capita China is still relatively low in the worldwide spectrum, although China's total domestic GDP ranks No. 1 in the world.

The factor with the least contribution is B3 (Food production per capita). China has a large planting area of crops with high and stable grain yield, so it has little impact on agricultural drought vulnerability.

4. Conclusions, Limitations, and Future Research

The paper uses entropy weight method, weighted comprehensive scoring method as well as k-means clustering algorithm to calculate and classify the vulnerability degree of agricultural drought. ArcGIS was used to show the spatial and temporal changes of agricultural drought vulnerability in China, then, using the contribution model to analyze the influencing factors and the degrees of agricultural drought vulnerability in China, the results show that:

(1) From 1978 to 2018, the vulnerability of agriculture to drought has been reduced and the numbers of China's highly vulnerable cities have declined. During the same time, there has been a trend appeared that high vulnerability cities have converted to the middle-level vulnerability cities while middle-level vulnerability cities have converted to mild-level or low-level vulnerability cities. The vulnerability towards agricultural drought disasters in China was generally at the middle and mild level in most regions while the vulnerability in Northwest China and Southwest China were more severe.

(2) China's agricultural drought vulnerability is mainly affected by sensitivity factors, among which multiple-crop index and the proportion of rural population have a higher contribution compared with other indicators. For resilience index, forest coverage rate and real GDP per capita carry a more important role.

In the data collection process of this paper, partially due to the wide time span selected, there is a lack of data from early years. Therefore, those crucial indicators that can be easily obtained with clean data have been selected for evaluation. Imperfection still exists although these selected indicators can truly reflect the vulnerability characteristics of agricultural drought in China. In the future, we will do some comparative studies on different evaluation methods to further optimize the research results.

Author Contributions: Conceptualization, H.G. and C.P.; Data curation, C.P. and J.C.; Formal analysis, H.G.; Funding acquisition, H.G.; Investigation, C.P. and J.C.; Methodology, H.G. and J.C.; Project administration H.G.; Resources, H.G.; Software, H.G. and J.C.; Supervision, C.P.; Validation, H.G. and J.C.; Visualization, H.G.; Writing—original draft H.G. and J.C.; Writing—review and editing, J.C. and C.P. All authors have read and agreed to the published version of the manuscript.

Funding: Social Science Fund Project of Jilin Province, China, grant No.2018BS33; MOE (Ministry of Education in China) Project of Humanities and Social Sciences, grant/award No.18YJC630128; Social Science Fund Project of "the 13th Five-Year" of Education Department of Jilin Province, China, grant No. JJKH20190736SK; and Jilin Province Education Planning Project, China, grant No. JJKH20190243SK.

Institutional Review Board Statement: Not applicable.

Informed Consent Statement: Not applicable.

Data Availability Statement: Publicly available datasets were analyzed in this study. This data can be found here: China Statistical Yearbook (http://www.stats.gov.cn/english/Statisticaldata/AnnualData/, accessed on 17 July 2020), China Rural Statistical Yearbook (https://data.cnki.net/trade/yearbook/single/n2019120190?z=z009, accessed on 23 July 2020), China Meteorological Administration Data Network (https://data.cma.cn/, accessed on 28 July 2020).

Conflicts of Interest: The authors declare no conflict of interests.

References

1. Sun, G.Z. Analysis of Natural disasters in China disaster and disaster reduction countermeasures. *Bull. Chin. Acad. Sci.* **1990**, *01*, 37–39.
2. Tian, Z.H.; Li, X.X. Drought Change Cycle in China's Major Grain Producing Areas Based On Disaster Situation. *J. Catastrophol.* **2020**, *35*, 40–45.
3. Duan, M. Ministry of Emergency Management: The Direct Economic Losses Caused by Natural Disasters in November Totaled 2.29 Billion Yuan. Available online: https://www.mem.gov.cn/xw/mtxx/201912/t20191205_341963.shtml (accessed on 15 December 2019).
4. Li, Z.Y. The Emergency Management Department Released the National Natural Disaster Situation in April 2020. Available online: https://www.mem.gov.cn/xw/bndt/202005/t20200508_352280.shtml (accessed on 8 May 2020).
5. Zarafshani, K.; Sharafi, L.; Azadi, H.; Van Passel, S. Vulnerability Assessment Models to Drought: Toward a Conceptual Framework. *Sustainability* **2016**, *8*, 588. [CrossRef]
6. Wu, J.S.; Lin, X.; Wang, M.J.; Peng, J. Assessing Agricultural Drought Vulnerability by a VSD Model: A Case Study in Yunnan Province, China. *Sustainability* **2017**, *9*, 918. [CrossRef]
7. Guan, Y.H.; Zheng, F.L.; Zhang, P.; Qin, C. Spatial and temporal changes of meteorological disasters in China during 1950–2013. *Nat. Hazards* **2015**, *75*, 2607–2623. [CrossRef]
8. IPCC. *Climate Change and Land*; Special Report; Intergovernmental Panel on Climate Change: Geneva, Switzerland, 2019.
9. IPCC. *AR5 Climate Change 2014: Impacts, Adaptation and Vulnerability*; Intergovernmental Panel on Climate Change: Geneva, Switzerland, 2014.
10. Murthy, C.S.; Yadav, M.; Ahamed, J.M.; Laxman, B.; Prawasi, R.; Sai, M.V.R.S.; Hooda, R.S. A study on agricultural drought vulnerability at disaggregated level in a highly irrigated and intensely cropped state of India. *Environ. Monit. Assess.* **2015**, *187*, 140. [CrossRef] [PubMed]
11. Yi, J.M. *Risk Assessment of Agricultural Drought in Dalian*; Liaoning Normal University: Dalian, China, 2010.
12. Yuan, X.C. *Climate Change Risk Assessement: Modeling and Applications*; Beijing Institute of Technology: Beijing, China, 2016.
13. Yan, L. A Study on Dynamic Risk Assessment of Maize Drought Disaster in Northwestern Liaoning Province. Master's Thesis, Northeast Normal University, Changchun, China, 2012.
14. Pang, Z.Y.; Dong, S.N.; Zhang, J.Q.; Tong, Z.J.; Liu, X.P.; Sun, Z.Y. Evaluation and regionalization of maize vulnerability to drought disaster in Western Jilin Province based on CERES-Maize model. *Chin. J. Eco Agric.* **2014**, *22*, 705–712.
15. Farhangfar, S.; Bannayan, M.; Khazaei, H.R.; Baygi, M.M. Vulnerability assessment of wheat and maize production affected by drought and climate change. *Int. J. Disaster Risk Reduct.* **2015**, *13*, 37–51. [CrossRef]
16. Liu, X.J.; Zhang, J.Q.; Ma, D.L. The Study of Vulnerability Assessment on Maize Drought in the Northwest of Liaoning Province Based on Modis. *Chin. J. Agric. Resour. Reg. Plan.* **2016**, *37*, 44–49.
17. Kim, S.-M.; Kang, M.-S.; Jang, M.-W. Assessment of agricultural drought vulnerability to climate change at a municipal level in South Korea. *Paddy Water Environ.* **2018**, *16*, 699–714. [CrossRef]
18. Lestari, D.R.; Pigawati, B. Drought disaster vulnerability mapping of agricultural sector in Bringin District, Semarang Regency. *IOP Conf. Ser. Earth Environ. Sci.* **2018**, *123*, 012031. [CrossRef]
19. Huang, J.; Fu, P.; Tong, J.P.; She, J.W.; Zhang, J.X. Evaluating the vulnerability of agricultural drought in Hetao Irrigation Area of Inner Mongolia Based on super efficiency DEA. *IOP Conf. Ser. Earth Environ. Sci.* **2019**, *330*, 032020. [CrossRef]
20. Frischen, J.; Meza, I.; Rupp, D.; Wietler, K.; Hagenlocher, M. Drought Risk to Agricultural Systems in Zimbabwe: A Spatial Analysis of Hazard, Exposure, and Vulnerability. *Sustainability* **2020**, *12*, 752. [CrossRef]
21. Das, R.; Das, P.K.; Bandyopadhyay, S.; Raj, U. Trends And Vulnerability Assessment of Meteorological and Agricultural Drought Conditions over Indian Region Using Time-Series (1982–2015) Satellite Data. *Int. Arch. Photogramm. Remote Sens. Spat. Inf. Sci.* **2019**, *XLII-3*, 453–459. [CrossRef]
22. Pei, H.; Wang, X.Y.; Fang, S.F. Study on Temporal-spatial Evolution of Agricultural Drought Vulnerability of China Based on DEA Model. *J. Catastrophol.* **2015**, *30*, 64–69.
23. Yuan, X.-C.; Wang, Q.; Wang, K.; Wang, B.; Jin, J.-L.; Wei, Y.-M. China's regional vulnerability to drought and its mitigation strategies under climate change: Data envelopment analysis and analytic hierarchy process integrated approach. *Mitig. Adapt. Strateg. Global Chang.* **2015**, *20*, 341–359. [CrossRef]
24. Ekrami, M.; Marj, A.F.; Barkhordari, J.; Dashtakian, K. Drought vulnerability mapping using AHP method in arid and semiarid areas: A case study for Taft Township, Yazd Province, Iran. *Environ. Earth Sci.* **2016**, *75*, 1039. [CrossRef]

25. Lakshmi, S.V.; Rakshith, R.K.; Manoj, N. Drought vulnerability assessment and mapping using Multi-Criteria decision making (MCDM) and application of Analytic Hierarchy process (AHP) for Namakkal District, Tamilnadu, India. *Mater. Today Proc.* **2020**, *43*, 1592–1599.
26. Sik, M.Y.; Ho, N.W.; Gi, J.M.; Joong, K.H.; Ku, K.; Chul, L.J.; Hyun, H.T.; Kwangya, L. Evaluation of Regional Drought Vulnerability Assessment Based on Agricultural Water and Reservoirs. *J. Korean Soc. Agric. Eng.* **2020**, *62*, 97–109.
27. Wu, H.; Qian, H.; Chen, J.; Huo, C. Assessment of Agricultural Drought Vulnerability in the Guanzhong Plain, China. *Water Resour. Manag.* **2017**, *31*, 1557–1574. [CrossRef]
28. Zagade, N.D.; Umrikar, B.N. Drought severity modeling of upper Bhima river basin, western India, using GIS–AHP tools for effective mitigation and resource management. *Nat. Hazards* **2020**, *105*, 1165–1188. [CrossRef]
29. Balaganesh, G.; Malhotra, R.; Sendhil, R.; Sirohi, S.; Maiti, S.; Ponnusamy, K.; Sharma, A.K. Development of composite vulnerability index and district level mapping of climate change induced drought in Tamil Nadu, India. *Ecol. Indic.* **2020**, *113*, 106197. [CrossRef]
30. Liu, L.-F.; Shi, Y.; Jin, J.-L.; Xiao, Z.-C.; Deng, M.-R. Assessment on Drought Vulnerability in Subtropical Rice Areas—A Case of Hengyang Basin in China. In Proceedings of the 2013 International Conference on Advanced in Earth Sciences (ICAES 2013), Pilani, India, 21–23 September 2013; Volume 8.
31. Guo, Y.; Wang, R.; Tong, Z.J.; Liu, X.P.; Zhang, J.Q. Dynamic Evaluation and Regionalization of Maize Drought Vulnerability in the Midwest of Jilin Province. *Sustainability* **2019**, *11*, 4234. [CrossRef]
32. Xu, H. Assessment of agricultural drought vulnerability and identification of influencing factors based on the entropy weight method. *Agric. Res. Arid Areas* **2016**, *34*, 198–205.
33. Yu, Z.L.; Yang, X.J.; Shi, Y.Z. Evaluation of Urban Vulnerability to Drought in Guanzhong Area. *Resour. Sci.* **2012**, *34*, 581–588.
34. Liu, Q.J.; Wei, M.D.; Li, Z.N.; Hu, Y.K. Study on vulnerability measurement and its differences of farmers in typically deep poverty-stricken regions based on household micro-data survey in Northwest Inland Arid Regions of China. *J. Chin. Agric. Mech.* **2020**, *41*, 215–223.
35. Addis, E. Agro-ecosystems' Vulnerability to Climate Change in Drought Prone Areas of Northeastern Ethiopia. *J. Nat. Sci. Res.* **2017**, *7*, 21–41.
36. Alamdarloo, E.H.; Khosravi, H.; Nasabpour, S.; Gholami, A. Assessment of drought hazard, vulnerability and risk in Iran using GIS techniques. *J. Arid Land* **2020**, *12*, 984–1000. [CrossRef]
37. Huang, L.; Yang, P.; Ren, S. The Vulnerability Assessment Method for Beijing Agricultural Drought. In *Computer and Computing Technologies in Agriculture VII*; Springer: Berlin/Heidelberg, Germany, 2014; p. 12.
38. Jain, V.K.; Pandey, R.P.; Jain, M.K. Spatio-temporal assessment of vulnerability to drought. *Nat. Hazards* **2015**, *76*, 443–469. [CrossRef]
39. Rojas, O.; Vrieling, A.; Rembold, F. Assessing drought probability for agricultural areas in Africa with coarse resolution remote sensing imagery. *Remote Sens. Environ.* **2010**, *115*, 343–352. [CrossRef]
40. Zhang, J.; Mu, Q.Z.; Huang, J.X. Assessing the remotely sensed Drought Severity Index for agricultural drought monitoring and impact analysis in North China. *Ecol. Indic.* **2016**, *63*, 296–309. [CrossRef]
41. Guo, H.; Zhang, X.M.; Lian, F.; Gao, Y.; Lin, D.G.; Wang, J.A. Drought Risk Assessment Based on Vulnerability Surfaces: A Case Study of Maize. *Sustainability* **2016**, *8*, 813. [CrossRef]
42. Chen, J.F.; Deng, M.H.; XIa, L.; Wang, H.M. Risk Assessment of Drought, Based on IDM-VFS in the Nanpan River Basin, Yunnan Province, China. *Sustainability* **2017**, *9*, 1124. [CrossRef]
43. Zeng, Z.Q.; Wu, W.X.; Li, Z.L.; Zhou, Y.; Huang, H. Quantitative Assessment of Agricultural Drought Risk in Southeast Gansu Province, Northwest China. *Sustainability* **2019**, *11*, 5533. [CrossRef]
44. Zarafshani, K.; Sharafi, L.; Azadi, H.; Hosseininia, G.; De Maeyer, P.; Witlox, F. Drought vulnerability assessment: The case of wheat farmers in Western Iran. *Glob. Planet. Chang.* **2012**, *98–99*, 122–130. [CrossRef]
45. Wu, J.J.; Geng, G.P. Global vulnerability to agricultural drought and its spatial characteristics. *Sci. Sin.* **2017**, *47*, 910–920. [CrossRef]
46. Bahareh Kamali, K.C.A.; Bernhard, W.; Yang, H. A Quantitative Analysis of Socio-Economic Determinants Influencing Crop Drought Vulnerability in Sub-Saharan Africa. *Sustainability* **2019**, *11*, 6135. [CrossRef]
47. Dai, Y.Q.; Wang, L.G.; Hu, B. Agricultural Drought Vulnerability Evaluation of Gansu Province. *J. Appl. Sci. Electron. Inf. Eng.* **2018**, *36*, 515–523.
48. Huang, L.M. Vulnerability Assessment of Agricultural Drought in Guizhou Province. Master's Thesis, Guizhou Normal University, Guiyang, China, 2017.
49. Ji, C.R. *Research on the Vulnerability Assessment of Agricultural Drought Disaster in Henan Province*; Henan University: Kaifeng, China, 2013.
50. Li, M.N. Evaluation of agricultural vulnerability to drought in Guanzhong Area. *Resour. Sci.* **2016**, *38*, 166–174.
51. Pei, W.; Fu, Q.; Liu, D.; Li, T.-X.; Cheng, K. Assessing agricultural drought vulnerability in the Sanjiang Plain based on an improved projection pursuit model. *Nat. Hazards* **2016**, *82*, 683–701. [CrossRef]
52. National Bureau of Statistics. Available online: http://www.stats.gov.cn/ (accessed on 17 July 2020).
53. China Meteorological Data Network. Available online: http://data.cma.cn/ (accessed on 28 July 2020).

54. Sun, R.; Liu, G.D.; Wang, C.D.; Zheng, S. Application of Combination Weighting Method to Agricultural Drought Vulnerability Assessment and Regionalization. *Environ. Sci. Technol.* **2015**, *38*, 374–378.
55. Wu, L.; Feng, L.P.; Li, Y.Z.; Wang, J.; Wu, L.H. A Yield-Related Agricultural Drought Index Reveals Spatio-Temporal Characteristics of Droughts in Southwestern China. *Sustainability* **2019**, *11*, 714. [CrossRef]
56. Xu, D.M.; Li, P.Y.; Wang, W.C.; Yin, P.B. Analysis on Vulnerability of Agricultural Drought Disaster Based on the Improved Grey Clustering Method. *J. Irrig. Drain.* **2016**, *35*, 87–91.
57. Ma, Y.L.; Guo, J.P.; Luan, Q.; Liu, W.P. Assessment of Agricultural Drought Vulnerability in Agro—Pastoral Ecotone in Northern Shanxi Province. *J. Catastrophol.* **2020**, *35*, 75–81.
58. Xie, J.Z.; Che, S.F.; Lin, Y. Vulnerabilidy Assessment and the Driving Force in the Management of Agricultultural Drought Hazard. *J. Southwest Univ.* **2017**, *43*, 03.
59. Wu, D.; Yan, D.H.; Yang, G.Y.; Wang, X.G.; Xiao, W.H.; Zhang, H.T. Assessment on agricultural drought vulnerability in the Yellow River basin based on a fuzzy clustering iterative model. *Nat. Hazards* **2013**, *67*, 919–936. [CrossRef]
60. Zhao, H.J.; Yu, F.W. Evaluation of Agricultural Green Development Level in Main Grain Producing Areas based on Entropy Method. *Reform* **2019**, *11*, 136–146.
61. Guo, H.; Xu, S.; Pan, C. Measurement of the Spatial Complexity and Its Influencing Factors of Agricultural Green Development in China. *Sustainability* **2020**, *12*, 9259. [CrossRef]
62. Xu, Y. Research and Application of K_means Algorithm and Swarm Intelligence Algorithm(PSO)Fusion. Master's Thesis, Inner Mongolia Agricultural University, Hohhot, China, 2019.
63. Guan, S.; Tang, G. *The Research of the Drought Loss Prediction of Distribution and Coping Strategies in China*; Yunnan University: Kunming, China, 2015.
64. Yang, Z.; Liu, Y. Spatiotemporal variation of drought in China from 1950 to 2010. In Proceedings of the Research on Water Governance and Sustainable Development in China, Kunming, Yunnan, China, 1–10 August 2012.
65. Zhou, J. Drought characteristics and economic loss assessment in China. *J. Catastrophol.* **1993**, *3*, 45–49.
66. Wang, Y.; Zhao, W.; Zhang, Q.; Yao, Y.-B. Characteristics of drought vulnerability for maize in the eastern part of Northwest China. *Sci. Rep.* **2019**, *9*, 964. [CrossRef]
67. Du, X.; Huang, S. Comprehensive assessment and zoning of vulnerability to agricultural drought in Tianjin. *Nat. Disasters* **2010**, *19*, 138–145.

International Journal of
Environmental Research and Public Health

Article

Rapid Estimation of Earthquake Fatalities in Mainland China Based on Physical Simulation and Empirical Statistics—A Case Study of the 2021 Yangbi Earthquake

Yilong Li [1], Zhenguo Zhang [1,2,3,*], Wenqiang Wang [1] and Xuping Feng [1]

[1] Department of Earth and Space Sciences, Southern University of Science and Technology, Shenzhen 518055, China; 11930700@mail.sustech.edu.cn (Y.L.); 11849528@mail.sustech.edu.cn (W.W.); 11930721@mail.sustech.edu.cn (X.F.)
[2] Southern Marine Science and Engineering Guangdong Laboratory (Guangzhou), Guangzhou 511458, China
[3] Guangdong Provincial Key Laboratory of Geophysical High-Resolution Imaging Technology, Southern University of Science and Technology, Shenzhen 518055, China
* Correspondence: zhangzg@sustech.edu.cn

Citation: Li, Y.; Zhang, Z.; Wang, W.; Feng, X. Rapid Estimation of Earthquake Fatalities in Mainland China Based on Physical Simulation and Empirical Statistics—A Case Study of the 2021 Yangbi Earthquake. *Int. J. Environ. Res. Public Health* **2022**, *19*, 6820. https://doi.org/10.3390/ijerph19116820

Academic Editors: Daisuke Sasaki and Mikio Ishiwatari

Received: 30 March 2022
Accepted: 31 May 2022
Published: 2 June 2022

Publisher's Note: MDPI stays neutral with regard to jurisdictional claims in published maps and institutional affiliations.

Abstract: At present, earthquakes cannot be predicted. Scientific decision-making and rescue after an earthquake are the main means of mitigating the immediate consequences of earthquake disasters. If emergency response level and earthquake-related fatalities can be estimated rapidly and quantitatively, this estimation will provide timely, scientific guidance to government organizations and relevant institutions to make decisions on earthquake relief and resource allocation, thereby reducing potential losses. To achieve this goal, a rapid earthquake fatality estimation method for Mainland China is proposed herein, based on a combination of physical simulations and empirical statistics. The numerical approach was based on the three-dimensional (3-D) curved grid finite difference method (CG-FDM), implemented for graphics processing unit (GPU) architecture, to rapidly simulate the entire physical propagation of the seismic wavefield from the source to the surface for a large-scale natural earthquake over a 3-D undulating terrain. Simulated seismic intensity data were used as an input for the fatality estimation model to estimate the fatality and emergency response level. The estimation model was developed by regression analysis of the data on human loss, intensity distribution, and population exposure from the Mainland China Composite Damaging Earthquake Catalog (MCCDE-CAT). We used the 2021 Ms 6.4 Yangbi earthquake as a study case to provide estimated results within 1 h after the earthquake. The number of fatalities estimated by the model was in the range of 0–10 (five expected fatalities). Therefore, Level IV earthquake emergency response plan should have been activated (the government actually overestimated the damage and activated a Level II emergency response plan). The local government finally reported three deaths during this earthquake, which is consistent with the model predictions. We also conducted a case study on a 2013 Ms7.0 earthquake in the discussion, which further proved the effectiveness of the method. The proposed method will play an important role in post-earthquake emergency response and disaster assessment in Mainland China. It can assist decision-makers to undertake scientifically-based actions to mitigate the consequences of earthquakes and could be used as a reference approach for any country or region.

Keywords: earthquake disaster; earthquake fatalities; rapid estimation; earthquake relief; disaster assessment; earthquake emergency response; numerical simulation; empirical method; Yangbi earthquake

1. Introduction

Earthquakes are among the most severely damaging natural disasters on Earth. This is especially true in China, where many catastrophic earthquakes have occurred in recent history, such as the 1976 Tangshan earthquake that caused ~242,000 deaths and the 2008 Wenchuan earthquake that caused ~69,000 deaths [1]. At present, no technology can accurately predict earthquakes. Therefore, facilitating timely, scientific judgments on the disaster situation can play an important role in relief after an earthquake. Rapidly estimating the number of fatalities and identifying the earthquake emergency response level can help decision-makers take scientifically-based actions to optimize the earthquake relief plan.

Current earthquake fatality estimation methods can be divided into two categories: analytical and empirical methods [2]. The former requires a variety of detailed data, including building inventories, building structure types, vulnerabilities, and building occupancy. After the earthquake, based on the seismic intensity distribution and the detailed list of the buildings on it, and further combining the damage vulnerability and human loss ratio of different structural types of buildings under different intensities to calculate the number of fatalities in all buildings [3–5]. In contrast, empirical methods use seismic parameters (e.g., seismic intensity, magnitude, or ground motion parameters) and demographic data (e.g., population density and number of destroyed buildings) to estimate the number of fatalities from previous earthquakes via statistical regression [2,6–10]. Generally, analytical methods are more accurate than empirical methods. However, implementing an accurate analytical process across entire Mainland China is not feasible. First, high-resolution databases of building characteristics (e.g., building inventory and types) do not currently exist nationwide. Second, Daniell et al. pointed out that >30% of all earthquake fatalities in the Asia Pacific are typically caused by non-structural damage (e.g., heart disease and fire) [11]. Third, the inherent dynamic characteristics of these databases (e.g., the inventory, type, and vulnerability of buildings, which change with societal development), lead to great uncertainty in rapid post-earthquake fatality estimation. Therefore, to rapidly obtain such an estimate after an earthquake in Mainland China, we opted to rely on an empirical method. Li et al. developed an earthquake fatality estimation model based on the Mainland China Composite Damaging Earthquake Catalog (MCCDE-CAT) [1,12]. Based on many earthquake events estimations, Li et al. verified that the model has an optimal evaluation ability for large earthquakes that resulted in tens of thousands of fatalities, such as the Wenchuan earthquake, and small earthquakes that resulted in relatively few fatalities [12].

This estimation model mainly relies on data of seismic intensity distribution and population exposure. Many high-precision population datasets are freely available on the Internet, such as CIESIN [13], Asiapop [14], and LandScan [15]. Three methods are commonly used to obtain seismic intensity data. The first relies on field surveys by earthquake administration. Professionals collect information on damages in a sampling survey unit. By calculating the average damage, the seismic intensity can be estimated according to the predefined Seismic Intensity Scale [16,17]. However, it relies on a limited number of sample sites. When investigators finally draw the seismic intensity distribution map, an elliptical distribution in space is generally assumed. This shape does not reflect the irregular pattern of seismic intensity resulting from differences in underground stress, rock lithology, site effects, and topography. In addition, the investigation process requires substantial human and material resources, and can last several days. Therefore, earthquake intensity cannot be used for the rapid estimation of post-earthquake fatalities. The second (empirical) method relies on the widely used Ground Motion Prediction Equations (GMPEs) [18–20]. However, it suffers from several limitations, such as the ergodic hypothesis, insufficient near-source observation data, and insufficient spatial correlation processing [21]. In addition, the statistical and approximate nature of this empirical method hinders appropriate characterization of heterogeneities in seismic intensity, as in the first method.

With the rapid development of computer technology, the third method, namely a large-scale parallel fast numerical calculation method based on physical simulation, has gained significant attention [22–24]. This method can simulate the entire physical process of seismic wavefield propagation from the source to the surface. It can calculate the influence of near-surface site effects, terrain, and other factors on the results, thus, producing a detailed spatial distribution of surface seismic intensity. This is a major improvement compared to the two aforementioned methods. Therefore, we selected this method to obtain the seismic intensity distribution, specifically using the three-dimensional (3-D) curved grid finite difference method (CG-FDM) [25–27] based on a parallel algorithm [24]. This method is characterized by high calculation speed and high accuracy and can reproduce complex undulating surfaces.

In the following sections of the paper, we first introduce the physically-based numerical simulation method and the earthquake fatality assessment model for Mainland China. Subsequently, considering the Ms 6.4 Yangbi earthquake that occurred in Yunnan Province on 21 May 2021 as an example, we present the results of a rapid assessment of the fatalities after the earthquake. Finally, the proposed method and its features are discussed, followed by concluding comments.

2. Methods

The proposed method combines the 3-D CG-FDM with an earthquake fatality estimation model based on regression analysis. The numerical simulation is solved by a GPU owing to its high computational performance.

2.1. Strong Ground Motion Numerical Simulation

As aforementioned, the 3-D CG-FDM [25,26] was selected as the strong ground motion numerical simulation method for calculating the seismic intensity. The method was selected considering its (1) high calculation accuracy, (2) high calculation efficiency, (3) flexible division of the calculation grid according to the undulating terrain, (4) use of the traction mirror method to calculate free surface, and (5) easy parallel computing. The method has been verified in seismic numerical simulations at various dimensions [25–28] and has been applied to strong ground motion numerical simulations and rapid seismic disaster analyses [29–32]. Considering that the theoretical derivation and application of this method have been explained comprehensively in the cited literature, here, we only provided a brief introduction.

The first-order velocity-stress equations for the wave equation in elasticity were as follows:

$$\rho v_{i,t} = \sigma_{ij,j} + f_i \text{ and} \tag{1}$$

$$\sigma_{ij,j} = \lambda \delta_{ij} v_{k,k} + \mu(v_{i,j} + v_{j,i}), \tag{2}$$

where ρ is density, f is the source term, v_i presents the velocity component, σ_{ij} presents the stress component, λ and μ are the Lamé constants, and δ_{ij} is the Kronecker tensor.

To simulate complex surface conditions, the physical quantities in Equations (1) and (2) were mapped from the curved grid in the physical space (x, y, z) to the uniform grid in the computing space (ξ, η, ζ), as shown in Figure 1.

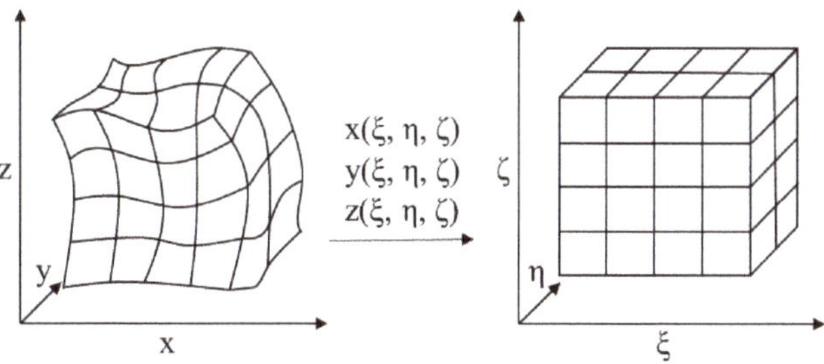

Figure 1. Mapping between (**left**) the curved grid in the physical space (x, y, z) and (**right**) the uniform grid in the computational space (ξ, η, ζ).

During the specific implementation of the numerical calculation, for integration in the time domain, we adopted the fourth-order Runge–Kutta integral form. When calculating the difference in the spatial domain, the forward and backward difference operators in the DRP/opt MacCormark scheme [33] were used alternately to solve the velocity-stress in Equations (1) and (2) in the Runge–Kutta time marching scheme. Taking the x-axis derivative as an example, the operator forms were as follows:

$$L_x^F(U)_i = \frac{1}{\Delta x} \sum_{n=-1}^{3} a_n U_{i+n} \text{ and} \tag{3}$$

$$L_x^B(U)_i = \frac{1}{\Delta x} \sum_{n=-1}^{3} -a_n U_{i-n}, \tag{4}$$

where U is a velocity-stress vector [$U = (v_x, v_y, v_z, \sigma_{xx}, \sigma_{yy}, \sigma_{zz}, \sigma_{xy}, \sigma_{xz}, \sigma_{yz})^T$]; L_x represents the spatial difference in the x-direction; superscripts F and B represent the forward and backward difference operators, respectively; i represents the grid index; and a_n are a set of difference coefficients. For a comprehensive derivation and application of the 3-D CG-FDM, see [25,26].

To rapidly calculate large-scale natural earthquakes, we must also adopt parallel calculation. The graphics processing unit (GPU) has strong parallel computing power, high memory access bandwidth, and low latency [34]. Wang et al. achieved a parallel operation of the CG-FDM and developed a platform for rapid response to earthquake disasters (CGFDM3D-EQR) based on a heterogeneous CPU/GPU architecture [24]. After the earthquake, the platform can rapidly estimate the seismic intensity distribution through parallel numerical simulation within 30 min. To verify the reliability and effectiveness of the platform, they did four numerical simulations and compared them with the instrument observation results, and obtained the expected results. For the case in this paper, we use this computing platform to carry out the strong ground motion numerical simulation.

2.2. Earthquake Fatality Estimation Model

We used the Mainland China earthquake fatalities estimation model developed by Li et al. [12], which is based on data of seismic intensity and losses for historical earthquakes. Clearly, the performance of this model depends on the completeness of data, and their integrity and consistency across events. Li et al. compiled an open-source MCCDE-CAT (see Data Availability Statement) from 1950 to 2018 based on six earthquake catalogs. They used a large number of previous studies, reports, and websites to verify and complement the information [1]. The catalog includes basic seismological data, social and economic loss data, population exposure data, and intensity distribution data for each earthquake event. Based on data recorded for the 377 earthquake events in the MCCDE-CAT (Figure 2), Li et al. used a logarithmic linear fatality ratio function with two free parameters to develop a set of earthquake fatality estimation models via regression analysis [12]. A large number of earthquake cases were used to verify that the model could estimate >85% of the values within an order of magnitude of the recorded values. A brief introduction to this method is provided below.

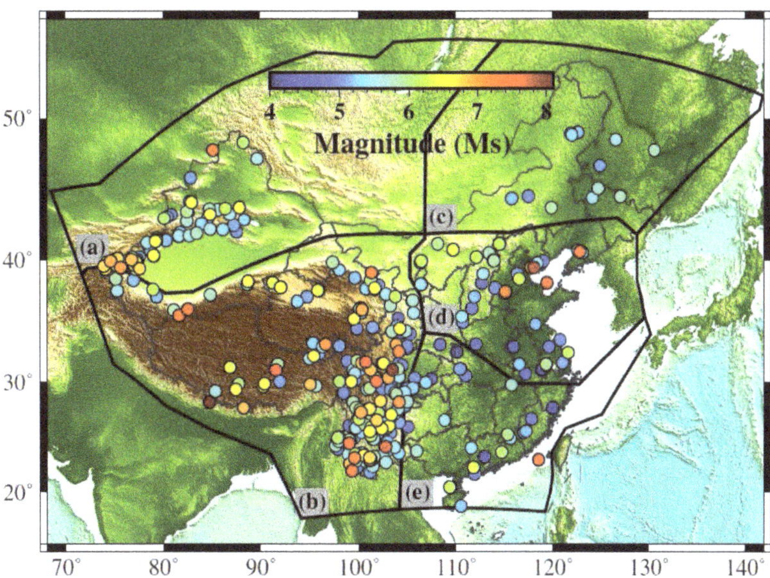

Figure 2. Location of the 377 earthquake epicenters in the Mainland China Composite Damaging Earthquake Catalog (MCCDE-CAT) used by Li et al. to develop the fatality assessment model [12]. The thick black lines divide Mainland China into the five sub-regions as follows (**a**): Xinjiang region, (**b**): Qinghai-Tibet Plateau region, (**c**): Northeast region, (**d**): North China region, and (**e**): South China region.

According to the characteristics of the geological structures, historical seismicity, focal mechanisms, and social and economic development in Mainland China, the MCCDE-CAT was divided into five sub-regions (as divided by thick black lines in Figure 2). Moreover, according to the differences in the earthquake fatality ratio with time, related to changes in the seismic resistance of buildings, public education, and other factors, the human development index (HDI) was used to evaluate changes in social development level over time to adjust the fatality ratio.

The earthquake-related fatalities were calculated as follows:

$$E = \sum_{I=V}^{XI} \left(r(I) \times \frac{HDI_{max-year}}{HDI_{event-year}} \right) \times Pe_{(region,\ intensity=I)} \tag{5}$$

where E represents the estimated value, $r(I)$ is the fatality ratio function (which is a logarithmic linear model, i.e., $\log(r) = \beta + \theta \times I$, with two free parameters (β, θ)), I is the seismic intensity. Intensity V-XI is the minimum and maximum intensities of all known seismic records of damaging earthquakes in Mainland China. *HDI* represents the human development index, which is the correction term for the fatality ratio, r, and $Pe_{(region,\ intensity=I)}$ represents the number of people exposed in regions with different seismic intensities, I.

The two parameters β and θ of the fatality function were obtained by minimizing the objective function as follows:

$$\varepsilon = \ln \left[\sqrt{\frac{1}{N} \sum_{i=1}^{N} (E_i - O_i)^2} \right] + \sqrt{\frac{1}{N} \sum_{i=1}^{N} \left[\ln \left(\frac{E_i}{O_i} \right) \right]^2}, \tag{6}$$

where O_i represents the recorded fatalities for the ith earthquake event, E_i represents the estimated fatalities in Equation (5), and N represents the number of earthquake events.

According to the relevant provisions in the National Earthquake Emergency Plan (NEEP) (see Data Availability Statement) for Mainland China, the Chinese government classifies the earthquake emergency response into four levels as follows: Level IV ($\leq$10 fatalities), Level III (10 < Number of fatalities $\leq$ 50), Level II (50 < Number of fatalities $\leq$ 300), and Level I (>300 fatalities). To rapidly evaluate the response level after an earthquake, we constructed a normal cumulative distribution function based on the estimated fatalities and fitting residual calculated using this method (its rationality has been verified using the normality test [12]). The probability P of a specific fatality range was calculated as follows:

$$P(a < \text{fatality} \leq b) = \Phi\left[\frac{\ln b - \ln(E_i)}{\zeta}\right] - \Phi\left[\frac{\ln a - \ln(E_i)}{\zeta}\right], \tag{7}$$

where Φ represents the normal cumulative distribution function with respect to the expected value, $\ln(E_i)$, and the log residual, ζ. For more details on the development and validation of this estimation model, see [12].

3. Case Study: 2021 Ms 6.4 Yangbi Earthquake

Using the above rapid earthquake fatality estimation method, we rapidly calculated the seismic intensity distribution, estimated a total of five deaths, and determined an earthquake emergency response of Level IV within 1 h after the 2021 Ms 6.4 Yangbi earthquake. Later, the government reported that there were three fatalities during this earthquake, consistent with our assessment expectations. This case study of the 2021 Yangbi earthquake is described in detail below.

3.1. Background and Data

At 21:48:34 on 21 May 2021, a Ms 6.4 earthquake occurred in Yangbi County, Dali Bai Autonomous Prefecture, Yunnan Province. The epicenter was located at 99.87° E, 25.67° N, and the focal depth was 8 km. The maximum intensity on the surface was VIII. The earthquake area was located near the southwest boundary of the Chuan-Dian block, where historically strong earthquakes have been relatively frequent. In addition, this region features complex mountains and gullies, rendering it difficult to perform post-earthquake relief work. Figure 3 shows the topographic distribution of the area near the Yangbi earthquake. The overall trend of the mountains in the area is northwest-southeast and the fault trend is similar. The black wireframe in Figure 3 represents the numerical simulation area. The blue wireframe is the projection of the fault on the surface while the red star is the projection of the epicenter onto the surface. The focal mechanism solution shows that this earthquake was a strike-slip earthquake with a large-dip angle (78.3° or 72.9°).

The 3-D terrain data were extracted from the Shuttle Radar Topographic Mission digital elevation dataset (see Data Availability Statement) provided by the Consultative Group for International Agricultural Research-Consortium for Spatial Information [35]. This dataset is based on the NASA data processed through an interpolation technique to obtain global seamless connection elevation data with a resolution of 90 m. The 3-D medium model used in the simulation was the Chinese crust-upper mantle seismic reference model [36] (see Data Availability Statement). This model is based on data from >2000 seismic stations obtained through environmental noise Rayleigh wave tomography and seismic tomography, with a lateral resolution of 0.5° × 0.5° and a maximum vertical depth of 150 km, covering the entire territory of China. The source data (see Data Availability Statement) were provided by Wang et al., the Institute of Qinghai-Tibet Plateau Research, and the Chinese Academy of Sciences [37].

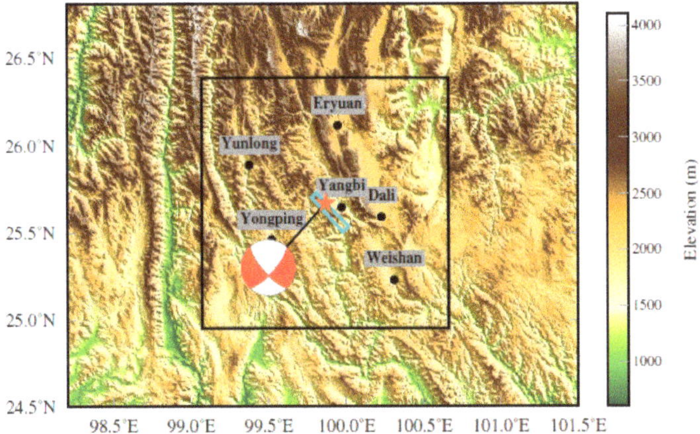

Figure 3. The terrain of the Yangbi earthquake area and its vicinity. The black wireframe represents the calculation region of the strong ground motion simulation. The blue wireframe represents the projection of the fault on the surface. The red star represents the projection of the epicenter onto the surface. The beach ball represents the focal mechanism solution.

3.2. Numerical Simulation Results

The Yangbi earthquake was numerically simulated using the GPU-based 3-D CD-FDM described in Section 2.1 and the input data described above. The grid size was set to 800 × 800 × 400 and the spatial resolution was 200 m × 200 m. Figure 4 shows four wavefield snapshots during the propagation of the seismic wavefield obtained via numerical simulation. The energy of this earthquake mainly spread outward along the fault strike and perpendicular to the fault strike. Energy in the other directions was weak, consistent with the relatively weak disaster situation in major cities to the east of the epicenter. In addition, the propagation of the seismic wavefield on the surface was complex. Abundant multiple reflection waves were generated near the peaks and ridges along the fault strike; coherent superposition enhanced the ground motion. These simulation results were consistent with those of previous studies on the impact of topography on seismic wavefield propagation [30,38–41].

Based on GPU parallel computing, we calculated the seismic intensity distribution according to the PGV-intensity relationship provided by the latest China Seismic Intensity Scale (GB/T 17742-2020) (see Data Availability Statement), as shown in Figure 5, where the base map represents the population distribution data (see Data Availability Statement) provided by LandScan [15]. The calculation results showed that the maximum seismic intensity was VIII, entirely concentrated in Yangbi County. The majority of the intensity VII area was distributed in Yangbi County and a small part extended to sparsely populated areas in Yongping County in the southwest. The densely populated areas of the counties and cities adjacent to Yangbi only suffered a maximum seismic intensity VI, among which the densely populated areas of Eryuan County in the north and Yongping County in the southwest only suffered a seismic intensity V. Damage caused by earthquakes with intensity V–VI are generally considered to be highly limited [42]. Therefore, we estimated that the overall damage due to this earthquake was relatively small. The intensity distribution began from the fault as the center, extended outward in an overall butterfly-like shape, and gradually decayed. The main reasons for this distribution pattern for the seismic intensity are that the fault rupture scale of this earthquake was small, and the fault slip was dominated by strike-slip. Therefore, strong S-wave energy was generated along the fault strike and perpendicular to the fault strike, finally presenting a distribution pattern similar to the double couple point source radiation pattern on the surface. In addition, the fault strike, dip angle and slip angle were about 138°, 73° and –163°, respectively. The fault hanging wall was located in the southwest. Owing to the hanging wall effect of the fault, the energy in the southwest direction was stronger than that in the footwall direction. The asymmetric shape of the specific details for the intensity also reflected the non-uniformity of the surface seismic intensity distribution under the actual complex conditions (i.e., the media and terrain).

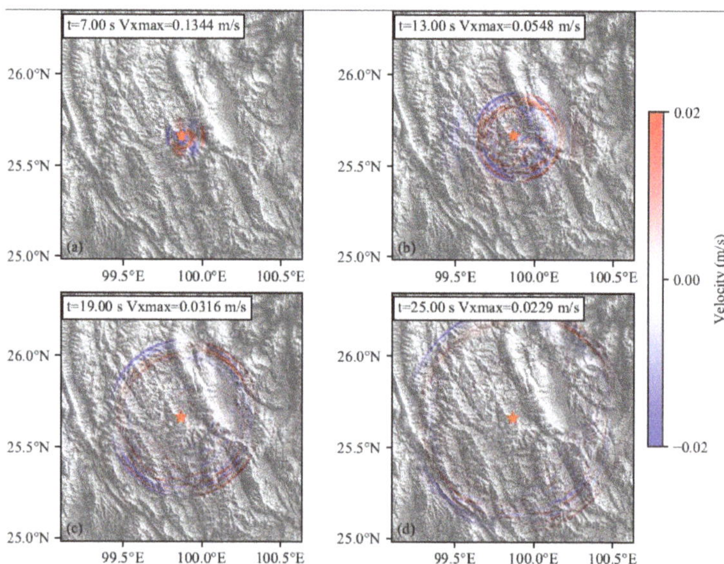

Figure 4. Evolution process of wavefield propagation (east-west velocity component, Vx) along the ground surface at the indicated time points for the Yangbi earthquake. (Note: the color range from −0.02 to 0.02 m/s is not the range of the Vx amplitude. The range was reduced to show the details of ground motion more clearly).

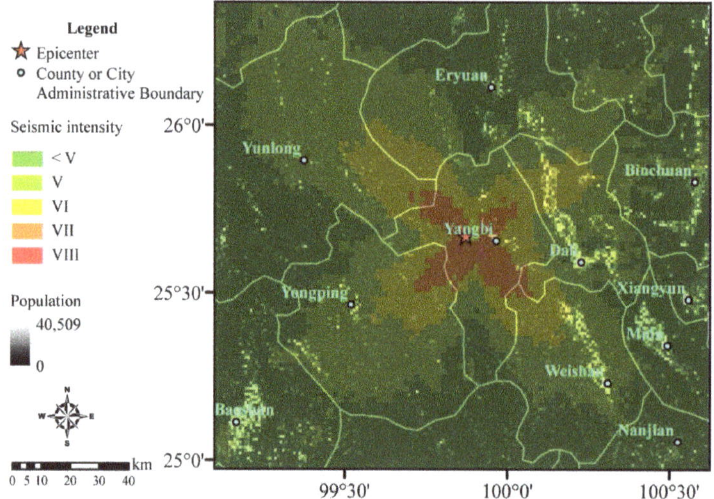

Figure 5. Seismic intensity distribution was obtained via rapid numerical simulations of the Yangbi earthquake. The base map represents the population distribution provided by LandScan [15].

3.3. Fatality Estimation Results

Based on the above seismic intensity distribution data, GIS software (Version 10.9) was used to calculate regional superposition using LandScan population distribution data to obtain population exposure values in different seismic intensity areas. Equations (5) and (7) are then used to estimate the fatalities related to the Yangbi earthquake and the probability of the emergency response levels, as shown in Figure 6. Five fatalities were estimated; the highest probability for the estimated fatality range was 0–10, with a value of 65.2%. According to the post-earthquake emergency response regulations in the NEEP, this event would be defined as a "General Earthquake Disaster Event" with a high probability, corresponding to a Level IV earthquake emergency response. According to the population density distribution, two relatively densely populated cities are located nearby the epicenter: Dali City in the east and Weishan County in the southeast. However, the densely populated areas only suffered from seismic intensity VI. Overall, we considered that the earthquake was relatively minor and did not cause major damage. Finally, we proposed the initiation of a Level IV emergency response plan, using Yangbi County as the main rescue area during earthquake relief. Weishan County, Dali City, Eryuan County, Yunlong County, and Yongping County were considered as secondary concerns. Among them, the northwest area in the two urban areas of Dali and Weishan is the key disaster relief area in the secondary disaster relief.

On 23 May, the Yunnan provincial government reported a final fatality count of three for the Yangbi earthquake (see Data Availability Statement). Even though the estimates deviated from the reported fatalities, the error remained within one order of magnitude, consistent with expectations. In addition, the reported fatalities also occurred within the 0–10 fatality range estimated in this study. In summary, the estimation results for the Yangbi earthquake were ideal, which shows that the fatality estimation method is sufficient to conform to the actual needs of the government for rapidly implementing earthquake emergency response decisions after an earthquake.

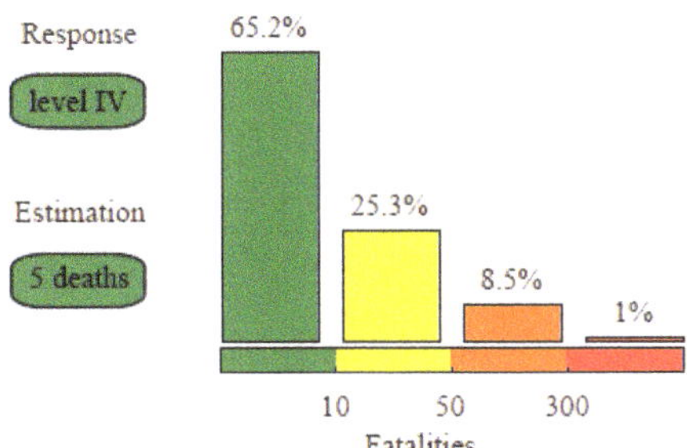

Figure 6. Rapid estimation of the fatalities and earthquake emergency response level for the Yangbi earthquake.

4. Discussion

Even though the Yangbi earthquake is a low-fatality earthquake event, it was selected as a case study here, because it was not among the earthquake events used in the original fatality estimation model. To further verify the effectiveness of this method, we selected the Ms 7.0 Sichuan Lushan earthquake on 20 April 2013 in our estimation model data. This earthquake caused great damage, which resulted in 196 deaths and direct economic losses of more than 10 billion US dollars. The source model used in this case was provided by Wang et al. [43]. The seismic intensity distribution was obtained by strong ground motion simulation, as shown in Figure 7a, where the maximum seismic intensity is VIII in this event. The hardest-hit areas of seismic intensity VII–VIII were concentrated in five cities near the epicenter (Lushan, Baoxing, Tianquan, Mingshan, and Ya An), in which Lushan suffered the most serious damage. The estimated high-intensity areas correspond to the most severely damaged areas reported after the earthquake (see Data Availability Statement). As shown in Figure 7b, the final estimated number of deaths is 104, and the fatality range with the largest probability of this earthquake is 50–300. According to the NEEP, a Level II earthquake emergency response plan should be initiated for this event. In this case, the expected and recorded fatalities have the same magnitude order, and the number of recorded deaths is in the estimated fatality range. If the plan was initiated according to our estimated earthquake emergency response level, it would fully meet the actual needs of this earthquake disaster relief. This reasonable result is significant for the earthquake relief and rescue effort.

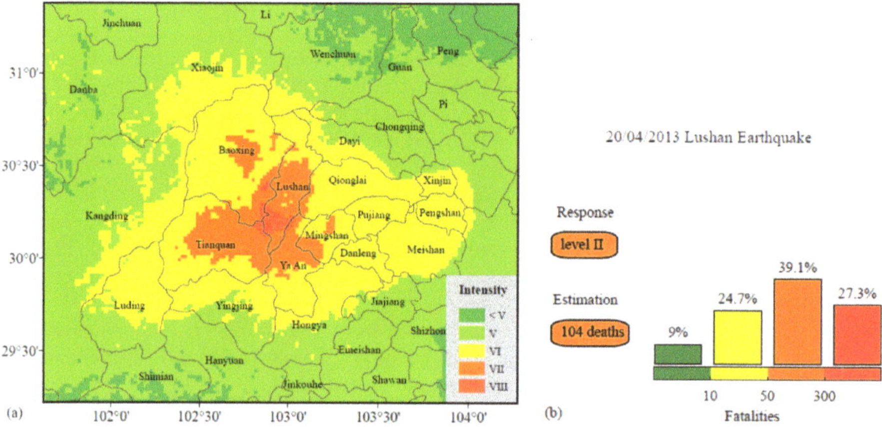

Figure 7. (**a**) The seismic intensity distribution calculated by numerical simulations for the 2013 Ms 7.0 Lushan earthquake. (**b**) The rapid estimation of the fatalities and earthquake emergency response level for the 2013 Ms 7.0 Lushan earthquake.

The numerical simulation results depended on the source model, velocity medium, terrain, and other data. At present, there are no unified results of other studies for the previously mentioned variables, except terrain, and determining the same is beyond the scope of this paper. Most source models are derived from the inversion of teleseismic waveform data; the maximum frequency of waveforms is usually lower than that of the near-field waveforms. Zhang et al. comprehensively discussed the influence of different frequencies on the numerical simulation results for strong ground motion, based on the Sunway TaihuLight supercomputer (Institute of Computing Technology, Chinese Academy of Sciences, China). The supercomputer is the world's first supercomputer with a peak performance exceeding 100 PFlops and 1.3 PB memory [44]. They found that the sensitivity of the surface seismic intensity to frequency change is low [45]. Therefore, the accuracy of the simulation results based on teleseismic data in this study was sufficient to meet the requirements of preliminary post-earthquake rapid estimation. In addition, the final fatality estimation results obtained in this study also demonstrated the reliability of the method. High-frequency waveforms generated in the near field could contain more details on sources. However, seismic stations are absent in some areas near epicenters, or the local government does not disclose the data, consequently, making it difficult for researchers to obtain near-field data at the first time. Teleseismic data not only ensure the implementation of this method rapidly after an earthquake but also produce timely estimation results. Further, the numerical intensity and estimated fatalities will be updated with continuous improvements to post-earthquake data collection (e.g., near-source data or more accurate teleseismic data).

The fatality estimation method is an empirical method based on regression analysis. Therefore, the estimated results cannot be used as an accurate prediction for the fatalities of an earthquake. However, the estimated value can be guaranteed to be within one order of magnitude of the actual number of fatalities [12]. In other words, the model can estimate the number of fatalities within a reasonable range. Therefore, the method is appropriate for rapid fatality estimations during a post-earthquake emergency response. Using this method to estimate the probability of each fatality range can provide a scientific quantitative basis to the government and related agencies to activate a specific earthquake emergency response plan based on the NEEP. Regarding the case study of the Yangbi earthquake on 21 May 2021, we determined the number of fatalities in the range of 0–10 (the calculated expected value was 5). Based on the quantitative estimated fatalities and NEEP, we evaluate that this earthquake should be initiated a Level IV earthquake emergency response plan. However, the local government actually initiated a Level II emergency response plan after the earthquake, which overestimated the damage degree of the earthquake and led to unnecessary wastage of resources. The emergency response level estimated by our method was more consistent with the actual mitigation requirements. In addition, the actual number of fatalities was within the estimated range and within an order of magnitude of the expected estimates. Therefore, it can be concluded that our estimation results were consistent with the expectations.

There are some limitations of this study, for example, the uncertainty of each parameter in the fatality estimation model needs to be further determined. The model is currently too simplified. In the future, some parameters that are highly correlated with the fatalities can be considered to further enhance the estimation accuracy. In addition, the far-field data carries fewer high-frequency components than the near-field. In the future, if we can cooperate with earthquake-related institutions in Mainland China to obtain near-field data as soon as possible, the accuracy of the final calculation results will be further increased.

The above result demonstrates the potential of the proposed method to assist the government in a rapid post-earthquake emergency response. In the future, if an earthquake occurs in Mainland China, the application of this method will facilitate rapid estimation of the earthquake fatalities and indicate the appropriate emergency response level. These estimations can provide scientific guidance to the government and relevant institutions for real-time disaster relief applications, such as rapid post-earthquake emergency response decision-making and appropriate relief and resource allocation. Finally, it can reduce additional losses and save more lives.

5. Conclusions

Combining a physics-based numerical simulation of earthquake intensity and a statistics-based fatality estimation model, we propose a method for rapidly estimating the number of fatalities and emergency response levels after an earthquake. We successfully applied it to the Ms6.4 Yangbi earthquake in Yunnan on 21 May 2021, and the calculated results are reasonable compared with the actual number of deaths. In the future, if an earthquake occurs in Mainland China, the application of this method will facilitate rapid estimation of the earthquake fatalities and indicate the appropriate emergency response level. These estimations can provide scientific guidance to the government and relevant institutions for real-time disaster relief applications, such as rapid post-earthquake emergency response decision-making and appropriate relief and resource allocation. Finally, it can reduce additional losses and save more lives.

Other similar countries or regions that have experienced many earthquakes can use the method proposed in this study as a reference. Based on the local historical human loss data, the fatality assessment model was constructed using regression analysis, and combined with the open-source GPU numerical simulation program; consequently, a method for timely assessment of fatalities was developed. In areas with few historical earthquakes, information on adjacent areas can be considered to evaluate the fatalities.

Author Contributions: Conceptualization, Y.L.; methodology, Y.L. and W.W.; software, Y.L. and W.W.; investigation, Z.Z.; data curation, Y.L.; writing—original draft preparation, Y.L.; visualization, X.F.; validation, Z.Z.; funding acquisition, Z.Z. All authors have read and agreed to the published version of the manuscript.

Funding: This research was funded by the National Natural Science Foundation of China (Grant No.41922024), Key Special Project for Introduced Talents Team of Southern Marine Science and Engineering Guangdong Laboratory (Guangzhou) (GML2019ZD0203), Guangdong Provincial Key Laboratory of Geophysical High-resolution Imaging Technology (2022B1212010002), and Shenzhen Science and Technology Program (KQTD20170810111725321). This work is supported by the Center for Computational Science and Engineering at the Southern University of Science and Technology.

Institutional Review Board Statement: Not applicable.

Informed Consent Statement: Not applicable.

Data Availability Statement: The data consulted for this article can be found from the following websites: (1) Mainland China Composite Damaging Earthquake Catalog available at https://zenodo.org/record/6514359#.Yoh2bhpByUm (accessed on 3 May 2022); (2) National Earthquake Emergency Plan (NEEP) for Mainland China available at http://www.gov.cn/yjgl/2012-09/21/content_2230337.htm (accessed on 21 May 2022); (3) Shuttle Radar Topographic Mission digital elevation dataset provided by the Consultative Group for International Agricultural Research-Consortium for Spatial Information available at https://cgiarcsi.community/data/srtm-90m-digital-elevation-database-v4-1/ (accessed on 21 May 2022); (4) Chinese crust-upper mantle seismic reference model available at http://chinageorefmodel.org (accessed on 13 January 2022); (5) China Seismic Intensity Scale (GB/T 17742-2020) available at http://www.gb688.cn/bzgk/gb/std_list?p.p1=0&p.p90=circulation_date&p.p91=desc&p.p2=GB/T%2017742 (accessed on 21 May 2022); (6) population distribution data provided by LandScan and available at https://landscan.ornl.gov/ (accessed on 13 January 2022); and (7) The Yangbi earthquake reported fatalities from the Yunnan Province government available at http://www.yn.gov.cn/ywdt/zsdt/202105/t20210523_222657.html (accessed on 13 January 2022); (8) The description of the Lushan earthquake disaster available at https://en.wikipedia.org/wiki/2013_Lushan_earthquake (accessed on 21 May 2022). (9) Figures 2 and 3 were created using Generic Mapping Tools [46]. (10) The seismic source data used in this study can be requested from Weimin Wang (personal communication).

Acknowledgments: The authors thank Weiming Wang from the Institute of Qinghai-Tibet Plateau, Chinese Academy of Sciences for providing the source data, and the China National Network Center for the epicenter information. The authors also thank Danhua Xin for technical assistance with the GIS software.

Conflicts of Interest: The authors declare no conflict of interest.

References

1. Li, Y.; Zhang, Z.; Xin, D. A Composite Catalog of Damaging Earthquakes for Mainland China. *Seismol. Res. Lett.* **2021**, *92*, 3767–3777. [CrossRef]
2. Tang, B.; Chen, Q.; Liu, X.; Liu, Z.; Liu, Y.; Dong, J.; Zhang, L. Rapid estimation of earthquake fatalities in China using an empirical regression method. *Int. J. Disast. Risk Reduct.* **2019**, *41*, 101306. [CrossRef]
3. FEMA. *HAZUS-MH 2.1 Technical Manual*; Federal Emergency Management Agency: Washington, DC, USA, 2006.
4. Porter, K. Cracking an Open Safe: HAZUS Vulnerability Functions in Terms of Structure-Independent Spectral Acceleration. *Earthq. Spectra* **2009**, *25*, 361–378. [CrossRef]
5. Ceferino, L.; Kiremidjian, A.; Deierlein, G. Probabilistic model for regional multiseverity casualty estimation due to building damage following an earthquake. *ASCE-ASME J. Risk Uncertain. Eng. Syst.* **2018**, *4*, 04018023. [CrossRef]
6. Kawasumi, H. Intensity and Magnitude of Shallow Earthquakes, Bureau Central Seism. *Intern. Ser. A Trav. Sci.* **1954**, *19*, 99–114.
7. Samardjieva, E.; Badal, J. Estimation of the expected number of casualties caused by strong earthquakes. *Bull. Seismol. Soc. Am.* **2002**, *92*, 2310–2322. [CrossRef]
8. Jaiswal, K.; Wald, D.J.; Hearne, M. *Estimating Casualties for Large Earthquakes Worldwide Using an Empirical Approach*; 2009-1136; US Geological Survey: Denver, CO, USA, 2009; p. 78.
9. Jaiswal, K.; Wald, D.J. An empirical model for global earthquake fatality estimation. *Earthq. Spectra* **2010**, *26*, 1017–1037. [CrossRef]
10. Firuzi, E.; Hosseini, K.A.; Ansari, A.; Izadkhah, Y.O.; Rashidabadi, M.; Hosseini, M. An empirical model for fatality estimation of earthquakes in Iran. *Nat. Hazards* **2020**, *103*, 231–250. [CrossRef]

11. Daniell, J.E.; Khazai, B.; Wenzel, F.; Vervaeck, A. The CATDAT damaging earthquakes database. *Nat. Hazards Earth Syst. Sci.* **2011**, *11*, 2235–2251. [CrossRef]
12. Li, Y.; Xin, D.; Zhang, Z. A rapid-response earthquake fatality estimation model for mainland China. *Int. J. Disast. Risk Re.* **2021**, *66*, 102618. [CrossRef]
13. (CIESIN), C.f.I.E.S.I.N.; (CIAT), C.I.d.A.T. Gridded Population of the World, Version 3 (GPWv3) Data Collection. Available online: http://sedac.ciesin.columbia.edu/data/collection/grump-v1 (accessed on 13 January 2022).
14. Tatem, A.; Linard, C. Population mapping of poor countries. *Nature* **2011**, *474*, 36. [CrossRef] [PubMed]
15. Rose, A.N.; McKee, J.J.; Urban, M.L.; Bright, E.A.; Sims, K.M. *LandScan 2018*, 2018th ed.; Oak Ridge National Laboratory: Oak Ridge, TN, USA, 2019.
16. Hu, Y. *Earthquake Engineering*; Seismological Press: Beijing, China, 1988.
17. Wang, D.; Wang, X.; Kou, A.; Ding, X. Primary study on the quantitative relationship between the typical building structures in western China. *Earthquake* **2007**, *27*, 105–110. [CrossRef]
18. Paolucci, R.; Gatti, F.; Infantino, M.; Smerzini, C.; Güney Özcebe, A.; Stupazzini, M. Broadband Ground Motions from 3D Physics-Based Numerical Simulations Using Artificial Neural Networks. *Bull. Seismol. Soc. Am.* **2018**, *108*, 1272–1286. [CrossRef]
19. Infantino, M.; Mazzieri, I.; Özcebe, A.G.; Paolucci, R.; Stupazzini, M. 3D Physics-Based Numerical Simulations of Ground Motion in Istanbul from Earthquakes along the Marmara Segment of the North Anatolian Fault. *Bull. Seismol. Soc. Am.* **2020**, *110*, 2559–2576. [CrossRef]
20. Stupazzini, M.; Infantino, M.; Allmann, A.; Paolucci, R. Physics-based probabilistic seismic hazard and loss assessment in large urban areas: A simplified application to Istanbul. *Earthq. Eng. Struct. Dyn.* **2021**, *50*, 99–115. [CrossRef]
21. Xin, D.; Zhang, Z. On the Comparison of Seismic Ground Motion Simulated by Physics-Based Dynamic Rupture and Predicted by Empirical Attenuation Equations. *Bull. Seismol. Soc. Am.* **2021**, *111*. [CrossRef]
22. Tromp, J.; Komatitsch, D.; Hjörleifsdóttir, V.; Liu, Q.; Zhu, H.; Peter, D.; Bozdag, E.; McRitchie, D.; Friberg, P.; Trabant, C. Near real-time simulations of global CMT earthquakes. *Geophys. J. Int.* **2010**, *183*, 381–389. [CrossRef]
23. Lee, S.-J.; Liang, W.-T.; Cheng, H.-W.; Tu, F.-S.; Ma, K.-F.; Tsuruoka, H.; Kawakatsu, H.; Huang, B.-S.; Liu, C.-C. Towards real-time regional earthquake simulation I: Real-time moment tensor monitoring (RMT) for regional events in Taiwan. *Geophys. J. Int.* **2014**, *196*, 432–446. [CrossRef]
24. Wang, W.; Zhang, Z.; Zhang, W.; Yu, H.; Liu, Q.; Zhang, W.; Chen, X. CGFDM3D-EQR: A Platform for Rapid Response to Earthquake Disasters in 3D Complex Media. *Seismol. Res. Lett.* **2022**. [CrossRef]
25. Zhang, W.; Chen, X. Traction image method for irregular free surface boundaries in finite difference seismic wave simulation. *Geophys. J. Int.* **2006**, *167*, 337–353. [CrossRef]
26. Zhang, W.; Zhang, Z.; Chen, X. Three-dimensional elastic wave numerical modelling in the presence of surface topography by a collocated-grid finite-difference method on curvilinear grids. *Geophys. J. Int.* **2012**, *190*, 358–378. [CrossRef]
27. Zhang, W.; Shen, Y.; Zhao, L. Three-dimensional anisotropic seismic wave modelling in spherical coordinates by a collocated-grid finite-difference method. *Geophys. J. Int.* **2012**, *188*, 1359–1381. [CrossRef]
28. Sun, Y.-C.; Ren, H.; Zheng, X.-Z.; Li, N.; Zhang, W.; Huang, Q.; Chen, X. 2-D poroelastic wave modelling with a topographic free surface by the curvilinear grid finite-difference method. *Geophys. J. Int.* **2019**, *218*, 1961–1982. [CrossRef]
29. Zhang, W.; Shen, Y.; Chen, X. Numerical simulation of strong ground motion for the Ms8.0 Wenchuan earthquake of 12 May 2008. *Sci. China Ser. D Earth Sci.* **2008**, *51*, 1673–1682. [CrossRef]
30. Zhu, G.; Zhang, Z.; Wen, J.; Zhang, W.; Chen, X. Preliminary results of strong ground motion simulation for the Lushan earthquake of 20 April 2013, China. *Earthq. Sci.* **2013**, *26*, 191. [CrossRef]
31. Zhang, Z.; Zhang, W.; Sun, Y.; Zhu, G.; Wen, J.; Chen, X. Preliminary simulation of strong ground motion for Yutian, Xinjiang earthquake of 12 February 2014, and hazard implication. *Chin. J. Geophys.* **2014**, *57*, 685–689. [CrossRef]
32. Zhang, Z.; Sun, Y.; Xu, J.; Zhang, W.; Chen, X. Preliminary simulation of strong ground motion for Ludian, Yunnan earthquake of 3 August 2014, and hazard implication. *Chin. J. Geophys.* **2014**, *57*, 3038–3041. [CrossRef]
33. Hixon, R. On Increasing the Accuracy of MacCormack Schemes for Aeroacoustic Applications. *AIAA Pap.* **1997**. [CrossRef]
34. Cheng, J.; Grossman, M.; McKercher, T. *Professional CUDA C Programming*; John Wiley & Sons: Hoboken, NJ, USA, 2014.
35. Reuter, H.; Nelson, A.; Jarvis, A. An Evaluation of Void-Filling Interpolation Methods for SRTM Data. *Int. J. Geogr. Inf. Sci.* **2007**, *21*, 983–1008. [CrossRef]
36. Shen, W.; Ritzwoller, M.H.; Kang, D.; Kim, Y.; Lin, F.-C.; Ning, J.; Wang, W.; Zheng, Y.; Zhou, L. A seismic reference model for the crust and uppermost mantle beneath China from surface wave dispersion. *Geophys. J. Int.* **2016**, *206*, 954–979. [CrossRef]
37. Wang, W.; He, J.; Hao, J.; Yao, Z. *Preliminary Result for Rupture Process of May. 21, 2021, M6.4 Earthquake, Dali, China*; Institute of Qinghai-Tibet Plateau Research, Chinese Academy of Sciences: Beijing, China, 2021.
38. Stidham, C.; Antolik, M.; Dreger, D.; Larsen, S.; Romanowicz, B. Three-dimensional structure influences on the strong-motion wavefield of the 1989 Loma Prieta earthquake. *Bull. Seismol. Soc. Am.* **1999**, *89*, 1184–1202.
39. Cárdenas-Soto, M.n.; Chávez-García, F.J. Regional path effects on seismic wave propagation in central Mexico. *Bull. Seismol. Soc. Am.* **2003**, *93*, 973–985. [CrossRef]
40. Komatitsch, D.; Liu, Q.; Tromp, J.; Suss, P.; Stidham, C.; Shaw, J.H. Simulations of ground motion in the Los Angeles basin based upon the spectral-element method. *Bull. Seismol. Soc. Am.* **2004**, *94*, 187–206. [CrossRef]

41. Khan, S.; Meijde, M.v.d.; Werff, H.v.d.; Shafique, M. The impact of topography on seismic amplification during the 2005 Kashmir earthquake. *Nat. Hazards Earth Syst. Sci.* **2020**, *20*, 399–411. [CrossRef]
42. *GB/T 17742-2020*; The Chinese Seismic Intensity Scale. General Administration of Quality Supervision, Inspection and Quarantine of the People's Republic of China (AQSIQ); Standardization Administration of the People's Republic of China (SAC): Beijing, China, 2020.
43. Wang, W.; Hao, J.; Yao, Z. Preliminary result for rupture process of Apr. 20, 2013, Lushan Earthquake, Sichuan, China. *Chin. J. Geophys.* **2013**, *56*, 1412–1417. [CrossRef]
44. Fu, H.; Liao, J.; Yang, J.; Wang, L.; Song, Z.; Huang, X.; Yang, C.; Xue, W.; Liu, F.; Qiao, F.; et al. The Sunway TaihuLight supercomputer: System and applications. *Sci. China Inf. Sci.* **2016**, *59*, 072001. [CrossRef]
45. Zhang, W.; Zhang, Z.; Fu, H.; Li, Z.; Chen, X. Importance of Spatial Resolution in Ground Motion Simulations With 3-D Basins: An Example Using the Tangshan Earthquake. *Geophys. Res. Lett.* **2019**, *46*. [CrossRef]
46. Wessel, P.; Smith, W.H.F. New improved version of Generic Mapping Tools released. *Eos Trans. Am. Geophys. Union* **1998**, *79*, 579. [CrossRef]

International Journal of
Environmental Research and Public Health

MDPI

Article

Examining the Indirect Death Surveillance System of The Great East Japan Earthquake and Tsunami

Xiang Zheng [1,*], Chuyao Feng [1] and Mikio Ishiwatari [1,2]

[1] Department of International Studies, University of Tokyo, Chiba 277-8561, Japan
[2] Japan International Cooperation Agency, Tokyo 151-0066, Japan
* Correspondence: zhengxiangemerge@gmail.com

Abstract: The long-term mortality risk of natural disasters is a key threat to disaster resilience improvement, yet an authoritative certification and a reliable surveillance system are, unfortunately, yet to be established in many countries. This study aimed to clarify the mechanism of post-disaster indirect deaths in Japan, to improve the existing disaster recovery evaluation system and support decision making in public policy. This study first investigated the definition of indirect deaths via a literature review before examining the observed number of indirect deaths via case study, census data from the Population Demographic and Household Surveys, other social surveys, and reports in the case of the Great East Japan Earthquake and Tsunami, which severely damaged northeastern Japan, especially the three prefectures, which are the target areas in this context (i.e., Fukushima, Iwate, and Miyagi). It was found that the reported number of indirect deaths was significantly underestimated. In total, 4657 indirect deaths were estimated to have occurred in the target prefectures. This was higher than the reported number, which was 3784. The overall statistics established via collaboration between local administrations and governments can be improved to provide better reference for researchers and policymakers to investigate the long-term effects of natural disaster.

Keywords: indirect death; long-term effects; excess mortality; earthquake fatalities; surveillance system

Citation: Zheng, X.; Feng, C.; Ishiwatari, M. Examining the Indirect Death Surveillance System of The Great East Japan Earthquake and Tsunami. *Int. J. Environ. Res. Public Health* **2022**, *19*, 12351. https://doi.org/10.3390/ijerph191912351

Academic Editor: William A. Toscano

Received: 2 August 2022
Accepted: 24 September 2022
Published: 28 September 2022

Publisher's Note: MDPI stays neutral with regard to jurisdictional claims in published maps and institutional affiliations.

1. Introduction

The Sendai Framework for Disaster Risk Reduction 2015–2030 was adopted at the Third United Nations World Conference on Disaster Risk Reduction during 2015 in Sendai, Japan, which includes four priorities and seven targets for the world [1,2]. Target A was to reduce global disaster mortality by 2030, for which the establishment of a mechanism and a mortality database was advocated [2]. The common classification of disaster mortality consists of direct and indirect deaths. Yet, it is still unclear how to count indirect death due to limited surveillance periods, as well as inappropriate definitions and criteria [3]. In fact, indirect death counting is still completely overlooked in some studies [4], as indirect loss data are generally not available [5]. To improve the utility of the Sendai Framework for Disaster Risk Reduction, it is necessary to address the issues of data collection and monitoring through identification and collective consideration, which holistically cope with the complexities associated with defining, reporting, and interpreting disaster mortality data regarding different challenges for different types of hazard events and subsequent disasters for this target [6].

A study of disaster-related deaths caused by Hurricane Andrew published in 1996 initially included indirectly related deaths as those resulting from any other disaster-related event, such as evacuation or cleanup [7]. In 2017, the Center for Disease Control and Prevention (CDC) defined it as any deaths that occur when unsafe or unhealthy conditions are present during any phase of a disaster (e.g., pre-event or preparing for the disaster, during the disaster event, or post-event during cleanup after a disaster) [8]. Although it is named disaster-related death (SaiGaiKanRenShi) in Japan, it is officially defined as death due to injuries aggravated by disaster hazards or illness caused by physical burden during evacuation, and it is recognized as such by the Law of Condolence Money (Act No. 82 of 1973) [9], which also recognizes such deaths that occur post disaster the same way [10,11].

The case of this study is the Great East Japan Earthquake and Tsunami (GEJE) in 2011 that triggered a tsunami and the Fukushima nuclear accident, which caused a total of 19,759 direct deaths [12] and 3784 reported indirect deaths [13]. This resulted in an extreme spike in annual crude death rate that has slowly progressed since 1987 (Figure 1) in the three targeted prefectures studied in this context, along with the general situation of Japan. Compared to direct deaths, indirect deaths are too subtle and intrinsic to be clearly seen or monitored. In Japan, the Reconstruction Agency publishes indirect official deaths based on analysis of death certifications, selection of condolence applications from municipal authorities, and hearings with government officers [14]. Although this surveillance system has been criticized for its biased screening process [15], they have still been monitoring indirect deaths resulting from natural disasters since 1995 and providing the country with authoritative statistics on the extent of the long-term effects of natural disasters.

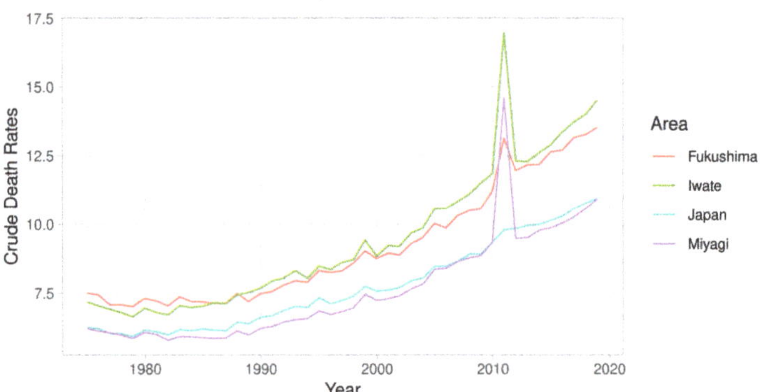

Figure 1. Annual all-cause crude death rates per 100,000 persons in Fukushima, Iwate, Miyagi, and Japan from 1975–2019.

To clarify the mechanism of post-disaster indirect deaths and assess Japan's surveillance system, a wide range of literature is reviewed to compare its definitions, and an excess mortality model was built to estimate the death toll of indirect mortality. Excess mortality is an epidemiological word defined as mortality above what would be expected based on the noncrisis mortality rate in the population of interest [16] and is calculated by subtracting the expected number of deaths from the observed number of deaths [17]. It is considered a better indicator for monitoring the scale of a pandemic such as COVID-19 [18], as well as investigating the long-term effects of natural disasters. After investigating 103 countries and territories, we found that in several of the worst-affected countries (Peru, Ecuador, Bolivia, Mexico) the excess mortality was above 50% of the expected annual mortality [19]. The pandemic not only caused internal and external tension in different countries [20], but also can be related to the dismal state of people's wellbeing due to the unplanned lockdown and subsequent period of socio-economic and health crisis [21], which is likely to contribute to indirect death. In this paper, the added value is that, more so than for direct death, the number of indirect deaths that cannot be easily collected is the initial focused in this methodology, which was developed for COVID-19. A constant value of excess death was first used in the field of natural disasters to explore deviations from baseline (expected death) in the case of the mid-July 1995 heat wave in Chicago [22]. In another instance, the temporal trends in the risk of excess mortality were assessed by constructing a Poisson regression using age, city, and year, which showed that indirect deaths that were impacted by physical health problems in Soma City and Minamisōma City were most severe in the first month after the disaster [23].

The hypothesis of this study is that post-disaster indirect mortality has been underestimated among prefectures, cities, towns, and villages. Thus, better accountability for indirect deaths is required so we can monitor the impact of disasters more comprehensively [24]. Using the case of the Great East Japan Earthquake and Tsunami, this study aims to inspect Japan's indirect post-disaster death surveillance system and clarify a mechanism to optimize its certification, which plays a crucial role in providing reliable statistics for recovery evaluation.

2. Data and Methods

This multidisciplinary study originally adopts the epidemiological methodology from a socioeconomic perspective by studying larger demographic data from severely damaged areas to examine the indirect surveillance system in Japan, and aims to provide reference for policymakers and researchers to improve this system in Japan or establish a better one in other countries. This paper is divided into two parts: the first part is a literature review intended to qualitatively analyze and investigate indirect deaths, as well as the pros and cons of the existing surveillance system; the second part uses a machine learning model to quantitatively analyze the gap between the expected and observed indirect death toll.

To firmly define indirect death, government reports from the Reconstruct Agency and Cabinet Office and research articles from the field of epidemiology, social science, and public policy are reviewed, in which the researchers performed their studies from selection of appropriate data to conducting analytic work and forming the concept. Regarding data selection, Japan's research related to indirect death focus mainly on domestic data, from central departments such as the Cabinet Office and local governments in disaster-prone areas such as Kobe [25], while global researchers turn to a wider range of information sources, including institutions such as the CDC. With regard to disasters, it is generally covered because among limited literature directly related to indirect death, some focus on storm hazards, some study seismic disasters, and others impose interests of all kinds. For example, government reports such as those from the Cabinet Office [9] and CDC applied their indirect death to all disasters globally, while Loris and Ueda kept an eye on the local area. Regarding the affected population, most of those studies were carried out in high-frequency areas where geological movements can cause more deaths that sometimes are thousands of people than any other type, costing dozens of lives. In terms of the time frame, it varies from study to study.

Since 2020, excess mortality has become a popular buzzword because it is considered the most objective indicator of the COVID-19 death toll [26,27]. Under the circumstances of the pandemic, officially reported deaths and demographic data are commonly used to calculate excess mortality. However, when it comes to natural disasters such as hurricanes and tsunamis, although the amount of death certifications and vital registration mortality data that can be utilized, respectively, to obtain observed deaths is limited, it is an ingenious and innovative way to predict death tolls by coping with the uneasy situation of confirming the true number of indirect deaths due to the lack of directly observed indirect deaths [19,28,29].

2.1. Data and Tools

Twenty-four official reports associated with the indirect deaths caused by the GEJE published every June and December from 2011 to 2021 by the Japan Reconstruction Agency were scrutinized, from which the number of observed indirect deaths was acquired. The observed number of direct deaths was obtained from the latest report published in 2021 by the Fire and Disaster Management Agency. The number of the registered population and all-cause deaths in 131 municipalities from 2000 to 2021 was extracted from resident registrations that are included in the Population, Demographic and Household Surveys based on the Basic Resident Ledger published every August by the Ministry of Internal

Affairs and Communications on the e-Stat Portal Site of Official Statistics of Japan. Since 2000, many municipalities have been merged into one (e.g., Yamagata village was combined into Kuji city in 2006). In this study, all the data from the despaired municipalities are added and counted into the latest merged municipalities. Moreover, geographic data of Geo JSON was obtained from the Ministry of Land, Infrastructure, Transport and Tourism of Japan. Although it is known from interviews that the Ministry of Health, Labor, and Welfare of Japan manages a database of death certifications, our requests for access to death certifications were rejected by both the local and national government. Thus, all the data used in this context are open access and available from official government websites. As for tools applied in this analysis, Python was used for geographical analysis and data cleaning by means of Jupyter Notebook and Google Colaboratory, while R was used for data organization, regression analysis, and plot export, using RStudio and RStudio Cloud.

2.2. Method

The mega disaster can be regarded as a colossal natural experiment [30,31], in which the control group (all municipalities) turned into an experimental group from 11 March 2011 due to the intervention of the GEJE. Morita's research quantified the excess mortality of two cities in Fukushima from 2015 to 2016 [23], while Uchimura estimated that the mortality ratio in the month post disaster, comparing the three prefectures with other prefectures was at 1.20 for those aged 60–69 years old [32]. Nevertheless, as an essential indicator of long-term environmental and health effects, the number of indirect deaths is still unknown [33]. Regarding almost all prefectures in Japan, the total reported indirect deaths on 31 March in 2019 reached 99.9% of the accumulated total reported indirect deaths in 2020, and dropped to zero and stopped growing after 31 March in 2020. The first death attributed to the COVID-19 pandemic in Japan occurred on 10 February 2020 [34], when it began to affect all-cause mortality and population. Thus, although we intended to observe the period 2000–2020, 2019 is selected as an end node for the long-term effects of GEJE. Population and all-cause death between 2000–2010 and 2018–2019 are deemed as the control group, while those occurring between 2011–2017 are deemed the experimental group. In order to create a control group for 2011–2017, data between 2000–2010 and 2018–2019 were used to train a Gaussian process model Equation (1) to estimate a set of plausible coefficients for calculating excess deaths in Equation (2). Then, Equation (3) was used to collect indirect deaths from different cohorts.

$$N_{m,y} = \alpha_m \cdot Y_{(2000-2010\&2018-2019)} + \beta_m + \epsilon, \ \epsilon \sim N\left(0, \sigma^2\right) \tag{1}$$

Regressing the Gaussian process with a Bayesian approach functions well on small time-series data sets capable of providing uncertainty measurements on the predictions [27] for each municipality employed as the smallest unit. In the generalized linear model Equation (1), N denotes the observed number of all-cause deaths in the municipality of m and the year of y. The constant α is a linear slope for the municipality of m across years. β refers to a separate intercept as fixed effects in each municipality of m from the difference in population structure, migration rate and other socioeconomic factors. Y is an independent variable ranging from the years 2000–2010 and 2018–2019, and ϵ is the Gaussian noise following the normal distribution fluctuating around zero.

$$E\left(N_{m,y}\right) = \widehat{\alpha_m} \cdot Y_{2011-2017} + \widehat{\beta_m} \tag{2}$$

$$I = \sum_j^C \{\sum_i^{2011-2017} [N_{j,i} - E(N_{j,i})] - D_j\} \tag{3}$$

In Equation (1), all α and β estimated from Equation (2) are used to calculate the expected number of counterfactual deaths in municipality m in year Y (2011–2017). Equation (3) was developed to predict the number of indirect deaths I by summing municipality j in cohort C after adding up the excess numbers of year i 2011–2017. By referring to a specific cohort, we can obtain the predicted number of indirect deaths by resetting C (e.g., prefectures, municipality, etc.).

3. Results

In our preliminary study, it was gauged by baseline death rate between 2010 and 2019 that 624,695 all-cause deaths would have occurred if the GEJE did not happen, compared to the 651,442 observed deaths that did occur according to Vital Statistics during 2010 and 2019. Correspondingly, after deducting 19,759 direct deaths that were investigated by the Fire and Disaster Management Agency from the 26,747 excess deaths, the number of indirect deaths should be 6988, which means approximately 3204 deaths were undercounted by the surveillance system.

Literature Review

In this session, government reports and research articles are reviewed in fields such as epidemiology, public policy, and social science to study indirect death and the surveillance system. We investigated different definitions and causes of indirect death as a background study and clarified the history, mechanism, advantages, and disadvantages of the surveillance system.

In Japan, the recognition of indirect deaths is closely related to the condolence policy following Act No. 82, launched in September 1973, which initially regulated condolence payments of up to JPY 500,000 to the immediate family of the victim who perished directly or indirectly from natural disasters [35]. After several rounds of legal amendments, the payment has been increased to up to JPY five million since 1991. When the concept of indirect death (Saigai Kanrenshi in Japanese) was first introduced in January 1995 (Table 1) when the Great Hanshin-Awaji Earthquake struck west Japan, 14.3% of 6405 total casualties were classified as indirect deaths [36,37].

Table 1. Literature review of definition of indirect death.

Reference	Terminology	Time frame	Definition/Description
Cabinet Office (2020) [9]	Disaster-related death	No implication	Death due to injuries aggravated by disaster hazards or illness caused by physical burden during evacuation
Loris et al. (2007) [3]	Indirect Death	Two weeks	Death resulting from suicide; fatal injury occurring during clean-up; post-disaster pulmonary embolism on account of sheltering in motor vehicles.
	Disaster-Triggered Deaths	Two weeks; one year	Death resulting from disruption of care for post-disaster chronic illness; suicide.
Asim et al. (2006) [38]; NWS/NOAA (2005) [39]	Indirect Death	No implication	Death that occurs in the vicinity of a hydro-meteorological event, or after it had ended, but were not directly caused by impact or debris from the event.
Ueda et al.(1996) [25]; Ueda (2014) [40]; Ueda (2016) [41]	Post-disaster-related death	72 h; two or three weeks depends on the scale; three months	Deaths due to indirect causes such as psychological shock and severe evacuation conditions, even if the disaster did not cause a wound.
Debra (1999) [7]; CDC (2017) [42]	Indirectly related disaster death	Any phase	Death that occurs when the unsafe or unhealthy conditions present during any phase of the disaster (i.e., pre-event, during the actual occurrence, or post-event).
Nagaoka (2004); MHLW (2011); Miyamoto (2013) [43]	Disaster-related death	One month; six months	Death that occurs due to abrupt change of environment or suicide caused by mental illness or stress from the disaster during six months post disaster.
Ehren B. (2001) [44]	Indirect death	Approximately two weeks following the event.	Deaths not primarily resulting from the initial and physical impact the hurricane.
Nishant et al. (2018) [45]	Indirect death	No implication	Deaths resulting from worsening of chronic conditions or from delayed medical treatments that may not be captured on death certificates.
Hyogo Prefecture, MHLW (1995) [36]	Disaster-related death	No implication	Deaths certified by the Disaster Condolence Grants Committee with a reasonable causal relationship to the earthquake disaster.

Based on the definitions in Table 1, indirect death is studied and described in various ways by classifications and causes. Two weeks are conditionally implied as a key node for indirect deaths despite the dissimilarity in the type and scale of disasters, regions, or emergency management. Ueda demonstrated a watershed of two weeks (up to three weeks in the case of a mega disaster such as GEJE) between the chronic phase and the acute phase, which can be further differentiated as the hyper-acute phase, acute phase, and sub-acute phase [25,40,41]. Since the elderly are commonly regarded as a vulnerable group, Ehren pointed out that 50% of those who died from cardiovascular causes post-disaster usually died within two weeks [44]. In Loris' study, two weeks is a general borderline to determine immediacy [3]. Since the end node of disaster effects depends on the disaster scale to a great extent, in GEJE the reported number of indirect deaths almost stopped increasing at 469 in 2018 for Iwate, 929 in 2020 for Miyagi, and 2,314 in 2020 for Fukushima [13], in which Sendai city is demonstrated by five wards (i.e., Aobaku, Izumiku, Miyaginoku, Taihakuku, and Wakabayashiku) (Figure 2).

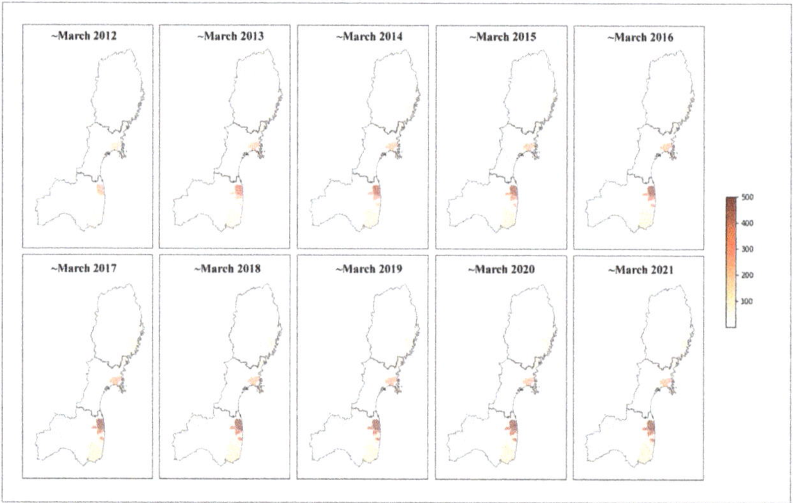

Figure 2. Annual geographical distribution of accumulated reported indirect death number of the Great East Japan Earthquake and Tsunami of 131 municipalities in Miyagi, Fukushima, and Iwate from March 2012–2021.

Thanks to infection control following GEJE [46], infectious illness is not a major cause of indirect death. The four leading causes comprise pneumonia, coronary heart disease, stroke, and cancer [23]. Among cities and towns hit by tsunami, a higher percentage of flooded households was associated with a higher risk of indirect death, lower expenditures on outpatient medical care, and lower expenditures on home care services [32]. It was reported that 50% of indirect death on average was due to physical and mental exhaustion of evacuating and living in a shelter, of which 60% was attributed to evacuees from Fukushima [14]. In the case of the 2004 Indian Ocean tsunami, indirect mortality was positively correlated with poor post-tsunami psychosocial health for males and loss of spouse for females [47].

To be recognized as a case of indirect death for individuals in Japan, someone (e.g., immediate family members) must submit a list of materials to the administrative authority or local government for application. After the screening process via the Disaster Condolence Grants Committee (consists of lawyers, doctors, administrative officers, and other professionals who are coordinated by the municipal/prefecture-level government), recognition and aid are then granted to the applicant. Since the amendment in 1991, if the victim is the breadwinner of the applicant's family, JPY 5 million is awarded. If the victim is not,

JPY 2.5 million yen is awarded instead. Accordingly, one-half of the payment is covered by the national government, one-fourth by the prefecture government, and one-fourth from local government of the municipality [48–50]. After reviewing reports and interviewing the government staff, it is clarified that the detailed reports from the Reconstruction Agency have been published twice a year since 2011 through continuous collection and analysis of data from death certifications, committee reports, and hearing from professionals and local governors, etc. [14]. Successful applications for condolence payments are consequently counted by committee reports and death certifications. Indirect deaths with no applications are only considered by scrutinizing other reports to a certain degree.

In Miyagi (Figure 3), it was observed that indirect deaths are under-reported in some municipalities, including Onagawamachi and Shiogamashi (in Japanese, the suffix of machi or cho refers to a town as a political district; shi refers to the city and mura or son refer to the village), while they were over-reported in municipalities such as Higashinatsushimashi and Kesennumashi etc. The rising rate in Sendaishi is stable, with a higher number of indirect deaths, due to its larger and younger population. Furthermore, the population in rural areas inclusive of Shichigashuku, Shikamamachi and Ohiramura was found to fluctuate dramatically due to the disaster's long-term effects.

Figure 3. All-cause death per 10,000 population of the municipality in Miyagi during 2000 and 2020. The bottom right gray number of each plot is the number of observed indirect deaths by March 2019 reported by the Reconstruction Agency; the bottom left black number of each plot is the estimated number of total excess deaths in the prefecture between January 2011 and December 2017. The top right blue number of each plot is the estimated indirect deaths calculated by subtracting reported direct deaths from total excess deaths; the top left red number of each plot is the under-counted number calculated by subtracting reported indirect deaths from estimated indirect deaths. * = $p < 0.05$, ** = $p < 0.001$, *** = $p < 0.001$.

In Iwate, indirect deaths are under-reported in municipalities including Ichinosekishi, Ichinohemachi, Ofunatoshi, and Okushushi, while they were over-reported in municipalities such as Minamisanrikumachi and Wtaricho (Figure 4). In some municipalities, the estimated number of indirect deaths turned out negative because excess mortality was lower than expected due to mortality selection that dominated scarring effects [47].

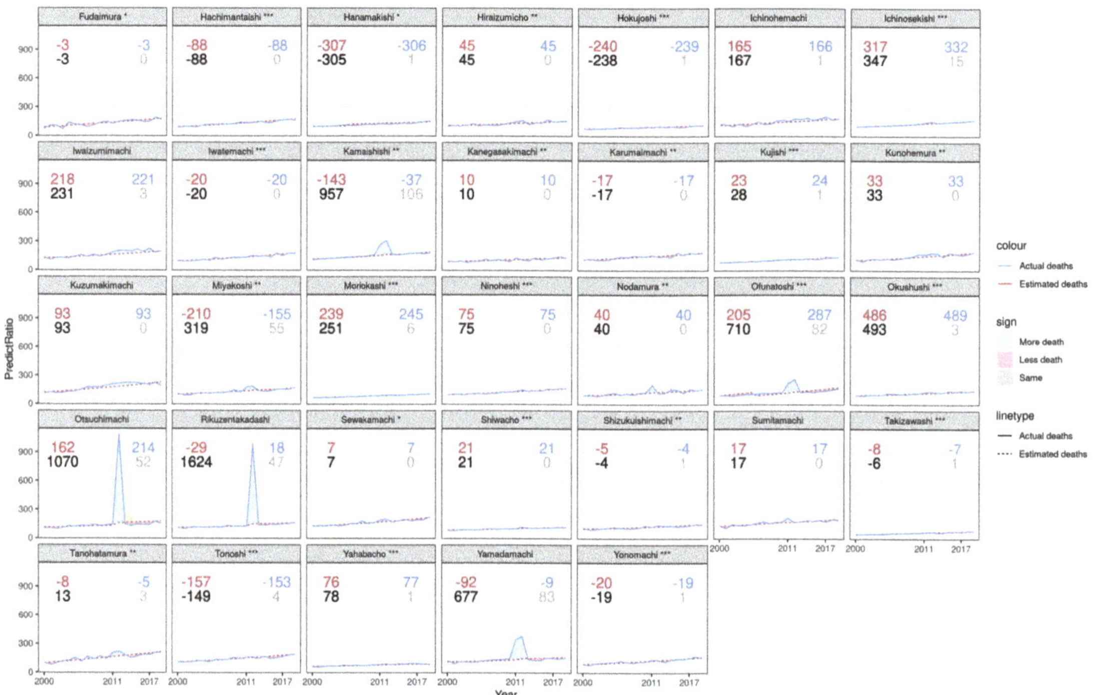

Figure 4. All-cause death per 10,000 population of municipality in Iwate during 2000 and 2020. The bottom right gray number of each plot is the number of observed indirect deaths by March 2019 reported by the Reconstruction Agency. The bottom left black number of each plot is the estimated number of total excess deaths in the prefecture between January 2011 and December 2017. The top right blue number of each plot is the estimated indirect deaths calculated by subtracting reported direct deaths from total excess deaths. The top left red number of each plot is the under-counted number calculated by subtracting reported indirect deaths from estimated indirect deaths. * = $p < 0.05$, ** = $p < 0.001$, *** = $p < 0.001$.

In Fukushima (Figure 5), where the nuclear disaster occurred at the Daiichi Nuclear Power Plant in Okumamachi, the number of displaced evacuees peaked at 165,000 in May 2012 and subsequently decreased to 37,000 in January 2021 [51]. Our analysis indicates that the mortality pattern in municipalities such as Showamura, Koorimachi and Shimogomachi destabilized after 2011. Indirect deaths in municipalities including Hanawamachi, Koorimachi, and Motomiyashi were under-reported, while they were over-monitored in municipalities such as Kawamuchimura and Narahamachi.

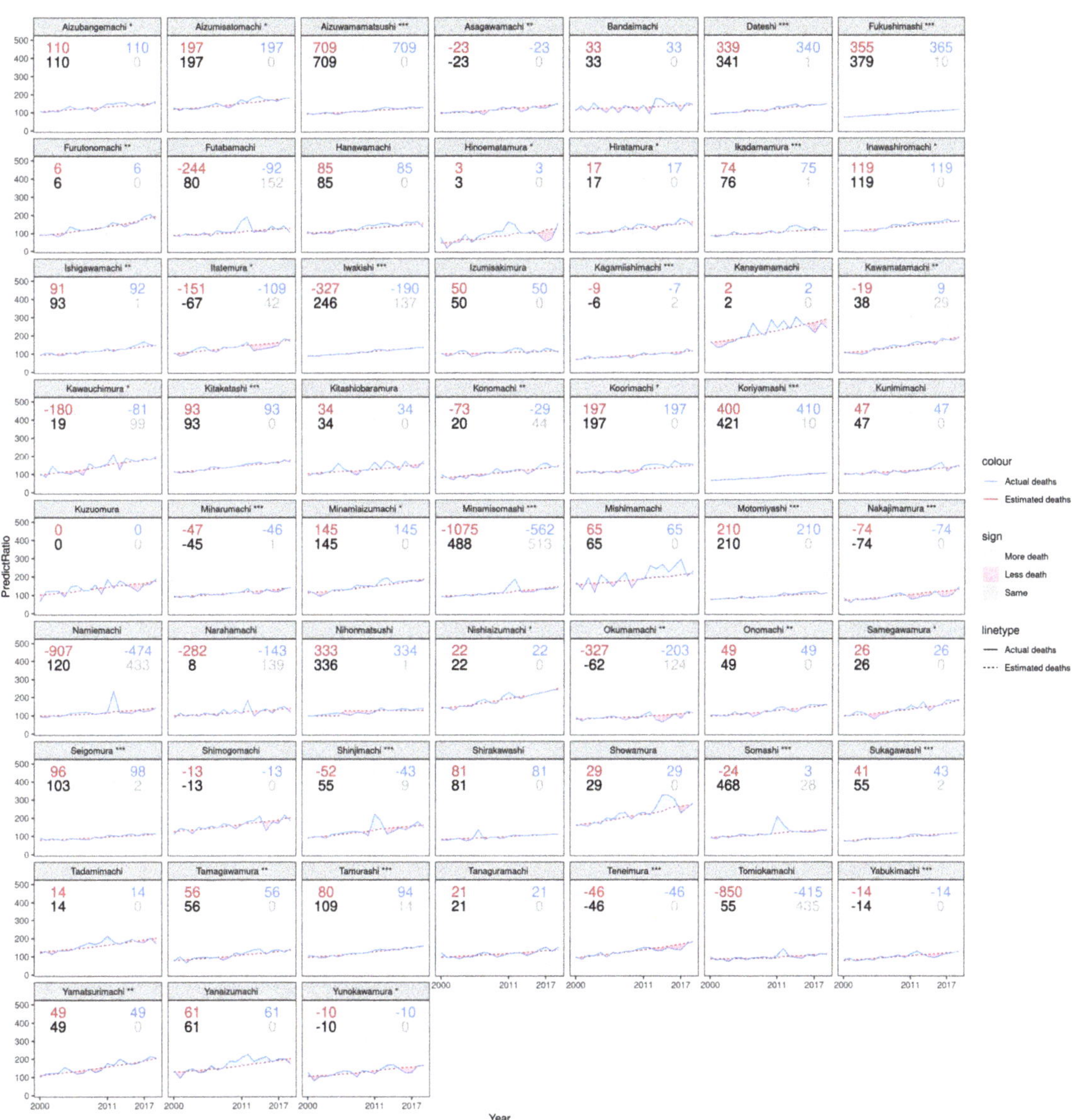

Figure 5. All-cause death per 10,000 population of municipality in Fukushima during 2000 and 2020. The bottom right gray number of each plot is the number of observed indirect deaths by March 2019 reported by the Reconstruction Agency; the bottom left black number of each plot is the estimated number of total excess deaths in the prefecture between January 2011 and December 2017. The top right blue number of each plot is the estimated indirect deaths calculated by subtracting reported direct deaths from total excess deaths; the top left red number of each plot is the undercounted number calculated by subtracting reported indirect deaths from estimated indirect deaths. * = $p < 0.05$, ** = $p < 0.001$, *** = $p < 0.001$.

4. Discussion

The impact of the megadisaster GEJE on the population was more profound than that reported by official statistics or news. A total of 4657 indirect deaths were estimated in

Fukushima, Iwate, and Miyagi prefectures, which is much higher than the reported number 3784, demonstrating a possible underestimation of indirect deaths in the case of the GEJE. To predict indirect deaths with more precision, all municipalities in the three prefectures were used as the smallest unit in the demographic panel data.

In Miyagi, it was estimated that 12,295 excess deaths occurred between January and December 2017, with 1734 possible indirect deaths. Subtracting the 925 reported direct deaths, we can confirm that 809 cases were ignored. In Iwate, 6457 excess deaths were captured by the model, which predicted 1350 indirect deaths compared to the 467 reported indirect deaths. Consequently, 883 cases were missed by the surveillance system. In Fukushima, the predicted 1573 indirect deaths fell below the reported 2229 indirect deaths, which means that 656 cases were overreported. By Figure 6, mortality selection can be identified on the plots of Miyagi and Iwate, but in Fukushima, the scarring effects slightly dominated the mortality selection.

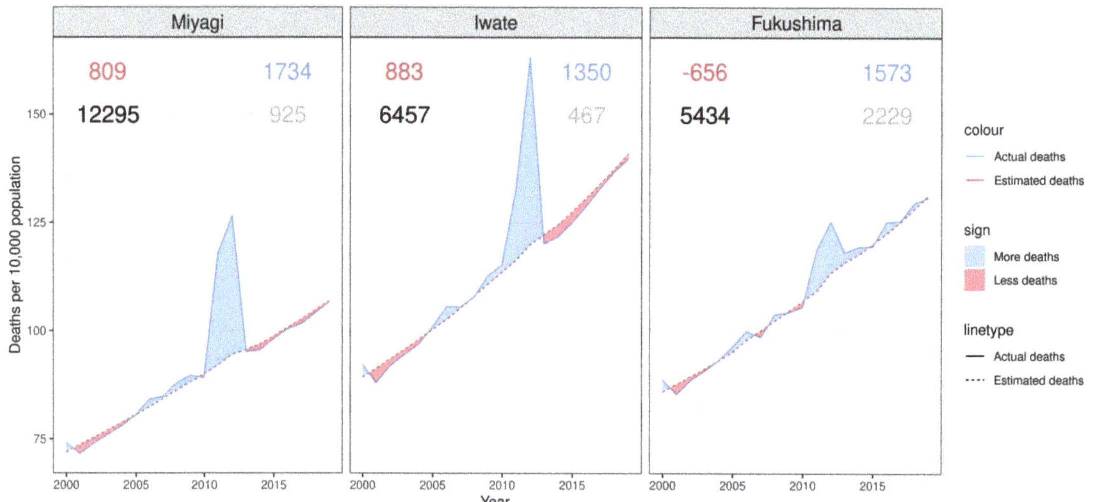

Figure 6. Actuality and estimation of all-cause crude death rate during 2000–2019 in three prefectures The bottom right gray number of each plot is the number of observed indirect deaths by March 2019 reported by the Reconstruction Agency. The bottom left black number of each plot is the estimated number of total excess deaths in the prefecture between January 2011 and December 2017. The top right blue numbers of each plot is estimated indirect death calculated by subtracting reported direct deaths from total excess deaths. The top left red number of each plot is the under-counted number calculated by subtracting observed deaths from estimated indirect deaths.

The concept of mortality selection originated from Darwin's theory of natural selection [52] and was later developed by Endler, who demonstrated it as a restatement of 'survival of the fittest' tautology [53]. Natural selection can be divided into sexual and non-sexual selection in a narrower sense; in the latter case, mortality selection and fecundity selection are included [53]. Mortality selection can be defined as a phenomenon where members with disadvantageous characteristics are more likely to die as the cohorts age [54,55]. Another similar term, selective mortality, is defined as a process in which disadvantaged individuals die at a younger age than their more advantaged peers [56]. It can be used to explain the drop in actual all-cause deaths after 2011 below the baseline death rate, because disadvantaged cohorts that are vulnerable in terms of physical, mental, and socioeconomic status in the targeted area died before when they would usually have died. On the contrary, scarring effects, which can be defined as a series of relatively negative effects on the post-disaster population, could lead to early death during their lifetime, as childhood adversities could lead to adult psychiatric disorders and eventually suicide [57]. Scarring effects are also regarded as the effects of a persistent change in beliefs about the probability of an extreme and negative shock [58]. Under several circumstances, the all-cause death rate can return to the baseline level plausibly, but only if there is a balance between mortality selection and scarring effects post the natural disaster. If the balance is not reached, the actual death rate fluctuates.

In the case of the GEJE that occurred on 11 March 2011, 17.0% out of 22,199 total post-disaster casualties [59,60] were classified as indirect deaths, signaling a worrying increase and a need to reconsider the underlying reasons as well as the effectiveness of disaster recovery, per se. Indirect death is a chronic and unignorable problem that can also be seen in the case of other natural disasters. For example, a review of 59 US tropical cyclones showed that the number of indirect deaths was just as large as the number of direct deaths, in which the main causes turned out to be cardiovascular failure, vehicle accident, evacuation failure, and power failure [61]. Moreover, underestimation occasionally happens in other cases of disasters as well, as demonstrated by related research articles. For instance, in the case of Hurricane Maria, it is estimated that there were 2975 excess deaths, in contrast with the sixty-seven deaths primitively reported by the Puerto Rico Department of Health [62].

Through unstructured interviews with residents of Ishinomaki and Sendai cities, six interviewees who lived in different places stated that they have great empathy for the suffering and death of their relatives or friends, but are not sure about condolence money or the concept of indirect death. It can be seen that their trauma from the past is connected to the notions of empathy and also to mental instability, which was reported to be one of the reasons for 13 suicides among the number of indirect deaths [14,63]. Understanding the mechanism of indirect death should take place from a complex system perspective in which indirect death correlates with social–cultural, natural, and physical environments [64].

The limitations of the study are the precision of the model and the volume of data, which can be improved by refining the model and expanding the dataset through further study. To provide more comprehensive evidence, more multidisciplinary studies are needed to explore causal relationships between socioeconomic effects and indirect death, as well as case studies investigating those municipalities where estimated indirect deaths significantly deviated from observed indirect deaths. Beyond improving the existing surveillance system for the future, overlooked deaths must also be counted with a more comprehensive methodology that considers migration and other variables. For greater generalization, more comparisons of different disasters should also be investigated.

5. Conclusions

The study estimated excess mortality in three prefectures severely damaged during the Great East Japan Earthquake and Tsunami. In total, 873 deaths were found to have not been reported as indirect deaths. Our methodology and results can be used as evidence to demonstrate that optimizing the existing surveillance system by improving death certification and standardization can provide better statistics in terms of disaster recovery evaluation for researchers and policy makers.

This paper makes a two-fold contribution to policy recommendations against the imperfections that exist in policies of the Japan condolence law and the Sendai framework. In total, 279 items of news reported by searching both keywords of indirect death and the Great East Japan Earthquake and Tsunami can be found via Google news by setting the time range between 11 March 2021 and 11 March 2022, which implies that indirect deaths caused by the GEJE are still a problem that receives attention from Japan society even after 10 years. However, how the number of indirect deaths in Japan is collected remains problematic, especially the immoderate reliance on condolence law, which cannot cover the perished who do not have any relative or spouse to apply for the condolence subsidy. Therefore, this study pointed out that there should be more policies to cooperate with the condolence law to maximize the accuracy of the statistics. On the other hand, to follow the status of whether and how the target A in the Sendai framework is being achieved, the number of deaths and missing persons attributed to disasters per 100,000 population of each country is collected by the Disaster Loss Data Collection Tool (called "DesInventar Sendai"), which is a subsystem of the Sendai Framework online monitoring tool. Despite the mention of attribution, death is applicably defined as the number of people who died during the disaster or directly after, as a direct result of the hazardous event in the Technical Guidance for monitoring and reporting on Progress in achieving the global targets of the Sendai Framework for Disaster Risk Reduction. This study addresses the importance of indirect death that can also be attributed to disasters and demonstrates the direction for developing a better methodology to monitor the number of deaths.

This study not only implies that a surveillance system monitoring long-term effects, such as the statistics of indirect death caused by natural disasters, can be built through methods inclusive of the existing policy and developing related policy, but also indicates the importance of upgrading the Data Collection tool serving the Sendai Framework by improving the criteria used to monitor countries' data and involving the number of indirect death that can also be attributed to disasters. In addition, social factors such as social networks that can underpin apparently spontaneous actions [65] should also be investigated to further develop the model in this study. Global demographic data can be collected to feed the model, which can be used more precisely to verify the exactness of the reported death toll or to estimate the number of indirect deaths in countries where the statistics for natural disasters are not yet well developed.

Author Contributions: Conceptualization, X.Z. and M.I.; methodology, X.Z.; software, X.Z. and C.F.; validation, X.Z. and M.I.; formal analysis, X.Z.; investigation, X.Z.; resources, X.Z.; data curation, X.Z.; writing—original draft preparation, X.Z.; writing—review and editing, X.Z.; visualization, X.Z. and C.F.; supervision, M.I.; project administration, X.Z.; funding acquisition, M.I. All authors have read and agreed to the published version of the manuscript.

Funding: This research was funded by the Japan Society for the Promotion of Science KAKENHI (grant number JP 21H03680).

Institutional Review Board Statement: Not applicable.

Informed Consent Statement: Not applicable.

Data Availability Statement: Please contact the corresponding author for the information of data availability.

Acknowledgments: The authors are thankful to Haruka Maeoka and Minglun Song for inspiration in the development of the algorithm.

Conflicts of Interest: The authors declare no conflict of interest.

Abbreviations

The following abbreviations are used in this manuscript:

CDC	Center for Disease Control and Prevention.
GEJE	Great East Japan Earthquake and Tsunami.
MHLW	Ministry of Health, Labour and Welfare of Japan.
NOAA	National Oceanic and Atmospheric Administration.
NWS	National Weather Service.
UNDRR	United Nations Office for Disaster Risk Reduction.
WHO	World Health Organization.

References

1. General Assembly. *Resolution Adopted by the General Assembly on 3 June 2015*; United Nations: New York, NY, USA, 2016.
2. United Nations. *The Sendai Framework for Disaster Risk Reduction*; United Nations: New York, NY, USA, 2015.
3. Uscher-Pines, L. "But for the Hurricane": Measuring Natural Disaster Mortality over the Long Term. *Prehospital Disaster Med.* **2007**, *22*, 149. [CrossRef] [PubMed]

4. McKinney, N.; Houser, C.; Meyer-Arendt, K. Direct and indirect mortality in Florida during the 2004 hurricane season. *Int. J. Biometeorol.* **2011**, *55*, 533–546. [CrossRef]

5. Wirtz, A.; Kron, W.; Löw, P.; Steuer, M. The need for data: Natural disasters and the challenges of database management. *Nat. Hazards* **2014**, *70*, 135–157. [CrossRef]

6. Green, H.K.; Lysaght, O.; Saulnier, D.D.; Blanchard, K.; Humphrey, A.; Fakhruddin, B.; Murray, V. Challenges with disaster mortality data and measuring progress towards the implementation of the Sendai framework. *Int. J. Disaster Risk Sci.* **2019**, *10*, 449–461. [CrossRef]

7. Combs, D.L.; Parrish, R.G.; McNABB, S.J.; Davis, J.H. Deaths related to hurricane Andrew in Florida and Louisiana, 1992. *Int. J. Epidemiol.* **1996**, *25*, 537–544. [CrossRef]

8. CDC. *A Reference Guide for Certification of Deaths in the Event of a Natural, Human-Induced, or Chemical/Radiological Disaster*; CDC: Atlanta, GA, USA, 2017.

9. Cabinet Office Policy Coordination Office; Disaster Prevention Counselor; DVA. *Disaster-Related Deaths*; Cabinet Office: London, UK, 2022.

10. Tsuboi, M.; Hibiya, M.; Tsuboi, R.; Taguchi, S.; Yasaka, K.; Kiyota, K.; Sakisaka, K. Analysis of disaster-related deaths in the Great East Japan Earthquake: A retrospective observational study using data from Ishinomaki City, Miyagi, Japan. *Int. J. Environ. Res. Public Health* **2022**, *19*, 4087. [CrossRef] [PubMed]

11. Hasegawa, A.; Ohira, T.; Maeda, M.; Yasumura, S.; Tanigawa, K. Emergency responses and health consequences after the Fukushima accident; evacuation and relocation. *Clin. Oncol.* **2016**, *28*, 237–244. [CrossRef] [PubMed]

12. Agency, J.M. *The 2011 off the Pacific coast of Tohoku Earthquake (Great East Japan Earthquake) (Report No. 162)*; OCHA: New York, NY, USA, 2021.

13. Agency, R. Number of Earthquake-Related Deaths in the Great East Japan Earthquake (Survey Results as of 30 September 2021). Reconstruction Agency, Cabinet Office, Fire and Disaster Management Agency. 2021. Available online: https://www.reconstruction.go.jp/topics/main-cat2/sub-cat2-6/20210630_kanrenshi.pdf (accessed on 30 August 2022).

14. Agency, R. Report on Earthquake-Related Deaths in the Great East Japan Earthquake. Reconstruction Agency. 2012. Available online: https://www.reconstruction.go.jp/topics/240821_higashinihondaishinsainiokerushinsaikanrenshinikansuruhoukoku.pdf (accessed on 30 August 2022).

15. Time, T.J. 'Indirect' Deaths from Disasters. The Japan Times. 2019. Available online: https://www.japantimes.co.jp/opinion/2019/03/23/editorials/indirect-deaths-disaster (accessed on 30 August 2022).

16. World Health Organization (WHO). The True Death Toll of COVID-19. 2021. Available online: https://www.who.int/data/stories/the-true-death-toll-of-covid-19-estimating-global-excess-mortality (accessed on 30 August 2022).

17. CDC. National Vital Statistics System. 2022. Available online: https://www.cdc.gov/nchs/nvss/potentially_excess_deaths.htm (accessed on 30 August 2022).

18. Statistics Canada. Estimation of Excess Mortality. 2022. Available online: https://www.statcan.gc.ca/en/statistical-programs/document/3233_D5_V1 (accessed on 30 August 2022).

19. Karlinsky, A.; Kobak, D. Tracking excess mortality across countries during the COVID-19 pandemic with the World Mortality Dataset. *Elife* **2021**, *10*, e69336. [CrossRef]

20. Creţan, R.; Light, D. COVID-19 in Romania: Transnational labour, geopolitics, and the Roma 'outsiders'. *Eurasian Geogr. Econ.* **2020**, *61*, 559–572. [CrossRef]

21. Azeez EP, A.; Negi, D.P.; Rani, A.; AP, S.K. The impact of COVID-19 on migrant women workers in India. *Eurasian Geogr. Econ.* **2021**, *62*, 93–112. [CrossRef]

22. Whitman, S.; Good, G.; Donoghue, E.R.; Benbow, N.; Shou, W.; Mou, S. Mortality in Chicago attributed to the July 1995 heat wave. *Am. J. Public Health* **1997**, *87*, 1515–1518. [CrossRef] [PubMed]

23. Morita, T.; Nomura, S.; Tsubokura, M.; Leppold, C.; Gilmour, S.; Ochi, S.; Ozaki, A.; Shimada, Y.; Yamamoto, K.; Inoue, M.; et al. Excess mortality due to indirect health effects of the 2011 triple disaster in Fukushima, Japan: A retrospective observational study. *J. Epidemiol. Community Health* **2017**, *71*, 974–980. [CrossRef] [PubMed]

24. United Nations Office for Disaster Risk Reduction (UNDRR). Human Cost of Disasters—An Overview of the Last 20 Years 2000–2019. 2020; ISBN 9789212320274. Available online: https://www.un-ilibrary.org/content/books/9789210054478/read (accessed on 30 August 2022).

25. Ueda, K. The Role of Medical Care and Welfare in Reducing Earthquake-Related Deaths from a Consideration of Estimation and Recognition of Earthquake-Related Deaths. Japan Society for Disaster Recovery and Revitalization. 2014. Available online: https://f-gakkai.net/wp-content/uploads/2014/06/10-1-2-1.pdf (accessed on 30 August 2022).

26. Helleringer, S.; Queiroz, B.L. Commentary: Measuring excess mortality due to the COVID-19 pandemic: Progress and persistent challenges. *Int. J. Epidemiol.* **2022**, *51*, 85–87. [CrossRef] [PubMed]

27. Staub, K.; Panczak, R.; Matthes, K.L.; Floris, J.; Berlin, C.; Junker, C.; Weitkunat, R.; Mamelund, S.-E.; Zwahlen, M.; Riou, J. Historically high excess mortality during the COVID-19 pandemic in Switzerland, Sweden, and Spain. *Ann. Intern. Med.* **2022**, *175*, 523–532. [CrossRef]

28. Santos-Burgoa, C.; Goldman, A.; Andrade, E.; Barrett, N.; Colon-Ramos, U.; Edberg, M.; Garcia-Meza, A.; Goldman, L.; Roess, A.; Sandberg, J.; et al. Ascertainment of the Estimated Excess Mortality from Hurricane Maria in Puerto Rico. The George Washington University. 2018. Available online: https://publichealth.gwu.edu/sites/default/files/downloads/projects/PRstudy/Acertainment%20of%20the%20Estimated%20Excess%20Mortality%20from%20Hurricane%20Maria%20in%20Puerto%20Rico.pdf (accessed on 30 August 2022).

29. Lloyd-Sherlock, P.; Ebrahim, S.; Martínez, R.; McKee, M.; Acosta, E.; Sempé, L. Estimation of All-Cause Excess Mortality by Age-Specific Mortality Patterns of COVID-19 Pandemic in Peru in 2020. *Lancet* **2021**, *2*, 3820553. [CrossRef]

30. Andrabi, T.; Daniels, B.; Das, J. Human capital accumulation and disasters: Evidence from the Pakistan earthquake of 2005. *J. Hum. Resour.* **2021**. [CrossRef]

31. Cavallo, E.; Galiani, S.; Noy, I.; Pantano, J. Catastrophic natural disasters and economic growth. *Rev. Econ. Stat.* **2013**, *95*, 1549–1561. [CrossRef]

32. Uchimura, M.; Kizuki, M.; Takano, T.; Morita, A.; Seino, K. Impact of the 2011 Great East Japan Earthquake on community health: ecological time series on transient increase in indirect mortality and recovery of health and long-term-care system. *J. Epidemiol. Community Health* **2014**, *68*, 874–882. [CrossRef]

33. Ranghieri, F.; Ishiwatari, M. *Learning from Megadisasters: Lessons from the Great East Japan Earthquake*; World Bank Publications: Washington, DC, USA, 2014.

34. Nikkei. First Japanese Death from New Pneumonia: A Woman in Her 80s in Kanagawa Prefecture. 2022. Available online: https://www.nikkei.com/article/DGXMZO55604310T10C20A2CC1000/ (accessed on 30 August 2022).

35. Justice, M. Act on Provision of Disaster Condolence Grants. Cabinet Office. 2021. Available online: https://elaws.e-gov.go.jp/document?lawid=348AC0100000082_20210520_503AC0000000030 (accessed on 30 August 2022).

36. Hyogo Prefecture. Investigation of the Hanshin-Awaji Earthquake. 2005. Available online: https://web.pref.hyogo.lg.jp\/kk42/pa20_000000016.html (accessed on 30 August 2022).

37. Mainichi Shinbun. Mom, I Miss You. I Can't Be a "Bereaved Family Member" of a Missing Person's Declaration of Disappearance. 2014. Available online: https://mainichi.jp/articles/20210911/k00/00m/030/197000c (accessed on 30 August 2022).

38. Jani, A.A.; Fierro, M.; Kiser, S.; Ayala-Simms, V.; Darby, D.; Juenker, S.; Storey, R.; Reynolds, C.; Marr, J.; Miller, G. Hurricane Isabel–related mortality—Virginia, 2003. *J. Public Health Manag. Pract.* **2006**, *12*, 97–102. [CrossRef]

39. National Weather Service. National Weather Service Instruction#10-1605. 2022. Available online: https://www.nws.noaa.gov/directives/sym/pd01016005curr.pdf (accessed on 30 August 2022).

40. Ueda, K. Post-disaster related death and its countermeasures. *Nippon Igaku Shinpo* **1996**, 40–44.

41. Ueda, K. Known in Q&A to Prevent Post-Disaster Related Deaths. Japanese Communist Party. Available online: https://cir.nii.ac.jp/crid/1522825129737600768 (accessed on 30 August 2022).

42. CDC. A Reference Guide for Certification of Deaths in the Event of a Natural, Human-Induced, or Chemical/Radiological Disaster. National Center for Health Statistics. 2022. Available online: https://stacks.cdc.gov/view/cdc/49294 (accessed on 30 August 2022).

43. Tomomi, M. On the Examination of Disaster-Related Deaths—From Iwate Prefecture's Efforts in the Great East Japan Earthquake. Arutesu Riberaresu. 2013. Available online: https://cir.nii.ac.jp/crid/1390290699641515008 (accessed on 30 August 2022).

44. Ngo, E.B. When disasters and age collide: Reviewing vulnerability of the elderly. *Nat. Hazards Rev.* **2001**, *2*, 80–89. [CrossRef]

45. Kishore, N.; Marqués, D.; Mahmud, A.; Kiang, M.V.; Rodriguez, I.; Fuller, A.; Ebner, P.; Sorensen, C.; Racy, F.; Lemery, J.; et al. Mortality in puerto rico after hurricane maria. *N. Engl. J. Med.* **2018**, *379*, 162–170. [CrossRef] [PubMed]

46. Ishiwatari, M.; Koike, T.; Hiroki, K.; Toda, T.; Katsube, T. Managing disasters amid COVID-19 pandemic: Approaches of response to flood disasters. *Prog. Disaster Sci.* **2020**, *6*, 100096. [CrossRef] [PubMed]

47. Frankenberg, E.; Sumantri, C.; Thomas, D. Effects of a natural disaster on mortality risks over the longer term. *Nat. Sustain.* **2020**, *3*, 614–619. [CrossRef] [PubMed]

48. Outline of Disaster Condolence Grants and Disaster Relief Payments. Cabinet Office. Available online: https://www.bousai.go.jp/taisaku/choui/pdf/siryo1-1.pdf (accessed on 30 August 2022).

49. Japan Federation of Bar Associations. Opinion on Disaster-Related Deaths. 2022.

50. Japan Federation of Bar Associations. Questionnaire on the Status of Screening for Disaster Condolence Grants. Available online: https://www.nichibenren.or.jp/library/ja/special_theme/data/condolence_money_questionnaire_2.pdf (accessed on 30 August 2022).

51. Fukushima Prefecture. Immediate Report on the Damage Caused by the 2011 Tohoku-Pacific Ocean Earthquake. Technical Report. Available online: https://www.pref.fukushima.lg.jp/site/portal/shinsai-higaijokyo.html (accessed on 30 August 2022).

52. Darwin, C. *On the Origin of Species, 1859*; Routledge: London, UK, 2004.

53. Endler, J.A. *Natural Selection in the Wild*; Princeton University Press: Princeton, NJ, USA, 1986.

54. Benjamin, E.J.; Blaha, M.J.; Chiuve, S.E.; Cushman, M.; Das, S.R.; Deo, R.; De Ferranti, S.D.; Floyd, J.; Fornage, M.; Gillespie, C.; et al. Heart disease and stroke statistics—2017 update: A report from the American Heart Association. *Circulation* **2017**, *135*, e146–e603. [CrossRef]

55. CDC. *Smoking-Attributable Mortality, Years of Potential Life Lost, and Productivity Losses–United States, 2000–2004*; CDC: Atlanta, GA, USA, 2008.

56. Zajacova, A.; Burgard, S.A. Healthier, wealthier, and wiser: A demonstration of compositional changes in aging cohorts due to selective mortality. *Popul. Res. Policy Rev.* **2013**, *32*, 311–324. [CrossRef]
57. Kessler, R.C.; Davis, C.G.; Kendler, K.S. Childhood adversity and adult psychiatric disorder in the US National Comorbidity Survey. *Psychol. Med.* **1997**, *27*, 1101–1119. [CrossRef]
58. Kozlowski, J.; Veldkamp, L.; Venkateswaran, V. *Scarring Body and Mind: The Long-Term Belief-Scarring Effects of COVID-19*; Technical Report; National Bureau of Economic Research: Cambridge, MA, USA, 2020. [CrossRef]
59. Shimbun, N.K. 15,899 Deaths, 9 Years after the Earthquake, National Police Agency Summary 2020. Available online: https://www.jstage.jst.go.jp/article/ijshs/20/0/20_202047/_pdf (accessed on 30 August 2022).
60. Corporation, J.B. 10 Years after the Great East Japan Earthquake, "Earthquake-Related Deaths" Certified at 3775.2021. Available online: https://www.theatlantic.com/photo/2021/03/photos-10-years-great-east-japan-earthquake/618243/ (accessed on 30 August 2022).
61. Rappaport, E.N.; Blanchard, B.W. Fatalities in the United States indirectly associated with Atlantic tropical cyclones. *Bull. Am. Meteorol. Soc.* **2016**, *97*, 1139–1148. [CrossRef]
62. Santos-Burgoa, C.; Sandberg, J.; Suárez, E.; Goldman-Hawes, A.; Zeger, S.; Garcia-Meza, A.; Pérez, C.M.; Estrada-Merly, N.; Colón-Ramos, U.; Nazario, C.M.; et al. Differential and persistent risk of excess mortality from Hurricane Maria in Puerto Rico: A time-series analysis. *Lancet Planet. Health* **2018**, *2*, e478–e488. [CrossRef]
63. Crețan, R.; Light, D.; Richards, S.; Dunca, A.M. Encountering the victims of Romanian communism: Young people and empathy in a memorial museum. *Eurasian Geogr. Econ.* **2019**, *59*, 632–656. [CrossRef]
64. Kircher, J.; Lee, S.; Jamal, T.; Donaldson, J.P. *Regenerating Tourism with an Ethic of Care and Empathy*; University of Massachusetts Amherst: Amherst, MA, USA, 2022; ISBN 978-0-692-46509-7. Available online: https://scholarworks.umass.edu/cgi/viewcontent.cgi?article=2833&context=ttra (accessed on 30 August 2022).
65. Crețan, R.; O'Brien, T. Corruption and conflagration: (in)justice and protest in Bucharest after the Colectiv fire. *Urban Geogr.* **2020**, *41*, 368–388. [CrossRef]

International Journal of
Environmental Research and Public Health

MDPI

Article

An Interdisciplinary Approach to Quantify the Human Disaster Risk Perception and Its Influence on the Population at Risk: A Case Study of Longchi Town, China

Shengnan Wu [1], Yu Lei [2,3,4,*] and Wen Jin [5]

[1] Chongqing Economic and Social Development Research Institute, Chongqing 400041, China
[2] Key Laboratory of Mountain Hazards and Earth Surface Processes, Institute of Mountain Hazards and Environment, Chinese Academy of Sciences, Chengdu 610041, China
[3] China-Pakistan Joint Research Center on Earth Sciences, Chinese Academy of Sciences-Higher Education Commission (CAS-HEC), Islamabad 45320, Pakistan
[4] University of Chinese Academy of Sciences, Beijing 100049, China
[5] National Disaster Reduction Center of China, Ministry of Emergency Management, Beijing 100084, China
* Correspondence: leiyu@imde.ac.cn

Abstract: Understanding disaster risk perception is vital for community-based disaster risk reduction (DRR). This study was set to investigate the correlations between disaster risk perception and the population at risk. To address this research question, the current study conducted an interdisciplinary approach: a household survey for measuring variables and constructed an Agent-based model for simulating the population at risk. Therefore, two correlations were defined, (1) between risk perception and willingness to evacuate, and (2) between willingness to evacuate and the population at risk. The willingness to evacuate was adopted as a mediator to determine the relationship between risk perception and the population at risk. The results show that the residents generally have a higher risk perception and willingness to evacuate because the study area frequently suffered from debris flow and flash floods. A positive correlation was found between risk perception and willingness to evacuate, and a negative correlation to the population at risk. However, a marginal effect was observed when raising public risk perception to reduce the number of the population at risk. This study provides an interdisciplinary approach to measuring disaster risk perception at the community level and helps policymakers select the most effective ways to reduce the population at risk.

Keywords: disaster risk reduction; disaster risk perception; the population at risk; agent-based modeling

Citation: Wu, S.; Lei, Y.; Jin, W. An Interdisciplinary Approach to Quantify the Human Disaster Risk Perception and Its Influence on the Population at Risk: A Case Study of Longchi Town, China. *Int. J. Environ. Res. Public Health* **2022**, *19*, 16393. https://doi.org/10.3390/ijerph192416393

Academic Editors: Mikio Ishiwatari and Daisuke Sasaki

Received: 29 September 2022
Accepted: 24 November 2022
Published: 7 December 2022

Publisher's Note: MDPI stays neutral with regard to jurisdictional claims in published maps and institutional affiliations.

1. Introduction

1.1. Background

Earthquakes, landslides, debris flows, and other types of disasters have posed significant threats to human safety, property, and critical infrastructures [1–3]. The Centre for Research on the Epidemiology of Disasters (CRED) reported that just in 2021, there were 432 disaster events around the world, accounting for 10,492 deaths, affecting 101.8 million people [4].

Global landmark agendas, including the Sendai Framework for Disaster Risk Reduction 2015–2030 (SFDRR), the 2030 Agenda for Sustainable Development (SDGs), and the Paris Agreement, highlight the importance of adopting disaster risk reduction (DRR) measures to help reduce the loss of life and property damage. With the deepening understanding of disaster risk, people's response strategy to disasters has undergone a conceptual change, and the focus of the strategy has gradually shifted from the traditional passive response to disasters to taking action for DRR actively. In this process, it has also been discovered that in addition to developing technology and techniques [5–9], it is important to consider human dynamic factors in DRR [10], such as risk perception, willingness to evacuate, and emergency behaviors [11–13].

Public risk-reducing behaviors and risk perception are interlinked to each other. People with higher risk perception are more likely to take actions to reduce or avoid disaster risks [14,15]. As one of the key factors that may influence public risk-reducing behaviors [16], risk perception has become an important research agenda in DRR and plays a pivotal role in ensuring the effectiveness of DRR measures [17–19]. However, to date, there is no unified standard for measuring disaster risk perception in the academic community. Many factors may affect public risk perception, and various methods have been developed and introduced to measure risk perception, such as disaster characteristics [11], distance to disasters [20], demographic factors [11,20,21], gender differences [22], and disaster experience [20]. Meanwhile, although mainstream schools believe that higher risk perception can encourage people to take DRR actions [23,24], it is difficult to quantify the impacts of risk perception on DRR. Therefore, the primary aim of this paper is to explore the correlation between risk perception and the population at risk, which may provide empirical evidence for how to reduce disaster risks by affecting public risk perception.

1.2. Research Question and Hypothesis

This study primarily set the research question as the correlation between public risk perception and the population at risk (Figure 1). Understanding the complexity of risk perception is vital, and different authors have measured risk perception in various ways. As the current study chose a Chinese town as a study site, we sorted out the existing research in risk perception measurement that set China as a case. We found that some scholars divide risk perception into disaster intensity, disaster loss, environmental sensitivity, and the characteristics of the affected population [25]. Some scholars also believe that disaster risk perception includes the likelihood, threat, knowledge, and attitude of resistance to disasters [13] or disaster characteristics, disaster experience, disaster knowledge, and risk communication [26]. Therefore, for reliability and validity of the questionnaire, exploring the existing studies in the same study area [13,26–28] and considering the actual situation in the study area, we classified disaster risk perception into four variables, namely (a) knowledge of disasters, (b) impacts of disasters, (c) participation in DRR, and (d) disaster experience, and measured the comprehensive risk perception of disasters by summing up four variables. Subsequently, this study sorted out the public willingness to evacuate and defined the possible correlations between the public willingness to evacuate and the four risk perception variables. Therefore, the research hypotheses *H1–H4* were proposed.

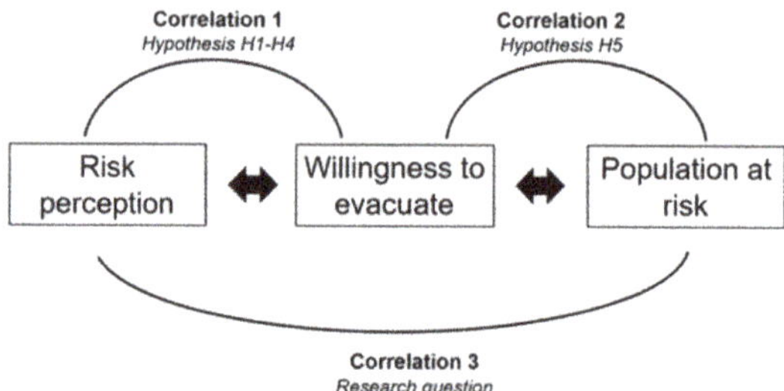

Figure 1. Research question and hypothesis.

(a) Knowledge of disasters: Knowledge of disasters refers to residents' knowledge of disasters, including the main types of local hazards, escape routes, and access to warning information. Many pieces of the literature confirmed the significance of knowledge for public risk perception [20,29]. The more the residents know about the major local hazards, the higher the risk perception they may have, and thus the more likely they are to take proactive measures to cope with the hazards [30]. Evacuation before disasters is one of the active measures to cope with disasters. Therefore, hypothesis *H1* is proposed:

H1: Knowledge of disasters had a significant positive effect on willingness to evacuate.

(b) Impacts of disasters: Impacts of disasters refer to how the residents perceive the level of disaster threat to their personal and local lives, property, production, and livelihood. Most studies have concluded that residents with stronger perceptions of disasters impacts of disasters tend to have a stronger willingness to evacuate [31–33]. For instance, people who live closer and may have a larger impact on the disaster may have higher risk perception [20,29,34]. However, some studies have also shown that disasters' impact levels and risk perception are not necessarily significantly related [35]. Based on this analysis, research hypothesis *H2* is proposed.

H2: Impacts of disasters had a significant positive effect on the willingness to evacuate.

(c) Participation in DRR: Participation in DRR refers to the willingness and cooperation of residents to participate in local disaster risk reduction practices. Literature reveals that government officials and residents vary greatly in risk perception [20]. Residents actively participating in local DRR activities tend to have higher risk perceptions [36]. Therefore, participation in DRR can be measured indirectly by measuring residents' participation in DRR and, thus, risk perception. Based on this analysis, research hypothesis *H3* is proposed.

H3: Participation in DRR had a significant positive effect on the willingness to evacuate.

(d) Disaster experience: Disaster experience refers to the number of disasters an individual has experienced. Disaster experience is one of the important factors affecting risk perception [20,26]. Many works of literature reveal that residents who have experienced disasters multiple times or have experienced large disasters tend to have higher perception levels [20,34,37]. Based on this analysis, research hypothesis *H4* was proposed.

H4: Disaster experience had a significant positive effect on willingness to evacuate.

Subsequently, this study developed an agent-based model to simulate the population at risk under different willingness to evacuate. It investigated the correlation between the population at risk and willingness to evacuate. The higher willingness to evacuate provides more time for evacuation and thus is more likely to reduce the overall number of local casualties due to disasters [13,33]. On the contrary, residents are less willing to evacuate and are unwilling to take the initiative to evacuate to a safe area immediately after receiving warning information. In this case, they can only passively escape from dangers when they encounter a disaster, which tends to have a lower survival rate. The population at risk refers to those residents who may be faced with dangers in disaster events. Therefore, hypothesis *H5* is proposed:

H5: Public risk perception had a significant positive effect on reducing the population at risk.

Once these two correlations, (1) between the willingness to evacuate and the risk perception and (2) between the public willingness to evacuate and the population at risk, were extracted, we can adopt the willingness to evacuate as a mediator to link risk perception and the population at risk. This study, therefore, explored the correlation between risk perception and the population at risk.

2. Materials and Methods

2.1. Data Preparation

2.1.1. Study Area

This study selected Longchi township as the study site (Figure 2). Longchi township is in Dujiangyan City, Sichuan Province, China, which is 80 km away from Chengdu. It is the largest township in Dujiangyan, with five communities, including Chaguan,

Yunhua, Liping, Nanyue, and Dongyue. Many natural and tourism resources, such as Dujiangyan National Forest Park, Giant Panda World Heritage Site, and Longxi-Hongkou National Nature Reserve, are located within Longchi town. According to statistics, before the 2018 Wenchuan earthquake, the output value of the primary and tertiary industries reached over 3,487,500 USD, and the fiscal revenue exceeded 159,030 USD.

Unfortunately, the 2018 Wenchuan earthquake struck the town and locked down its industries. During the earthquake, Longchi was devastatingly hit, with 36 people dead, 17 missing, and 1558 injured due to it occurring only three kilometers from the epicenter, Yingxiu town. Meanwhile, its roads, electricity, and communications were shut down, and the tourist resort suddenly became isolated. After the earthquake, due to the post-earthquake fractured rock and loose soil on the steep slopes [38], there were a total of 217 seismic hazard sites, 137 threatening farming sites, 77 threatening road safety sites, and 4239 threatening people [26]. These disasters after the earthquake repeatedly hit its pillar industry, tourism, as the National Forest Park was closed to the public.

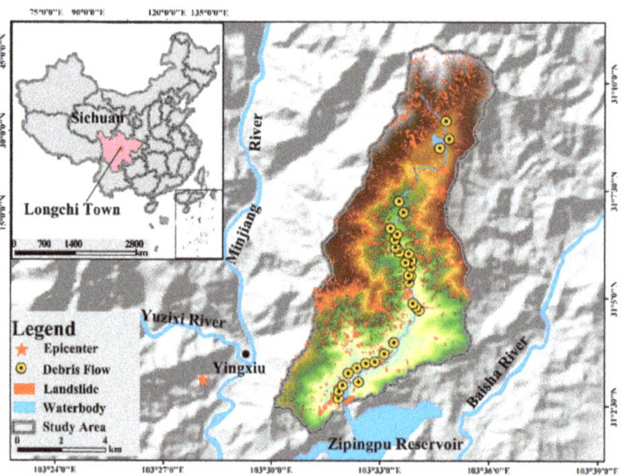

Figure 2. Location of Longchi town.

Among the secondary disaster events after the earthquake, the debris flow and flash flood on 13 August 2010 (the "813" disaster event) was one of the most destructive and typical disasters. More than 50 gullies in the area were affected by flash floods and debris flows, which destroyed many houses and roads and caused 495 casualties and huge economic losses. Thus, the current study takes the "813" disaster event as a case to simulate the population at risk and test the hypothesis.

2.1.2. Survey Design

This study used questionnaires and semi-structured interviews to collect data. Questions were asked about the four variables of risk perception (knowledge of disasters, impacts of disasters, participation in DRR, disaster experience) and willingness to evacuate, considering the study area's actual situation and the study participants. Additionally, respondents who had experienced the "813" disaster event were asked about their actual emergency behaviors at that time. In addition, the questionnaire included personal information about the residents, such as gender, age, and occupation.

In this study, we first conducted a focus group interview with 13 residents to gain an in-depth understanding of their risk perception variables, willingness to evacuate, and behaviors on the "813" disasters to adjust the survey questions and layout. The follow-up survey was conducted in May 2020 with a total of 166 respondents in the Longchi township. Each question survey lasted for half an hour. The targeted samples were taken separately from the whole area of Longchi township, which covered nearly 50% of resident households, and were widely distributed in locations, professions, gender, and age.

The basic information of the sample is shown in Table 1, in which the proportion of male and female residents in the sample was equal, with slightly more males than females. The age of the sample was predominantly middle-aged and elderly (59.26%), between 41 and 60 years old. They were mainly engaged in self-employment (37%) and agriculture and forestry (38.9%). The sample structure is consistent with the basic situation of Longchi town and is representative.

Table 1. Sample distribution.

Items	Categories	Proportion (%)
Gender	Male	60
	Female	41
Age	18–25	2
	26–30	7
	31–40	6
	41–50	26
	51–60	33
	Over 60	26
Occupation	Business	37
	Farming	39
	Others	24

2.2. Measuring Variables

2.2.1. Risk Perception

As discussed in Section 1.2, Research Question and Hypothesis, for reliability and validity of the questionnaire, exploring the existing studies in the same study area [13,26–28] and considering the actual situation in the study area, we classified disaster risk perception into four variables, namely (a) knowledge of disasters, (b) impacts of disasters, (c) participation in DRR, and (d) disaster experience. The sum of the four variables was calculated to measure the public disaster risk perception, and each variable was measured separately (Table 2). Meanwhile, the value of Cronbach's alpha for each variable is 0.6 on average, which indicates a good agreement between entries.

Table 2. Index for measuring public risk perception.

	Variables	No.	Variable Description and Definition
a	knowledge to disasters	A1	I know the main types of disasters in my community.
		A2	I know how to escape from these disasters.
		A3	I have access to disaster information, including early warnings, warning signs, evacuation routes.
b	Impacts of disasters	B1	Disasters can harm me.
		B2	Disasters can have a serious impact on my properties.
		B3	Disasters can have a serious impact on Longchi town.
c	Participation in DRR	C1	I would like to participate in local DRR activities.
		C2	I have participated in many DRR activities.
		C3	If I receive an early warning, I am willing to cooperate with the community for DRR.
d	Disaster experience	D1	I have experienced disasters many times.

Note: The answers to the questionnaire were measured using a 5-point Likert scale: 1 = strongly disagree, 2 = disagree, 3 = neutral, 4 = agree, 5 = strongly agree.

2.2.2. Willingness to Evacuate

For measuring the public willingness to evacuate in the face of a disaster, a short questionnaire was designed to ask participants to rate how strongly they agreed with the statement: Would you be willing to evacuate when a disaster threatens your place? The answers to the questionnaire were measured using a 5-point Likert scale: 1 = strongly disagree, 2 = disagree, 3 = neutral, 4 = agree, 5 = strongly agree.

2.3. Simulating Population at Risk

2.3.1. Agent-Based Modeling Design

Agent-based modeling (ABM) is a microscopic to macroscopic modeling approach based on the systems theory. With a multidisciplinary intersection background, it is widely used in mathematics, physics, biology, sociology, and other fields [39]. It has also been introduced to disaster risk assessment recently [10,40,41], especially for behavioral decisions on disaster response that fully consider the diversity of the population.

The current study takes advantage of the ABM modeling approach and relies on the Netlogo platform language programming to develop a model for simulating at-risk populations. The model inputs the real geospatial environment and generates many different types of agents (resident agents, disaster agents, building agents, etc.) (Figure 3).

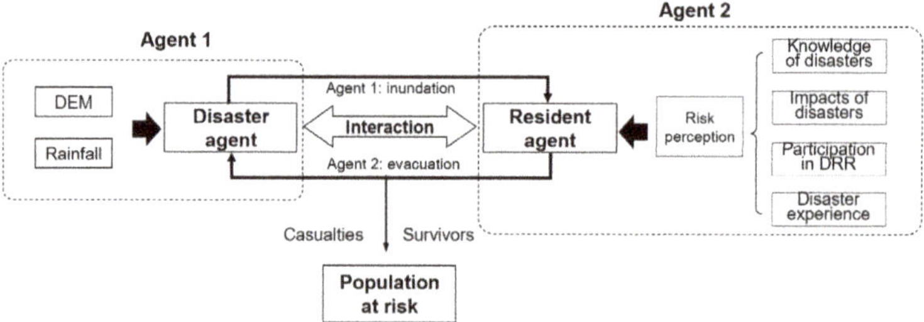

Figure 3. Interactions of agents in the ABM model.

Resident agents: This type of agent is set to simulate different risk-reducing behaviors of each resident. Each agent can display autonomously in the system and has its own physical attributes (age, gender, vision, stamina, safety) and mental attributes (four variables of risk perception). Data are obtained from the Sixth National Population Census of the People's Republic of China [42] and the survey investigation.

Disaster agents: This agent refers to the flash floods and debris flows in the "813" disaster event and can simulate the inundation areas dynamically. Their attributes include locations and debris flow volumes.

All agents are distributed in a 2D space according to established rules. They can act autonomously according to the set rules, perform activities, and interact with other agents and the surrounding environment (patches). The behaviors of all subjects occur in parallel. They are updated asynchronously, through which it is possible to simulate the behaviors of various types of residents and other agents in the event of a disaster. The model can measure the population at risk under different scenarios of willingness to evacuate by reproducing the entire emergency response process from the bottom up.

2.3.2. Experiment Process

The Netlogo platform (Figure 4) was used to reproduce the "813" disaster event in Longchi town according to the natural and socioeconomic conditions and to visually assess the population at risk under different willingness to evacuate scenarios. For each willingness scenario, 50 independent repeated experiments were conducted. Before starting the scenario simulation, the model was validated by using a questionnaire to obtain the willingness to evacuate through a survey of respondents who experienced the "813" disaster event and inputting the data into the simulation platform to verify the accuracy of the model. The repeated simulation results were projected to be 481 casualties on average, which was consistent with the real situation (486 casualties) and indicated that the risk assessment model has good reliability.

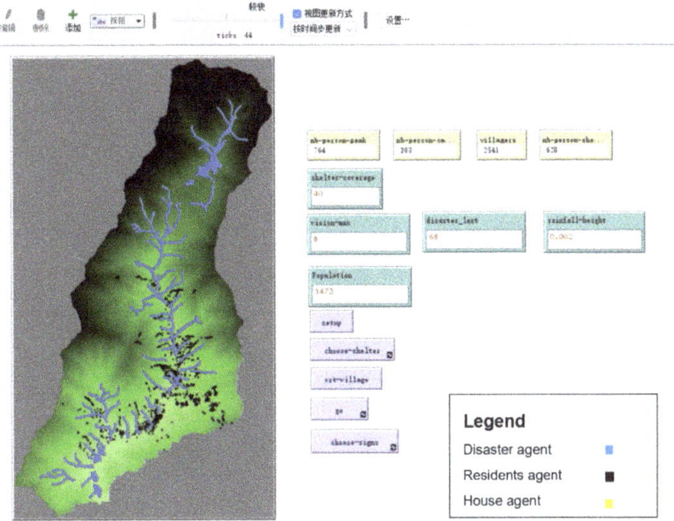

Figure 4. Screenshot of simulation on Netlogo platform.

3. Results

3.1. Descriptive Statistics

3.1.1. Risk Perception

As mentioned in Section 2, Materials and Methods, the four variables of public risk perception, namely, (a) knowledge of disasters, (b) impacts of disasters, (c) participation in DRR, and (d) disaster experience, were added to calculate the overall risk perception. The results are shown in Figure 5. The risk perception values were in the range of 0–20. The value of Cronbach's alpha for each variable is 0.6 on average, which indicates a good agreement between entries. The overall level of public risk perception is moderately high, and 50% of the respondents perceived the risk as between 13 and 17. Specifically, the maximum value is 20, the minimum value is 7, the mean value is 15, and the variance is 8.01. The results of each variable are displayed in Figure 6

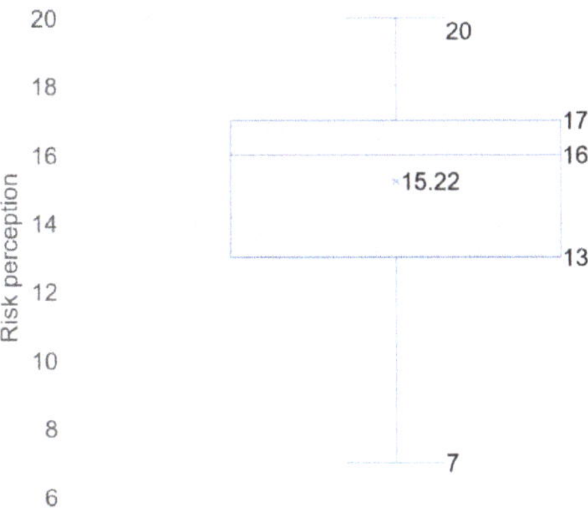

Figure 5. Measurement of risk perception.

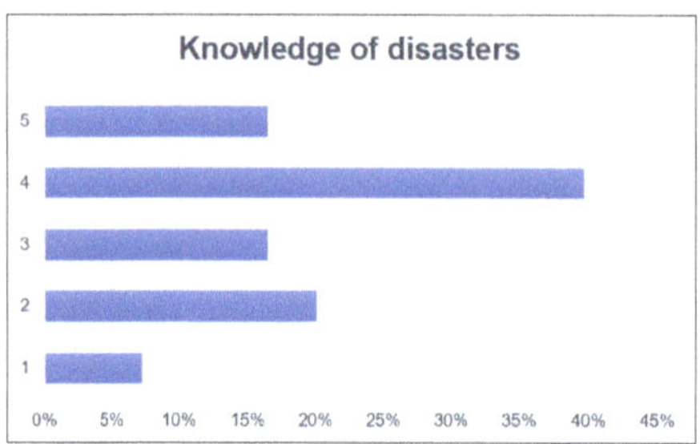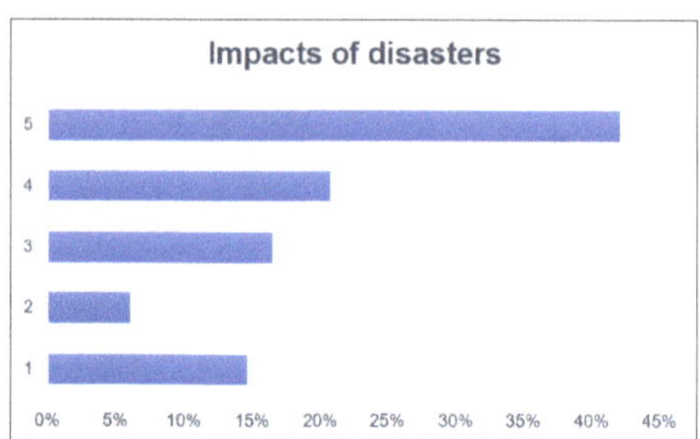

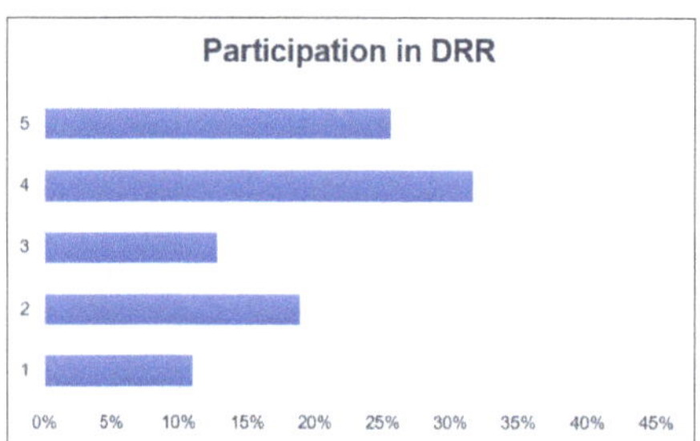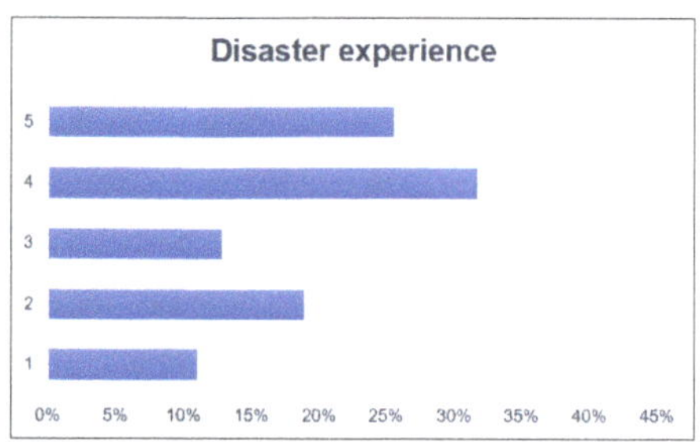

Figure 6. Statistic distribution of each variable of risk perception (1 = strongly disagree, 2 = disagree, 3 = neutral, 4 = agree, 5 = strongly agree).

Knowledge of disasters: The survey results show that overall public knowledge of disasters is high. Among them, 56% of the respondents are relatively or very well informed. However, 27% of the respondents still have insufficient knowledge of disasters, especially about how to escape from disasters.

Impacts of disasters: Respondents believe disasters will have a greater impact on their lives and property, as 63% of the respondents thought the level of impact was very high or relatively high. In contrast, only 21% thought the impact level was very low or relatively low, and 16% remained neutral. Meanwhile, 76% of respondents believe disasters impact the township more than individuals.

DRR participation: The survey results show that residents' participation in DRR is high, with 58% of respondents having relatively high or very high participation in DRR and 30% having low or very low participation in DRR.

Disaster experience: The survey results show that most residents have experienced multiple large-scale disasters. Among them, 98% of the respondents have experienced the Wenchuan earthquake and secondary disasters, and only three have ever experienced disasters. This result is consistent with how the Wenchuan earthquake affected the local area.

3.1.2. Willingness to Evacuate

From the frequency distribution of public willingness to evacuate (Figure 7), it can be seen that Longchi town residents are strongly willing to evacuate when disaster strikes. Specifically, 66% of the 164 respondents said they were very willing to evacuate in the event of a disaster, and 23% were willing to evacuate. In contrast, only 6% of residents said they were reluctant to evacuate, and 5% of residents were neutral about whether to evacuate immediately in the event of a disaster.

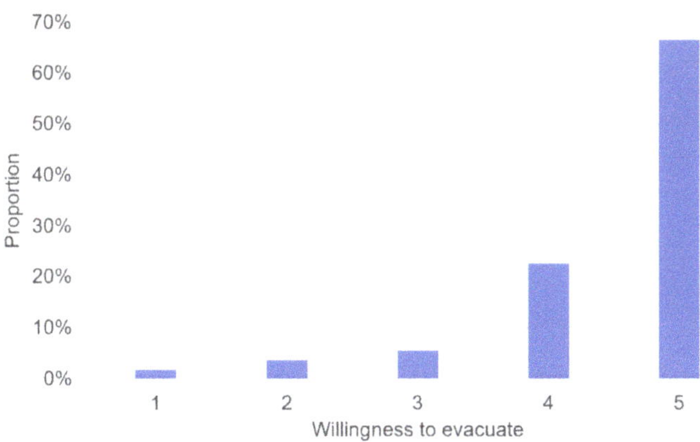

Figure 7. Statistic distribution of willingness to evacuate (Question: Would you be willing to evacuate when a disaster threatens your place? Answer: 1 = strongly disagree, 2 = disagree, 3 = neutral, 4 = agree, 5 = strongly agree).

3.2. Defining the Correlations

3.2.1. Between Risk Perception and Willingness to Evacuate

In this study, regression analysis was used to reveal the correlation between risk perception (including four variables respectively) and willingness to evacuate by using SPSS26, by which to verify the research hypothesis *H1–H4* mentioned in Section 1.2. The ANOVA results showed that the overall significance test statistic was at the 1% level, which indicated that a follow-up analysis could be conducted.

The regression analysis results of risk perception and willingness to evacuate are shown in Table 3. Meanwhile, their correlation is displayed in Figure 8. Resident risk perception was positively and significantly correlated with their willingness to evacuate (f = 52.542, $p < 0.001$). Therefore, the higher the disaster risk perception, the stronger their willingness to evacuate. More specifically, for each 1-unit increase in risk perception, the willingness to evacuate increases by an average of 0.157 units. Among the four variables of risk perception, three variables were all positively and significantly correlated with willingness to evacuate, namely (a) knowledge of disasters, (b) impacts from disasters, and (c) participation in DRR. Specifically, for each unit of increase in knowledge of disasters and participation in DRR, the public willingness to evacuate increased by 0.193 and 0.275 units, respectively. For each unit increase in impacts of disasters, there is a corresponding increase of 0.086 units in their willingness to evacuate. Therefore, the higher the value of these three variables the residents may have, the more willing they are to evacuate. Meanwhile, there was one surprising variable of disaster risk perception, namely (d) disaster experience, which was not significantly related to willingness to evacuate.

Table 3. Regression analysis between risk perception and willingness to evacuate.

Variables	Willingness to Evacuation				
	β	Standard Error	F	R^2	Adjusted R^2
risk perception	0.157 ***	0.022	52.542	0.245	0.240
knowledge to disasters	0.193 ***	0.051			
Impacts of disasters	0.086 **	0.042	22.478	0.361	0.345
Participation in DRR	0.275 ***	0.046			
Disaster experience	−0.142	0.080			

Note: *** means significant at less than 1% probability ** means significant at less than 5% probability.

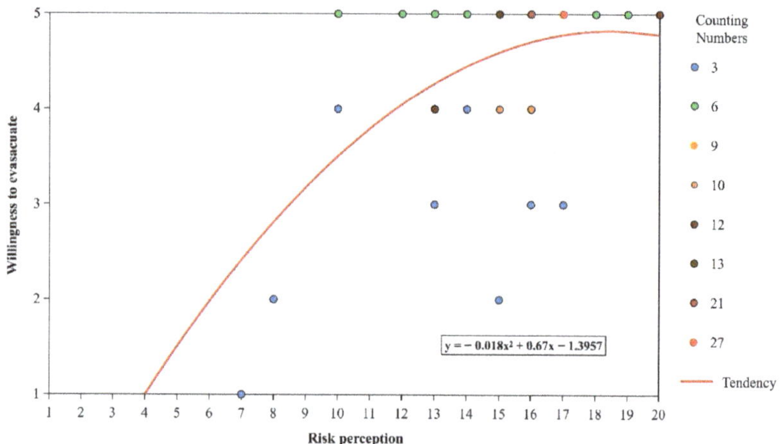

Figure 8. Correlation between risk perception and willingness to evacuate.

3.2.2. Between Willingness to Evacuate and the Population at Risk

The ABM model projected the correlation between willingness to evacuate and the population at risk (Section 2.3). Figure 9 presents the population at risk under different scenarios of willingness to evacuate, which can be indicated that with successive increases in the willingness to evacuate, the population at risk gradually decreased. Specifically, the average number of casualties decreased from 384.48 (willingness = 1) to 354.11 (willingness = 3) and 322.41 (willingness = 5). The number of casualties decreased by 8% versus 16%.

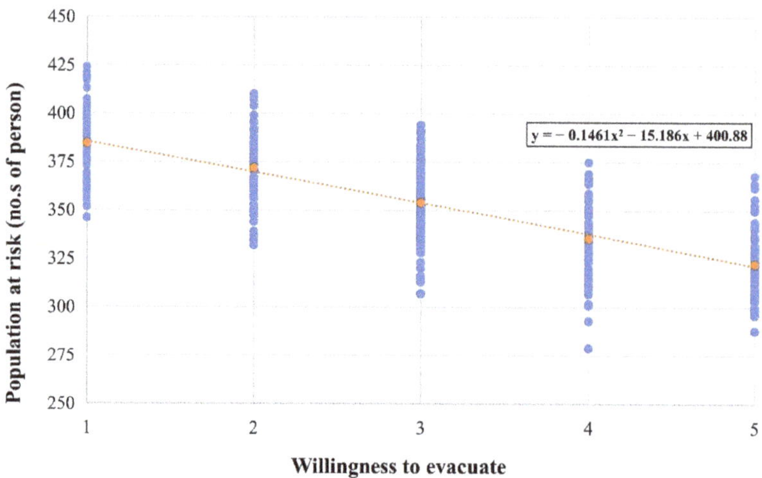

Figure 9. Correlation between willingness to evacuate and the population at risk.

3.2.3. Between Risk Perception and the Population at Risk

Willingness to evacuate playing as a mediator, the correlation between risk perception and the population at risk is investigated from two correlations obtained previously in Sections 3.2.1 and 3.2.2. Figure 10 provides the scatter diagram of the relationship between risk perception and the population at risk. It shows that there has been a gradual decline in populations at risk at the rate of 21.2%. Specifically, the number of populations at risk has been reduced from 412 to 325, with a gradual increase in risk perception.

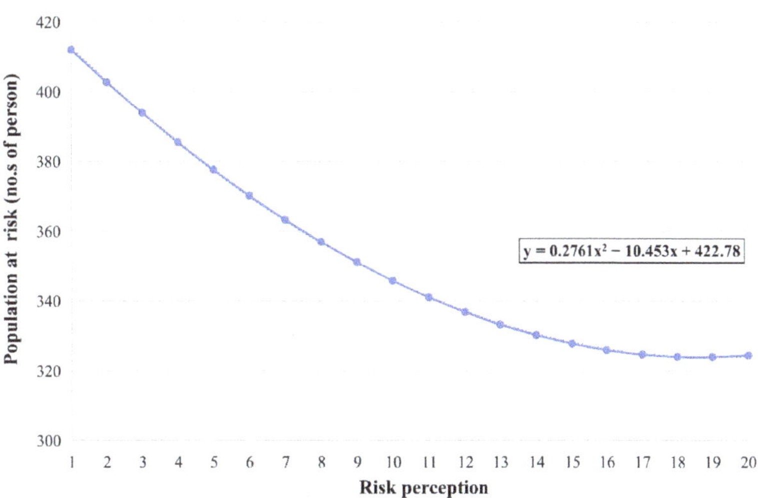

Figure 10. Correlation between risk perception and the population at risk.

In addition, although there is an overall trend of decline in the number of populations at risk, the rate of decline is different in three stages. When the risk perception increased from 1 to 6, the decline rate was over 2%. Subsequently, the rate drops to around 1% and less than 1% as the risk perception reaches 13.

4. Discussion

The results of the study show that most residents were willing to evacuate before disasters (Figure 7). One possible reason for the high level of willingness is owed to the government's efforts over the years after the Wenchuan earthquake in 2008. The local government has recently attached great importance to DRR [26]. After the Wenchuan earthquake, Longchi town gradually established a good mechanism for dealing with local geohazards. Longchi town organizes various DRR activities every year, for example, holding a town-wide flood prevention and geohazard mobilization meeting at the beginning of the year and conducting flash flood and geohazard warning training and emergency drills. These activities allow all stakeholders to actively participate in the local disaster prevention and mitigation efforts, including officials, local DRR practitioners, hospitals, schools, and individuals. In other words, this may be evidence that the efforts of community governments are useful in raising the public willingness to willingness and may provide practitioners with more confidence and motivation to accomplish community DRR.

Subsequently, this study was set to define three correlations. The correlation between risk perception and willingness to evacuate was set to demonstrate the *H1–H4* (Section 1.2). Among them, the findings are consistent with the research hypotheses *H1* and *H3*, where (a) knowledge of disasters and (c) participation in DRR strongly correlate with their willingness to evacuate (Figure 8). The possible reason for this is that after the Wenchuan earthquake, residents were regularly trained in disaster prevention and mitigation knowledge and skills and were motivated to participate in local disaster prevention and mitigation efforts through incentives. These DRR measurements brought people knowledge and awareness about disasters, consequently making them more willing to evacuate when faced with disasters. Several studies also indicated that knowledge [29,30,36] and participation in community DRR [20,36] have positive impacts on public risk perception.

Meanwhile, though it is found that (b) impacts of disasters are consistent with the research hypotheses H2, their correlation seems slightly less (Figure 8). This result is slightly different from the previous studies as most believed the impacts of disasters, such as distance to disasters, had strong significance to risk perception and associated actions [20,29,43].

One possible reason is related to the fact that most respondents may believe that the degree of impact of disasters on the Longchi township is higher than that of themselves. Thus, although residents are aware of the dangers of disasters, the correlation of this dimension with a willingness to evacuate is relatively lower than that of other dimensions because they do not perceive the impacts of disasters to be higher for themselves.

However, only one variable of disaster risk perception, namely (d) disaster experience, was not significantly related to willingness to evacuate. This result is inconsistent with research hypothesis *H4*. Although much previous literature [20,27,34,37] reveals that public past disaster experience is positively associated with risk perception and related actions, our study obtains an opposite finding in this variable. The possible reason is that this study was conducted with the residents of Longchi town as the study population. Longchi is adjacent to the epicenter of the Wenchuan earthquake, and most respondents have experienced the Wenchuan earthquake and the series of secondary disasters after the earthquake. For almost all interviewees, similar disaster experiences left a consistent and deep psychological feeling. Additionally, the study was designed to investigate the correlation between willingness to evacuate and the at-risk population. The result in Figure 9 was deployed by the ABM model and showed that the increasing willingness to evacuate could significantly reduce the population at risk, supporting hypothesis 5.

For the research question, this study aimed to understand the correlation between risk perception and the at-risk population. The result in Figure 10 shows an overall trend of decline in the number of populations at risk when the public risk perception increases, which provides scientific evidence for policymakers that the public disaster risk perception should be raised. Improving disaster risk perception can enable residents to be more willing to evacuate, thus effectively reducing casualties during the disaster. Meanwhile, raising public disaster risk perception is relatively low-cost compared with engineering measurement [36]. For example, low-cost engineering measures, such as disaster education and community DRR activities, can effectively improve public risk perception. Therefore, this approach to improving risk perception can be replicated in those communities in developing countries threatened by disasters. In terms of specific DRR measures, the government can carry out a variety of education and training, including the knowledge on the prevention and treatment of main disaster types and capacities to avoid and save themselves. In addition, a multi-stakeholder DRR participation system can be established to increase the risk perception of all relevant stakeholders, such as schools, hospitals, and NGOs [44]. Residents can improve their risk perception and reduce population risk by participating in community DRR activities.

Meanwhile, one interesting finding is that there is a marginal effect of raising public risk perception on reducing population risk. This phenomenon may be because when most of the residents have a high level of risk perception, other factors can also influence the risk of the population, such as the time of early warning, evacuation patterns, and different vulnerabilities. Since there is a marginal effect for raising risk perception in DRR when most residents have a high level of risk perception, the government can dedicate a portion of the investment to other aspects that are likely to affect public emergency behavior. For example, there are differences in vulnerability between populations. Disadvantaged populations, such as the disabled, pregnant women, and the elderly, are more vulnerable during disasters [12,45]. They usually need to take a longer time for evacuation or even need help from the government to evacuate to safe places in a timely manner. In this case, the government may use partial investment to tailor DRR measures for disadvantaged populations after most people have a higher risk perception.

5. Conclusions

This study has tried to combine interdisciplinary methods from both social sciences and natural sciences, including survey questions, statistical analysis, and ABM models, to establish a correlation between disaster risk perception and the population at risk. This attempt may provide some methodologies and ideas for subsequent interdisciplinary studies. In this paper, the following conclusions were obtained by measuring risk perception, willingness to evacuate, and the population at risk and exploring the correlation between the above.

Residents generally have a higher risk perception and willingness to evacuate. The results showed that 50% of the respondents had risk perception levels between 13 and 17 (maximum 20), and 66% of the residents indicated they were very willing to evacuate in the event of a disaster. The increased risk perception and willingness are inseparable from local investments in disaster education and other efforts in recent years. This result, to some extent, reflects the effectiveness of local government investment in DRR in recent years.

A positive correlation was found between risk perception and willingness to evacuate. For every 1 unit increase in risk perception, willingness to evacuate increased by an average of 0.157 units. Regarding each risk perception variable, three variables, namely (a) disaster risk perception, (b) knowledge of disasters, and (c) participation in DRR, have a significant positive correlation with the willingness to evacuate, supporting hypotheses *H1*, *H2*, and *H3*. However, the results between (d) disaster experience and the willingness to evacuate were not statistically significant, which does not correspond to hypothesis *H4*.

The ABM simulation results indicated that as the willingness to evacuate rises, the population at risk decreases rapidly (up to 16%). The correlation between willingness to evacuate and the population at risk can support hypothesis H5.

The correlation between risk perception and the population at risk also provided a positive significance. The observations have indicated a serious decline in the population at risk from 412 to 325 when their risk perception rises from 1 to 20. However, the rate of decline becomes slower after their risk perception reaches 13, which indicates a marginal effect of raising public risk perception on reducing the risk of population

These findings have significant implications for understanding public risk perception and how to reduce disaster risks by changing public risk perception. Firstly, the public risk perception should be increased since research findings show that increased public risk perception is associated with a higher willingness to evacuate, which can effectively reduce the population at risk. Moreover, raising risk perceptions is often low-cost and affordable for most communities. Therefore, it is recommended that most communities can conduct education to increase public risk perception and thus reduce the number of casualties at the community level to some extent. Secondly, although it is advocated from all levels to strengthen community disaster mitigation, it is nevertheless not requiring communities to invest heavily all at once. A large investment without a goal may not necessarily achieve very good results. Therefore, at the community level, what needs to be carried out is policy optimization, using the results of scientific research to provide more effective disaster reduction measures based on a combination of local natural, social, and economic conditions.

Author Contributions: Conceptualization, S.W. and Y.L.; Methodology, S.W. and Y.L.; Formal analysis, S.W. and Y.L.; Data curation, S.W.; Writing—original draft, S.W., Y.L. and W.J.; Writing—review & editing, S.W., Y.L. and W.J.; Visualization, S.W., Y.L. and W.J. All authors have read and agreed to the published version of the manuscript.

Funding: This research was funded by National Natural Science Foundation of China: 41941017; Chongqing Social Science Foundation: 2021SZ32; Sichuan Science and Technology Program: 2021YFH0009; Chinese Academy of Sciences Presidents International Fellowship Initiative: 2020FYC0004.

Institutional Review Board Statement: Not applicable.

Informed Consent Statement: Not applicable.

Data Availability Statement: The data presented in this study are available on request from the corresponding author. The data are not publicly available due to funding project policy.

Conflicts of Interest: The authors declare no conflict of interest.

References

1. Wang, W.; Yang, S.; Stanley, H.E.; Gao, J. Local floods induce large-scale abrupt failures of road networks. *Nat. Commun.* **2019**, *10*, 2114. [CrossRef] [PubMed]
2. Lei, Y.; Gu, H.; Cui, P. Vulnerability assessment for buildings exposed to torrential hazards at Sichuan-Tibet transportation corridor. *Eng. Geol.* **2022**, *308*, 106803. [CrossRef]
3. GAR. Global Assessment Report on Disaster Risk Reduction 2022. Back Page. Available online: https://www.undrr.org/publication/global-assessment-report-disaster-risk-reduction-2022 (accessed on 1 September 2022).
4. CRED. The Centre for Research on the Epidemiology of Disasters. 2021 Disasters in Numbers—World. ReliefWeb. 2022. Available online: https://reliefweb.int/report/world/2021-disasters-numbers (accessed on 3 September 2022).
5. Fell, R.; Ho, K.K.S.; Lacasse, S.; Leroi, E. A framework for landslide risk assessment and management. In *Landslide Risk Management*; CRC Press: Boca Raton, FL, USA, 2005; pp. 13–36.
6. Büchele, B.; Kreibich, H.; Kron, A.; Thieken, A.; Ihringer, J.; Oberle, P.; Merz, B.; Nestmann, F. Flood-risk mapping: Contributions towards an enhanced assessment of extreme events and associated risks. *Nat. Hazards Earth Syst. Sci.* **2006**, *6*, 485–503. [CrossRef]
7. Noferini, L.; Pieraccini, M.; Mecatti, D.; Macaluso, G.; Atzeni, C.; Mantovani, M.; Marcato, G.; Pasuto, A.; Silvano, S.; Tagliavini, F. Using GB-SAR technique to monitor slow moving landslide. *Eng. Geol.* **2007**, *95*, 88–98. [CrossRef]
8. Qiu, H.; Zhu, Y.; Zhou, W.; Sun, H.; He, J.; Liu, Z. Influence of DEM resolution on landslide simulation performance based on the Scoops3D model. *Geomat. Nat. Hazards Risk* **2022**, *13*, 1663–1681. [CrossRef]
9. Liu, Z.; Qiu, H.; Zhu, Y.; Liu, Y.; Yang, D.; Ma, S.; Zhang, J.; Wang, Y.; Wang, L.; Tang, B. Efficient Identification and Monitoring of Landslides by Time-Series InSAR Combining Single- and Multi-Look Phases. *Remote Sens.* **2022**, *14*, 1026. [CrossRef]
10. Aerts, J.C.J.H.; Botzen, W.J.; Clarke, K.C.; Cutter, S.L.; Hall, J.W.; Merz, B.; Michel-Kerjan, E.; Mysiak, J.; Surminski, S.; Kunreuther, H. Integrating human behaviour dynamics into flood disaster risk assessment. *Nat. Clim. Chang.* **2018**, *8*, 193–199. [CrossRef]
11. Ho, M.C.; Shaw, D.; Lin, S.; Chiu, Y.C. How do disaster characteristics influence risk perception? *Risk Anal. Int. J.* **2008**, *28*, 635–643. [CrossRef]
12. Nakanishi, H.; Black, J.; Suenaga, Y. Investigating the flood evacuation behaviour of older people: A case study of a rural town in Japan. *Res. Transp. Bus. Manag.* **2019**, *30*, 100376. [CrossRef]
13. Wu, S.; Lei, Y.; Yang, S.; Cui, P.; Jin, W. An Agent-Based Approach to Integrate Human Dynamics into Disaster Risk Management. *Front. Earth Sci.* **2022**, *9*, 53. [CrossRef]
14. Lepesteur, M.; Wegner, A.; Moore, S.A.; McComb, A. Importance of public information and perception for managing recreational activities in the Peel-Harvey estuary, Western Australia. *J. Environ. Manag.* **2008**, *87*, 389–395. [CrossRef] [PubMed]
15. Slovic, P. Perception of risk. *Science* **1987**, *236*, 280–285. [CrossRef] [PubMed]
16. Gotham, K.F.; Campanella, R.; Lauve-Moon, K.; Powers, B. Hazard experience, geophysical vulnerability, and flood risk perceptions in a postdisaster city, the case of New Orleans. *Risk Anal.* **2018**, *38*, 345–356. [CrossRef] [PubMed]
17. Baker, E.J. Hurricane evacuation behavior. *Int. J. Mass Emerg. Disasters* **1991**, *9*, 287–310.
18. Whitehead, J.C.; Edwards, B.; Van Willigen, M.; Maiolo, J.R.; Wilson, K.; Smith, K.T. Heading for higher ground: Factors affecting real and hypothetical hurricane evacuation behavior. *Glob. Environ. Chang. Part B Environ. Hazards* **2000**, *2*, 133–142. [CrossRef]
19. Huang, S.-K.; Lindell, M.K.; Prater, C.S.; Wu, H.-C.; Siebeneck, L.K. Household evacuation decision making in response to hurricane Ike. *Nat. Hazards Rev.* **2012**, *13*, 283–296. [CrossRef]
20. Qasim, S.; Khan, A.N.; Shrestha, R.P.; Qasim, M. Risk perception of the people in the flood prone Khyber Pukhthunkhwa province of Pakistan. *Int. J. Disaster Risk Reduct.* **2015**, *14*, 373–378. [CrossRef]
21. Shaw, D.G.; Chen, S.H.; Lin, S.Y.; Tsai, M.F.; Huang, H.H.; Huang, T.S. *The Executive Report of the Survey of Social-Economic Impacts and Risk Perception of Floods and Landslides in 2004*; National Science and Technology Center for Disaster Reduction: New Taipei City, Taiwan, 2006.
22. Gustafsod, P.E. Gender Differences in risk perception: Theoretical and methodological perspectives. *Risk Anal.* **1998**, *18*, 805–811. [CrossRef]
23. Brewer, N.T.; Weinstein, N.D.; Cuite, C.L.; Herrington, J.E. Risk perceptions and their relation to risk behavior. *Ann. Behav. Med.* **2004**, *27*, 125–130. [CrossRef]
24. Conner, M.; Norman, P. *EBOOK: Predicting and Changing Health Behaviour: Research and Practice with Social Cognition Models*; McGraw-Hill Education: London, UK, 2015.
25. Tang, J.; Yang, S.; Liu, Y.; Yao, K.; Wang, G. Typhoon Risk Perception: A Case Study of Typhoon Lekima in China. *Int. J. Disaster Risk Sci.* **2022**, *13*, 261–274. [CrossRef]
26. Chen, X.; Zhang, L.; Ke, X. *Fangfanhuajie Zhongdaanquanfengxian yu Yingjiwenhuajianshe*; Southwestern University of Finance and Economics Press: Chengdu, China, 2020; ISBN 9787550444218.
27. Chen, Y.; Li, X.; He, L.; Xu, Y.; Xue, N. Factors Influencing People's Perception of Dual Risks in Population Resettlement. *J. Southwest Pet. Univ. (Soc. Sci. Ed.)* **2021**, *23*, 24–33. [CrossRef]
28. Wu, J.; Deng, X.; Xu, D. Disaster risk perception, evacuation and relocation willingness of farmers in earthquake-stricken areas. *Mt. Res.* **2021**, *39*, 552–562.
29. Botzen, W.J.; Aerts, J.C.J.H.; van den Bergh, J.C. Dependence of flood risk perceptions on socioeconomic and objective risk factors. *Water Resour. Res.* **2009**, *45*, 10. [CrossRef]

30. Gregg, C.; Houghton, B.; Johnston, D.; Paton, D.; Swanson, D. The perception of volcanic risk in Kona communities from Mauna Loa and Hualālai volcanoes, Hawai'i. *J. Volcanol. Geotherm. Res.* **2004**, *130*, 179–196. [CrossRef]
31. Houts, P.S.; Lindell, M.K.; Hu, T.W.; Cleary, P.D.; Tokuhata, G.; Flynn, C.B. Protective action decision model applied to evacuation during the three mile island crisis. *Int. J. Mass Emerg. Disasters* **1984**, *2*, 27–39.
32. Riad, J.K.; Norris, F.H.; Ruback, R.B. Predicting evacuation in two major disasters: Risk perception, social influence, and access to resources. *J. Appl. Soc. Psychol.* **1999**, *29*, 918–934. [CrossRef]
33. Xu, D.; Peng, L.; Liu, S.; Su, C.; Wang, X.; Chen, T. Influences of Sense of Place on Farming Households' Relocation Willingness in Areas Threatened by Geological Disasters: Evidence from China. *Int. J. Disaster Risk Sci.* **2017**, *8*, 16–32. [CrossRef]
34. Siegrist, M.; Gutscher, H. Flooding risks: A comparison of lay people's perceptions and expert's assessments in Switzerland. *Risk Anal.* **2006**, *26*, 971–979. [CrossRef]
35. Lazo, J.K.; Bostrom, A.; Morss, R.E.; Demuth, J.L.; Lazrus, H. Factors affecting hurricane evacuation intentions. *Risk Anal.* **2015**, *35*, 1837–1857. [CrossRef]
36. Wu, S.-N.; Lei, Y.; Cui, P.; Chen, R.; Yin, P.-H. Chinese public participation monitoring and warning system for geological hazards. *J. Mt. Sci.* **2020**, *17*, 1553–1564. [CrossRef]
37. Zhu, X.; Jiang, L.; Dong, C.; Jin, W.; Wang, Y. Study on Risk Perception and Early warning Mechanism of Severe Disaster Events Such as Typhoon etc. *J. Catastrophology* **2012**, *27*, 62–66. [CrossRef]
38. Zhang, G.; Cui, P.; Yin, Y.; Liu, D.; Jin, W.; Wang, H.; Yan, Y.; Ahmed, B.N.; Wang, J. Realtime monitoring and estimation of the discharge of flash floods in a steep mountain catchment. *Hydrol. Process.* **2019**, *33*, 3195–3212. [CrossRef]
39. Green, D.G.; Sadedin, S. Interactions matter—Complexity in landscapes and ecosystems. *Ecol. Complex.* **2005**, *2*, 117–130. [CrossRef]
40. Huang, H.; Fang, Y.; Yang, S.; Li, W.; Guo, X.; Lai, W.; Wang, H. A multi-agent based theoretical model for dynamic flood disaster risk assessment. *Geogr. Res.* **2015**, *34*, 1875–1886. [CrossRef]
41. Wang, W.; Yang, S.; Gao, J.; Hu, F.; Zhao, W.; Stanley, H.E. An Integrated Approach for Assessing the Impact of Large-Scale Future Floods on a Highway Transport System. *Risk Anal.* **2020**, *40*, 1780–1794. [CrossRef]
42. National Bureau of Statistics of the People's Republic of China. Sixth National Population Census of the People's Republic of China. 2010. Available online: http://www.stats.gov.cn/ztjc/zdtjgz/zgrkpc/dlcrkpc/ (accessed on 12 January 2020).
43. Miceli, R.; Sotgiu, I.; Settanni, M. Disaster preparedness and perception of flood risk: A study in an alpine valley in Italy. *J. Environ. Psychol.* **2008**, *28*, 164–173. [CrossRef]
44. Zeng, Z.; Wang, Z.; Ding, S. Behaviours and Risk prevention of local government in the construction of characteristic towns. *J. Hangzhou Norm. Univ. (Humanit. Soc. Sci.)* **2021**, *43*, 125–136. [CrossRef]
45. Cutter, S.L. Vulnerability to environmental hazards. *Prog. Hum. Geogr.* **1996**, *20*, 529–539. [CrossRef]

International Journal of
*Environmental Research
and Public Health*

MDPI

Article

People's Response to Potential Natural Hazard-Triggered Technological Threats after a Sudden-Onset Earthquake in Indonesia

Fatma Lestari [1,2], Yasuhito Jibiki [3,*], Daisuke Sasaki [4], Dicky Pelupessy [2,5], Agustino Zulys [2,6] and Fumihiko Imamura [4]

[1] Occupational Health & Safety Department, Faculty of Public Health, Universitas Indonesia, Depok, Java Barat 16424, Indonesia; fatma@ui.ac.id or fatmalestari@icloud.com
[2] Disaster Risk Reduction Center, Universitas Indonesia, Depok, West Java 16424, Indonesia
[3] Next Generation Volcano Researcher Development Program, Graduate School of Science, Tohoku University, Sendai, Miyagi 980-8577, Japan
[4] International Research Institute of Disaster Science, Tohoku University, Sendai, Miyagi 980-8577, Japan; dsasaki@irides.tohoku.ac.jp (D.S.); imamura@irides.tohoku.ac.jp (F.I.)
[5] Faculty of Psychology, Universitas Indonesia, Depok, West Java 16424, Indonesia; dickypsy@ui.ac.id
[6] Chemistry Department, Faculty of Mathematics and Natural Sciences, Universitas Indonesia, Depok, West Java 16424, Indonesia; zulys@ui.ac.id
* Correspondence: yasuhito.jibiki.e3@tohoku.ac.jp

Citation: Lestari, F.; Jibiki, Y.; Sasaki, D.; Pelupessy, D.; Zulys, A.; Imamura, F. People's Response to Potential Natural Hazard-Triggered Technological Threats after a Sudden-Onset Earthquake in Indonesia. *Int. J. Environ. Res. Public Health* **2021**, *18*, 3369. https://doi.org/10.3390/ijerph18073369

Academic Editor: Paul B. Tchounwou

Received: 19 February 2021
Accepted: 20 March 2021
Published: 24 March 2021

Abstract: (1) Background: We aim to examine whether people activate initial protection behavior, adopt evacuation behavior, worry about the possibility of a tsunami, and consider natural hazard-triggered technological (Natech) situations in a sudden-onset earthquake. The literature suggests that risk perception is a significant predictor of people's response to potential Natech threats. We aim to empirically verify the variables relating to people's responses. (2) Methods: We conducted a household survey following a January 2018 earthquake in Indonesia. (3) Results: Immediately after the earthquake, almost 30% of the respondents assembled at the evacuation point. However, sequential steps of people's response were not observed: evacuation immediately after the earthquake was due to worry about the possibility of a tsunami, but this worry was not related to Natech damage estimation. The relevant factors for evacuation behavior were information access, worry about the possibility of a tsunami, and knowledge of groups and programs related to disaster risk reduction (DRR). The survey location (two villages), perceived earthquake risk, and DRR activity participation are less relevant to the behavior of assembling at the evacuation point. (4) Conclusions: Contrary to the existing literature, our results do not support that higher risk perception is associated with evacuation behavior, or that immediate evacuation is related to foreseeing cascading sequential consequences.

Keywords: natural hazard-triggered technological (Natech); risk perception; protective actions; evacuation; household survey; Cilegon; Indonesia

1. Introduction

1.1. Policy Background on Natech

The description "natural hazard-triggered technological" (Natech) for threats or accidents is an emerging technical term for comparison with familiar natural hazard types. Recently, it has become recognized globally in the context of the Sustainable Development Goals (SDGs) and Sendai Framework for Disaster Risk Reduction (SFDRR). The related SDG target addresses specific aspects of Natech in "3.D National and global health risks," "11.3 Urbanization," "12 Consumption and production," and "13 Climate-related hazards." Paragraph 15 of the SFDRR notes that its scope includes "related environmental, technological and biological hazards and risks."

According to an Organisation for Economic Co-operation and Development survey [1] (p. 35, para. 59), there is low visibility of Natech events in risk communication systems. For example, 6 respondents out of 17 samples (14 countries and 3 institutions representing science and industry) stated that information had been "provided to the public in case of emergencies due to chemical accidents" [1] (p. 36). This result indicates that people do not find information provision to be sufficient and this lack of information may not cause preferable behavior for safety in a crisis.

The United Nations Office for Disaster Risk Reduction Asia-Pacific Science Technology and Academia Advisory Group [2] recommends the need for an early warning mechanism, awareness, and training as important activities for Natech risk management. The United Nations Office for Disaster Risk Reduction [3,4] emphasizes almost the same consideration. These policy papers implicitly indicate that warning development, awareness raising, and training opportunities have not been well organized so far. The Izmit Earthquake (also known as the Kocaeli Earthquake) that struck Turkey in 1999, causing a massive fire at the Tupras Izmit refinery and an acrylonitrile spill at the Aksa acrylic fiber production plant, was showcased by Girgin [5] to describe the complexities and great difficulties involved in the evacuation process. More than 20 years have passed since the Izmit Natech incident, and yet still the world has not developed well-prepared methods for people to react to Natech threats.

1.2. Case Description of a Potential Natech Threat

Even though policies have not yet been well standardized, in reality, industrial parks are exposed to potential Natech threats. As a case study, we adopt a city in Indonesia, Cilegon city, to analyze how local people respond to Natech threats.

Cilegon is located on the western edge of the Java islands and is known as one of the most well-known and significant heavy industrial areas in Indonesia [6–11]. The government of Indonesia and the Association of Southeast Asian Nations (ASEAN) have paid close attention to the potential threats of Natech scenarios in Cilegon. After the Indian Ocean Tsunami of 2004, the Indonesian government conducted a national tsunami simulation for Cilegon in 2007 taking full account of potential Natech threats [12]. In addition, the Cilegon city government developed a tsunami early warning system [13]. These efforts finally culminated in a large-scale preparation exercise in November 2018, the ASEAN Regional Disaster Emergency Response Simulation Exercise (ARDEX 2018), which was conducted in Cilegon [14]. For this reason, Cilegon was selected for our study.

While Jibiki et al. [15] revealed that the community surrounding the industrial facilities in Cilegon was aware of Natech risks, Pelupessy et al. [16] clarified that such awareness was not necessarily connected with organized behaviors in the case of the Anak Krakatau eruption and tsunami, which occurred on the night of Saturday 22 December 2018. Even though no huge tsunami reached the coastal areas in Cilegon in the Anak Krakatau case, great confusion and social disorder were observed [16].

Cilegon experienced an event prior to the Anak Krakatau case. An earthquake of magnitude 6.4 was recorded on Tuesday 23 January 2018, 13:34:50 (local time), and people felt the ground shaking in Cilegon. According to the Indonesian Metrological Agency (Badan Meteorologi Klimatologi dan Geofisika, BMKG), the nearest observation station of our survey area detected "MMI IV (Modified Mercall Intensity)" [17]. This intensity means that the perceived shaking is light and there is no potential damage [17]. However, the local media reported that hundreds of employees felt the shock at one of the major petrochemical factories in Cilegon and evacuated out of the building [18]. The earthquake did not cause a tsunami, and no tsunami warning was issued. We focus on this earthquake in the present study.

Considering the geographical and socioeconomic characteristics of Cilegon, a preferable action in relation to Natech seems to be identified. When shaking is felt in Cilegon, the most preferable action is evacuation to higher grounds after the initial protection behavior (drop, cover, and hold). Since it is difficult to determine whether the epicenters of earthquakes are located inland or are megathrust, it seems preferable to consider the likelihood of tsunamis associated with earthquakes. A warning may be issued, but it is better to save time for evacuation without waiting for a warning. In addition, Natech issues need to be considered. Such a sequential relationship of response is desirable for those living near the industrial park in coastal areas in Cilegon.

1.3. Literature Review

While the policy settings have not been well synthesized to achieve the desired evacuation behavior, as noted earlier, we can refer to some literature on the factors generating evacuation behavior specifically for Natech events. Yu et al. [19] used a case study of a fire at a refinery caused by the tsunami triggered by the Great East Japan Earthquake and analyzed factors influencing evacuation behavior. Yu et al. [19] stated that only a few studies [20,21] have examined risk perception of and protective actions against technological threats. According to Yu et al.'s [19] logistic regression analysis to predict households' immediate evacuation, the first significant predictor is "respondents' direction to the industrial park" and the second is perceived severity of the Natech threat once they had perceived that a Natech accident would occur. For other factors, Yu et al. [19] stated that households were more likely to evacuate immediately if they felt that their lives or property would be impacted by the Natech accident to a very great extent when they perceived its occurrence. Furthermore, with reference to some studies [20,22–24], they pointed out that demographic variables have weak and inconsistent correlations with risk perception and protective responses.

Although Lindell et al. [25] did not deal directly with Natech events, they comprehensively examined the immediate behavioral responses to earthquakes. They concluded that risk perceptions matter for immediate responses to earthquakes, but no previous studies appear to have addressed this matter [25]. Earthquake information and emergency preparedness were associated with lower levels of negative emotions and maladaptive behavior, as well as with increased levels of adaptive behavior; this is one of their most important findings because it supports the effectiveness of pre-impact training activities [25]. Furthermore, they argued that fear was positively related to immediate evacuation. Lindell et al. [25] connected this point to past research and theorized that fear does not necessarily produce loss of control or non-rational flight [26–28].

While Yu et al. [19] and Lindell et al. [25] paid attention to psychological aspects, there is also relevant literature in the discipline of safety science. Feng et al. [29] studied post-earthquake evacuation using verbal protocol analysis in immersive virtual reality. The results of their experiments show that participants had wait-or-flight responses in post-earthquake evacuation. They also revealed that people's decision making tended to be driven, at least partially, by what those around them were doing in the greatest numbers [30]. In addition, Feng et al. [29] found that participant behavior was particularly influenced by those who appeared to be in authority positions, which has been observed in real-life evacuation cases [31–33]. Nascimento and Alencar [34] conducted a systematic review of the literature on Natech events, but their study provided few insights on people's responses. A systematic literature review by Suarez-Paba et al. [35] identified that only 6.1% of the total studies analyzed dealt with risk communication and risk perception. Yu et al. [19] and Yu and Hokugo [36] highlighted the fact that inhabitants' risk perception triggers their protective behavior (e.g., time to evacuate their house) during a disaster and that this is influenced by such parameters as location, demographic characteristics, and age.

1.4. Research Question and Hypothesis

As stated in Section 1.2, we primarily examine the sequential relationship of response as the research question, even though it is quite a hypothetical assumption: the evacuation action is required after the initial protection behavior (drop, cover, and hold). Worry about the possibility of a tsunami is also important, and such emotion needs to be linked with damage estimation, which could be induced by Natech events.

As summarized in Table 1, the existing literature provides relevant factors that seem to influence behavior. We employ proxy variables to verify whether we can obtain similar results to those in the existing literature. Our variables are set from a household-basis questionnaire survey, which we explain in detail in the next section. For the factor of "Direction," we use "Village location" because almost all the villages in Cilegon are located to the east of the industrial facilities due to the topographical characteristics of Cilegon. As an alternative variable, we test whether the differences in villages may affect behavior (the two villages are coded as binary data). Regarding demographic data, we do not include individuals' gender and age since we use a household survey in which each respondent provides answers on behalf of the household.

Table 1. Classification of variables influencing behavior.

Relevant Variables from the Existing Literature	Variables in the Present Study
Risk perception	Proneness of risk as perceived risk
Emotion	Worry about the possibility of a tsunami
Direction	Village location
Impact estimation	Natech damage estimation
Information seeking	Information access, including social media and
Reference to authority positions	government agencies
Preparedness	Disaster risk reduction activity participation, knowledge

The rest of the paper is structured as follows. Section 2 introduces the survey design and data. Following the Method section, we verify our hypothetical assumption (sequential steps of people's response) in Section 3.1. In addition, we examine whether risk perception plays a significant role in accordance with earlier works in Section 3.2. In addition to risk perception, other aspects are analyzed to clarify whether they are related to Natech damage estimation (Section 3.3). In contrast to the existing literature, village location, information access, and preparedness are investigated (Section 3.3, Section 3.4, Section 3.5). Section 4 summarizes the results of our analysis, concludes whether we find similar findings to the earlier works, and states the limitations of our study.

2. Method

2.1. Survey Design

A household survey was carried out at two sites in Cilegon city. The first survey was conducted in Lebak Gede village ("village" here is translation of "kelurahan" in the local language and context) during 14–18 February 2018, while the second survey was conducted in Gunung Sugih village from 6 to 9 March 2018. The sample size was determined in proportion to the total number of households in Lebak Gede (2907 households) and those in Gunung Sugih (1945 households). In order to collect the sufficient number that well represents the total households (4852) of the two villages, we initially planned to gather 500 samples. The samples were proportionally distributed in accordance with the number of households at the neighborhood association level (Rukun Tetangga, or RT) in each village. It was possible to specifically identify the number of households in each RT, and we calculated the composition ratio for each RT. For example, one RT in Gunung Sugih village has 107 households. The composition ratio was calculated as follows: the denominator was 4852 and the numerator was 107. Then, the composition ratio was 0.022052762. We multiplied the composition ratio in the RT by 500 samples (as the expected total number of samples), and we could determine the sample number (11.02638087) in the said RT. Technically, the calculated value was rounded off to the first decimal place, and we finally obtained 11 samples in this example. Finally, 497 samples (299 samples from Lebak Gede and 198 samples from Gunung Sugih) were collected. The enumerators (survey interviewers) were nominated and trained from Indonesian Red Cross volunteers and local residents.

The questionnaire used in this survey was constructed through an elicitation process learned from the interviews (including group interviews) with community leaders and members prior to the survey. Face validity of the questionnaire was acquired through enumerator training in which one session in the training was dedicated to obtaining feedback from the enumerators regarding the readability of the questionnaire items. Enumerator training was conducted to ensure that the enumerators followed the survey protocol and maintained its reliability.

The total number of the question items was 54 in our questionnaire. In the elicitation process for developing our questionnaire, we referred to the relevant literature for each factor. In risk perception, as one of the core factors in our study, we referred to Yu et al. [19] and Lindel et al. [25]. For evacuation behavior, we mainly referred to Yu et al. [19], Lindel et al. [25], and Feng et al. [29]. Insights from Jibiki et al. [15] and Yu et al. [19] were helpful for considering question items regarding Natech damage estimation.

2.2. Data Set

In our questionnaire, we included questions asking whether the respondents felt the shaking of the earthquake and whether they stayed in their villages. A total of 380 respondents (76.5% of the total samples; 231 from Lebak Gede, and 149 from Gunung Sugih) answered that they felt the shaking and that they stayed in the village when the earthquake occurred. Based on the chi-square test, we did not find a statistical difference between the two survey locations, and thus there was no need to deal with the data separately. After analyzing the data, we focused on the selected respondents (380 respondents) and those who experienced the earthquake. All respondents did not necessarily answer all questions. In analysis, the total number of the respondents does not reach 380 in some results due to the deficit values.

For reference, we introduced income information relating to our research target locations. Our targeted households in Cilegon city are located in Banten Province. Table 2 shows "Average of net wage/salary per month of formal employee by province and main occupation." The income level of Banten Province in 2018 was approximately the same as the average level and lower than that of Jakarta Special Province.

Table 2. Average of net wage/salary per month of formal employee by province and main occupation (Indonesia rupiahs).

	Year 2018	Year 2020
Banten Province	3,468,768	3,693,411
Jakarta Special Province	4,523,453	4,224,720
Average by Province	3,592,501	2,756,345

Source: BPS-Statistics Indonesia [37,38]. Note: As our survey was implemented in 2018, we introduce data of Year 2018. Data of Year 2020 are the latest information, but the value seems to be influenced by COVID-19.

For the data analysis, we used IBM SPSS Statistics 22 (IBM Cooperation, New York, the United States). Our study adopted a level of less than 5% to assess the statistical significance of the analysis.

2.3. Logistic Regression Analysis

In Section 3.3, we conduct a logistic regression analysis to examine whether some variables may be related to Natech damage estimation. In the analysis, the cases in which respondents answered "I don't know" are listwise deleted. First, we input the following 16 independent variables: worry about the possibility of a tsunami, evacuation behavior, access to information sources (6 variables), participation in DRR activities, and PMI activities (7 variables). Subsequently, we identify the best model using the stepwise method based on the Akaike information criterion (AIC).

3. Results and Interpretations

3.1. Sequential Steps of People's Response

As stated in Section 1.4, the primary purpose of this study is to verify whether sequential steps of the response can be observed. First, 29.2% (111 households) of the total respondents assembled at the evacuation point immediately after the earthquake. Among those who assembled at the evacuation point, 43.2% were "very worried" and 51.4% were "worried" about the possibility of a tsunami (see Table 3). Compared with those who did not assemble at to the evacuation point, those who assembled at the evacuation point had smaller proportions of being slightly worried (2.7%) and not worried (2.7%) than those who did not assemble at the evacuation point. The chi-square test showed a statistically significant relationship between evacuation and worry about a tsunami.

Table 3. Worry about the possibility of a tsunami.

	Worry about the Possibility of a Tsunami			
	Very Worried	**Worried**	**Slightly Worried**	**Not Worried**
Assembled at evacuation point (n = 111)	43.2%	51.4%	2.7%	2.7%
No (n = 267)	56.6%	35.6%	4.5%	3.4%
Sum (N = 378)	52.6%	40.2%	4.0%	3.2%
χ^2 (3, N = 378) = 8.233, $p < 0.05$.				

Next, we investigate the relationship between worry about a tsunami and Natech damage estimation. Our survey asked the following question about Natech damage estimation, "Do you think that a tsunami (2–3 m high) would damage industrial facilities?" The simple tabulation demonstrates that 53.7% answered "Yes," 16.9% responded "No," and 29.4% selected "I don't know." We employed a detailed cross-tabulation to investigate the relationship between worry about the possibility of a tsunami and damage estimation caused by a tsunami by dividing the sample into two groups (those who assembled at the evacuation point and those who did not; see Table 4). In both groups, we find no statistically significant relationship between worry about the possibility of a tsunami and the estimation of the damage. The respondents of both groups selected "I don't know [the damage estimation]," even if they were worried about a tsunami. These results imply that it was difficult for the respondents to imagine the cascading consequences at the time the earthquake occurred.

Table 4. Natech damage estimation.

			Were You Worried about Whether the Earthquake Would Generate a Tsunami?			
			Very Worried	**Worried**	**Slightly Worried**	**Not Worried**
Assembled at evacuation point	Do you think that a tsunami (2–3 metres in height) would cause damage?	Yes (n = 59)	42.4%	54.2%	1.7%	1.7%
		No (n = 23)	47.8%	39.1%	4.3%	8.7%
		I don't know. (n = 28)	39.3%	57.1%	3.6%	0.0%
	Sum (N = 110)		42.7%	51.8%	2.7%	2.7%

Table 4. *Cont.*

			Were You Worried about Whether the Earthquake Would Generate a Tsunami?			
			Very Worried	**Worried**	**Slightly Worried**	**Not Worried**
		χ^2 (6, N = 110) = 5.677, *n.s.*				
No (did not assemble at evacuation point)	Do you think that a tsunami (2–3 metres in height) would cause damage?	Yes (n = 144)	59.0%	33.3%	3.5%	4.2%
		No (n = 41)	70.7%	26.8%	0.0%	2.4%
		I don't know. (n = 81)	44.4%	44.4%	8.6%	2.5%
	Sum (N = 266)		56.4%	35.7%	4.5%	3.4%
		χ^2 (6, N = 266) = 12.413, *n.s.*				
Total	Do you think that a tsunami (2–3 metres in height) would cause damage?	Yes (n = 203)	54.2%	39.4%	3.0%	3.4%
		No (n = 64)	62.5%	31.3%	1.6%	4.7%
		I don't know. (n = 109)	43.1%	47.7%	7.3%	1.8%
	Sum (N =376)		52.4%	40.4%	4.0%	3.2%
		χ^2 (6, N = 376) = 11.650, *n.s*				

These chi-square tests do not verify the sequential steps of the response from evacuation to Natech damage estimation.

3.2. Risk Perception

As many relevant studies (e.g., (19,25,35)) suggest that risk perception is one of the significant predictors for people's response, we analyzed whether our survey shows similar findings to the literature. We asked respondents whether they considered their village to be prone to earthquakes. The simple tabulation demonstrates that earthquake risk perception is not very high (see Table 4). The sum of "not prone" (24.9%) and "slightly prone" (42.8%) is bigger than that of "very prone" (8.7%) and "prone" (23.7%). Such a low perception indicates that people felt the earthquake occurred suddenly, and thus we can identify the earthquake in our case study as a sudden-onset disaster.

There is no statistical significance between the two groups (those who assembled at the evacuation point and those who did not) for earthquake risk perception. This indicates that the perceived earthquake risk does not have a significant relationship with evacuation behavior (see the upper part of Table 5). In addition, we find no significant relationship between risk perception and Natech damage estimation (see the middle part of Table 5). However, statistical significance is detected in the relationship between risk perception and worry about a tsunami (see the lower part of Table 5). Our results do not clearly show that risk perception has a significant impact on people's responses.

Table 5. Relationship between perceived earthquake risk and people's response.

	How Prone Is Your Village to an earthquake?			
	Very Prone	Prone	Slightly Prone	Not Prone
Assembled at evacuation point (n = 47)	8.5%	17.0%	48.9%	25.5%
No (n = 126)	8.7%	26.2%	40.5%	24.6%
Sum (N = 173)	8.7%	23.7%	42.8%	24.9%
	χ^2 (3, N = 173) = 1.801, *n.s.*			

Table 5. *Cont.*

	How Prone Is Your Village to an earthquake?			
	Very Prone	Prone	Slightly Prone	Not Prone
Natech damage estimation—Yes (n = 103)	10.7%	25.2%	35.9%	28.2%
Natech damage estimation—No (n = 20)	15.0%	25.0%	35.0%	25.0%
Natech damage estimation—I don't know (n = 50)	2.0%	20.0%	60.0%	18.0%
Sum (N = 173)	8.7%	23.7%	42.8%	24.9%
χ^2 (6, N = 173) = 10.654, *n.s.*				

Worry about the possibility of a tsunami	How prone is your village to an earthquake?			
	Very prone	Prone	Slightly prone	Not prone
Very worried. (n = 84)	10.7%	35.7%	41.7%	11.9%
Worried (n = 74)	8.1%	10.8%	41.9%	39.2%
Slightly worried (n = 8)	0.0%	37.5%	37.5%	25.0%
Not worried (n = 7)	0.0%	0.0%	71.4%	28.6%
Sum (N = 173)	8.7%	23.7%	42.8%	24.9%
χ^2 (9, N = 173) = 27.5945 $p < 0.01$				

3.3. Significant Variables Related to Natech Damage Estimation

We perform chi-square tests to identify significant variables listed in Table 1 relating to Natech damage estimation. As a result, access to the website of the National Disaster Management Agency (Badan Nasional Penanggulangan Bencana, or BNPB), access to the BMKG website, and online news access are significant (see Table 6). Village location has no significant relationship with Natech damage estimation (see Table 7). Regarding preparedness, we employ the following three types of preparedness: disaster risk reduction (DRR) activity participation; knowledge of the Indonesian Red Cross (Palang Merah Indonesia, PMI) activities; and knowledge of DRR-related groups and programs. None of the three preparedness variables have a significant relationship with Natech damage estimation (see Table 8).

Table 6. Natech damage estimation and information access (N = 378).

	1. BNPB [*1] Website Accessed	No	2. BMKG [*2] Website Accessed	No
Natech damage estimation—Yes (n = 203)	7.4%	92.6%	7.4%	92.6%
Natech damage estimation—No (n = 64)	20.3%	79.7%	20.3%	79.7%
Natech damage estimation—I don't know (n = 111)	8.1%	91.9%	8.1%	91.9%
	χ^2 (2, N = 378) = 9.706, $p < 0.01$		χ^2 (2, N = 378) = 9.706 $p < 0.01$	
	3. TV news program accessed	No	4. Radio accessed	No
Natech damage estimation—Yes (n = 203)	78.3%	21.7%	0.5%	99.5%
Natech damage estimation—No (n = 64)	78.1%	21.9%	1.6%	98.4%
Natech damage estimation—I don't know (n = 111)	82.0%	18.0%	1.8%	98.2%
	χ^2 (2, N = 378) = 0.658, *n.s*		χ^2 (2, N = 378) = 1.362, *n.s*	

Table 6. *Cont.*

	5. Online news accessed	No	6. Social media*3 accessed	No
Natech damage estimation—Yes (n = 203)	4.4%	95.6%	14.3%	85.7%
Natech damage estimation—No (N = 64)	20.3%	79.7%	21.9%	78.1%
Natech damage estimation—I don't know (n = 111)	0.9%	99.1%	20.7%	79.3%
	χ^2 (2, N = 378) = 28.860, p < 0.01		χ^2 (2, N = 378) = 3.104, *n.s*	

*1 Badan Nasional Penanggulangan Bencana (National Disaster Management Agency in English) *2 Badan Meteorologi Klimatologi dan Geofisika (Meteorological, Climatological, and Geophysical Agency in English) *3 In the survey, the authors explained to the respondents that social media means Facebook, Twitter, WhatsApp, Line, Path, and Instagram.

Table 7. Natech damage estimation and village location (N = 378).

	Lebak Gede (n = 230)	Gunung Sugih (n = 148)
Natech damage estimation—Yes (n = 203)	58.1%	41.9%
Natech damage estimation—No (n = 64)	73.4%	26.6%
Natech damage estimation—I don't know (n = 111)	58.6%	41.4%
	χ^2 (2, N = 378) = 0.116, *n.s*	

Table 8. Natech damage estimation and preparedness (N = 378).

	DRR*1 Activities Participation—Yes	No
Natech damage estimation—Yes (n = 202)	35.1%	64.9%
Natech damage estimation—No (n = 63)	27.0%	73.0%
Natech damage estimation—I don't know (n = 111)	31.5%	68.5%
	χ^2 (2, N = 376) = 1.554, *n.s*	
	PMI*2 activities—Known	No
Natech damage estimation—Yes (n = 202)	31.0%	69.0%
Natech damage estimation—No (n = 63)	40.6%	59.4%
Natech damage estimation—I don't know (n = 111)	28.8%	71.2%
	χ^2 (2, N = 378) = 2.787, *n.s*	
	DRR-related groups and programs—Known	No
Natech damage estimation—Yes (n = 202)	21.2%	78.8%
Natech damage estimation—No (n = 63)	34.4%	65.6%
Natech damage estimation—I don't know (n = 111)	26.1%	73.9%
	χ^2 (2, N = 378) = 4.666, *n.s*	

*1 Disaster Risk Reduction *2 Palang Merah Indonesia (Indonesian Red Cross in English).

Furthermore, we conducted a logistic regression analysis to examine whether other variables may be related to Natech damage estimation. The value of AIC in the model that adopted all of the 16 independent variables was 290.72, while the smallest value of AIC among the examined models was 275.56. As it is considered that the smaller value of AIC indicates a more suitable fit in terms of the statistical model, the authors determined the latter model as the best model. The best model showed that BNPB website access and online news access were significant at the 5% and 1% levels, respectively, while neither worry about the possibility of a tsunami nor evacuation behavior appeared to be related to the Natech damage estimation (see Table 9). The results indicate that the respondents were likely not to estimate Natech damages if they had accessed the BNPB website or online news. This result can be interpreted as follows. Information on Natech might not have been available on both the BNPB website and online news because a tsunami was not generated in the earthquake of our case study. Therefore, it is understandable that those who accessed the BNPB website or online news did not estimate Natech damage.

Table 9. Results of logistic regression analysis.

	B	(SE)
(Constant)	1.493 **	(0.1710)
BNPB website access	−1.304 *	(0.5478)
Online news access	−1.492 **	(0.4903)
PMI activity [Risk mapping with local community participation]	−1.079 †	(0.6482)
N	265	
Akaike Information Criterion	275.56	
Nagelkerke R^2	0.125	

** $p < 0.01$, * $p < 0.05$, † $p < 0.1$.

3.4. Village Location

The chi-square test reveals that there is no statistically significant difference between village location and people's responses (see Table 10). The result for Natech damage estimation is shown in Table 7.

Table 10. Evacuation behavior by village location.

	Lebak Gede (n = 231)	Gunung Sugih (n = 149)
Assembled at evacuation point (n = 111)	59.5%	40.5%
No (n = 269)	61.3%	38.7%
χ^2 (1, N = 380) = 0.116, *n.s*		
	Lebak Gede (n = 230)	Gunung Sugih (n = 148)
Worry about the possibility of a tsunami—Very worried (n = 199)	62.8%	37.2%
Worried (n = 152)	55.3%	44.7%
Slightly worried (n = 15)	80.0%	20.0%
Not worried (n = 12)	75.0%	25.0%
χ^2 (3, N = 378) = 5.631, *n.s*		

3.5. Information Access

Almost 80% (79.3%) of respondents accessed a TV news program after the earthquake, and this percentage was remarkably the highest (see Table 11). The chi-square test showed statistical significance for this behavior.

Table 11. Information access after the earthquake (N = 380).

1. BNPB *1 Website	Accessed	No
Assembled at evacuation point (n = 111)	17.1%	82.9%
No (n = 269)	1.1%	98.9%
χ^2 (1, N = 380) = 36.889, $p < 0.01$		
2. BMKG*2 website	Accessed	No
Assembled at evacuation point (n = 111)	21.6%	78.4%
No (n = 269)	4.8%	95.2%
χ^2 (1, N = 380) = 25.200, $p < 0.01$		
3. TV news programs	Accessed	No
Assembled at evacuation point (n = 111)	79.3%	20.7%
No (n = 269)	79.2%	20.8%
χ^2 (1, N = 380) = 0.000, *n.s*		
4. Radio	Accessed	No
Assembled at evacuation point (n = 111)	2.7%	97.3%
No (n = 269)	0.4%	99.6%
χ^2 (1, N = 380) = 4.099, *n.s*		
5. Online news	Accessed	No
Assembled at evacuation point (n = 111)	11.7%	88.3%
No (n = 269)	3.7%	96.3%
χ^2 (1, N = 380) = 8.831, $p < 0.01$		
6. Social media*3	Accessed	No
Assembled at evacuation point (n = 111)	24.3%	75.7%
No (n = 269)	14.9%	85.1%
χ^2 (1, N = 380) = 4.836, $p < 0.05$		

*1 Badan Nasional Penanggulangan Bencana (National Disaster Management Agency in English) *2 Badan Meteorologi Klimatologi dan Geofisika (Meteorological, Climatological, and Geophysical Agency in English) *3 In the survey, the authors explained to the respondents that social media means Facebook, Twitter, WhatsApp, Line, Path, and Instagram.

Except for TV news program and radio access, we find a statistical significance in each information source based on the chi-square tests. Those who assembled at the evacuation point tended to access information sources more and this pattern is quite clear.

It is notable that access to social media (24.3%) is greater than access to BNPB (17.1%) and BMKG (21.6%) websites. These results seem to reflect the current Indonesian context. During the interviews, we exemplified to the respondents that social media refers to Facebook, Twitter, WhatsApp, Line, Path, and Instagram. Although we need to pay attention to fake news (locally often described as "hoax problems"), social media enables people to observe what people close to them are doing and to gather unassessed information quickly.

Compared with the results of Yu et al. [19], information access in our survey was more active. In Yu et al. [19], 12% of the respondents tried to search for information. Their study dealt with the Great East Japan Earthquake and many affected areas faced a shortage of electricity supply. By contrast, there was no black out in our case study. We consider that this is the reason we have a big difference in information access between the two surveys. In addition to the findings of Feng et al. [29], people accessed information provided by the BNPB and BMKG as authority positions and got to know other people's responses through social media.

Based on a hypothetical assumption that information may cause worries about the possibility of a tsunami, although chi-square tests are different to cause–effect analysis, three types of information access are found to be related to the worry: the BMKG website access, TV news program access, and social media access (see Table 12). Those who accessed these information sources tended to be less worried about the possibility of a tsunami. Regarding the relationship between information access and Natech damage estimation, we demonstrate the results in Table 6.

Table 12. Information access and worry about the possibility of a tsunami (N = 378).

Worry about the Possibility of a Tsunami	1. BNPB*[1] Website Accessed	No	2. BMKG*[2] Website Accessed	No
Very worried (n = 199)	4.5%	95.5%	6.5%	93.5%
Worried (n = 152)	7.2%	92.8%	10.5%	89.5%
Slightly worried (n = 15)	13.3%	86.7%	26.7%	73.3%
Not worried (n = 12)	0.0%	100.0%	33.3%	66.7%
	χ^2 (3, N = 378) = 3.454, *n.s*		χ^2 (3, N = 378) = 14.855, $p < 0.01$	
Worry about the possibility of a tsunami	3. TV news program accessed	No	4. Radio accessed	No
Very worried (n = 199)	78.9%	21.1%	1.0%	99.0%
Worried (n = 152)	83.6%	16.4%	1.3%	98.7%
Slightly worried (n = 15)	60.0%	40.0%	0.0%	100.0%
Not worried (n = 12)	58.3%	41.7%	0.0%	100.0%
	χ^2 (3, N = 378) = 8.330, $p < 0.05$		χ^2 (3, N = 378) = 0.390, *n.s*	
Worry about the possibility of a tsunami	5. Online news accessed	No	6. Social media*[3] accessed	No
Very worried (n = 199)	7.5%	92.5%	19.1%	80.9%
Worried (n = 152)	3.9%	96.1%	12.5%	87.5%
Slightly worried (n = 15)	6.7%	93.3%	53.3%	46.7%
Not worried (n = 12)	8.3%	91.7%	16.7%	83.3%
	χ^2 (3, N = 378) = 2.065, *n.s*		χ^2 (3, N = 378) = 16.153, $p < 0.01$	

*1 Badan Nasional Penanggulangan Bencana (National Disaster Management Agency in English) *2 Badan Meteorologi Klimatologi dan Geofisika (Meteorological, Climatological, and Geophysical Agency in English) *3 In the survey, the authors explained to the respondents that social media means Facebook, Twitter, WhatsApp, Line, Path, and Instagram.

3.6. Preparedness

In the survey, we asked whether the respondents had ever participated in any type of drill/simulation/exercise/activities on DRR. About 30% respondents stated that they had (32.8%) (see Table 13a). However, such participation does not have a statistical relationship with evacuation immediately after an earthquake. We find no statistical significance between DRR activity participation and worry about the possibility of a tsunami (see Table 13b).

In addition, we clarify whether the respondents knew about DRR-related groups and programs. At the survey sites, we identified that the local actors, such as PMI, DRR Forum, Youth Group for Disaster Preparedness (Taruna Siaga Bencana, Tagana), Disaster Response Village (Desa Tanggap Bencana), Disaster Prepared Village (Kampung Siaga Bencana), Alert Village (Desa Siaga), and Sultan Ageng Tirtayasa University, had been implementing DDR activities. Among them, the PMI was the most active. When we consider the group of respondents who assembled at the evacuation point, 42.3% knew about PMI activities (see the left part of Table 14), which was a significantly higher proportion than those who did not assemble at the evacuation point. For the other six groups and programs, the results show the same tendency (see the right part of Table 14). Those who assembled at the evacuation point knew more about DRR activities. Regarding the relationship with worry about the possibility of a tsunami, awareness of PMI activities showed statistical significance (see the left part of Table 15).

Table 13. Preparedness (disaster risk reduction (DRR) activity participation) and (a) immediate evacuation (N = 378) and (b) worry about the possibility of a tsunami (N = 376).

(a) Immediate evacuation	Yes	No
Assembled at evacuation point (n = 110)	40.0%	60.0%
No (n = 268)	29.9%	70.1%
χ^2 (1, N = 378) = 3.644, *n.s*		
(b) Worry about the possibility of a tsunami	Yes	No
Very worried (n = 198)	31.8%	68.2%
Worried (n = 151)	35.1%	64.9%
Slightly worried (n = 15)	40.0%	60.0%
Not worried (n = 12)	16.7%	83.3%
χ^2 (3, N = 376) = 2.207, *n.s*		

Table 14. Preparedness (knowledge) and immediate evacuation (N = 380).

	PMI[*1] Activities		DRR[*2]-Related Groups and Programs	
	Known	**No**	**Known**	**No**
Assembled at evacuation point (n = 111)	42.3%	57.7%	37.8%	62.2%
No (n = 269)	27.9%	72.1%	19.7%	80.3%
χ^2 (1, N = 380) = 7.539, *p* < 0.05			χ^2 (1, N = 380) = 13.783, *p* < 0.01	

*1 Palang Merah Indonesia (Indonesian Red Cross in English) *2 Disaster Risk Reduction.

Table 15. Preparedness (knowledge) and worry about the possibility of a tsunami (N = 378).

	PMI[*1] Activities		DRR[*2]-Related Groups and Programs	
Worry about the Possibility of a Tsunami	**Known**	**No**	**Known**	**No**
Very worried (n = 199)	32.7%	67.3%	21.6%	78.4%
Worried (n = 152)	31.6%	68.4%	28.3%	71.7%
Slightly worried (n = 15)	53.3%	46.7%	46.7%	53.3%
Not worried (n = 12)	0.0%	100.0%	8.3%	91.7%
χ^2 (3, N = 378) = 8.835, *p* < 0.05			χ^2 (3, N = 378) = 7.655, *n.s*	

*1 Palang Merah Indonesia (Indonesian Red Cross in English) *2 Disaster Risk Reduction.

Lindell et al. [25] addressed the effectiveness of pre-impact training activities, while Nakaya et al. [39] provided evidence to advocate the administration of tsunami drills in seaside communities to enhance evacuation behavior immediately after the disaster onset. Our study does not show such a strong implication. It is not easy to interpret why activity participation is not significant, but knowledge is. One possible interpretation is that local people witnessed advertising banners of relevant activities even though they did not attend them. For more detailed analysis for identifying effects of activity participation and knowledge, qualitative methods, such as in-depth interviews and focused group discussions, would be complementary to quantitative approaches.

Moreover, aspects of preparedness in the present study were limited, and further analysis is required. Theoretical frameworks such as the Social-Cognitive Model (SCM) [40,41] and the Protective Action Decision Model (PADM) [42] argued, in detail, a variety of factors influencing the adoption of preparedness. The SCM analyzed preparedness by setting a three-stage reasoning process (motivation to prepare, forming intentions to prepare, and their conversion into actual preparation). Paton et al. [41] clarified that the mechanism of "Intentions to Prepare" and "Intentions to Seek Information" are qualitatively different and stressed "This distinction has significant implications for conceptualising the preparedness process." Findings from the PADM research support encouraging emergency preparedness during the continuing hazard phase (the time between incidents). These arguments enable the present study to extend a more general understanding of preparedness.

4. Conclusions

4.1. Summary

Immediately after the earthquake, almost 30% of the respondents assembled at the relevant evacuation point. We explored, in detail, whether their evacuation action was related to worries about a tsunami and Natech damage estimation. In addition, to compare the key findings provided by the relevant literature, we assessed differences in location, information access, perceived risk, and preparedness.

Based on the survey results, no sequential steps of people's response were observed: evacuation right after the earthquake was related to worry about the possibility of a tsunami, but not to Natech damage estimation. The factors relevant to evacuation behavior were information access (except TV news program and radio), worry about the possibility of a tsunami, and knowledge of the DRR-related groups and programs. Meanwhile, the survey location (two villages), perceived earthquake risk, and DRR activity participation were less relevant to the behavior of assembling at the evacuation point. The survey results do not support the finding from the existing literature that higher risk perception is associated with evacuation behavior, nor do they support the assumption that a life-saving quick action is related to foreseeing cascading sequential consequences. At least based on our survey results, we consider that the importance of risk perception should not be excessively emphasized and needs to be further empirically evaluated.

4.2. Implications

In the earthquake that we used as a case study, shaking intensity was not so strong, and people did not face building collapse, at least in Cilegon. In addition, the earthquake occurred during the day. In this situation, people were able to access several information sources and seemed to be able to decide whether to take actions or not. However, at the same time, they needed to process and digest a lot of information, and they were likely to have been confused because few people anticipated the earthquake (see the upper part of Table 5). Although further clarification is necessary, some information raised or mentioned concerns about a tsunami, which is why people were worried about it. The results of Natech damage estimation imply that the content of information provided after disasters needs to be improved to easily make people understand that Natech situations need to be considered.

Some studies suggested to utilize the critical moment right after the hazard event occurrence for science communication and education campaigns [43,44]. The affected people urgently look for vital information as long as the communication line works, while families and friends of the potentially affected people also urgently search relevant information. For those who do not have anyone to care for in the affected zones, it seems very effective to provide learning opportunities. Specifically considering the country contexts in Indonesia, there have been growing concerns about the fake news problems [45,46]. Possibilities of large-scale information dissemination after the incidents should be carefully considered.

Our key finding is that sequential steps of people's response are not easily organized. However, it solely relied on one local study. For further verification, more studies are needed. There is a port area in Sendai city (Miyagi Prefecture, Japan), and the characteristics of the port area are similar to the case of the present study: the port area has some industrial facilities, and the residential area is located near the port. In the Great East Japan Earthquake (GEJE), the area experienced the Natech after the strong shaking and tsunami arrival [19]. After GEJE, at least, the Sendai Port area experienced two relatively bigger earthquakes. The first earthquake occurred in November 2016. Another earthquake happened in February 2021. While the epicenters of these two earthquakes were both located off Fukushima Prefecture, relatively strong shaking was observed around the Sendai Port area. We assume that higher risk perception and worries about a tsunami would be expected in the Sendai Port area based on its own experience. Preparedness might be a key factor for determining the Natech damage estimation, since a variety of practices have been carried out in that area.

4.3. Limitations

Our analysis was mainly limited to simple tabulations and statistical analysis using the chi-square test, except one trial of logistic regression analysis. In addition, we explored only the relationship between variables and did not analyze multiple variables concurrently (e.g., multi-variable analysis techniques). The reason that we did not conduct multi-variable analysis is that some answers to our questions are binary and categorical, and they are not suitable for multi-variable analysis. For example, when asking about information seeking, we simply asked if the respondents accessed the government agencies (Yes/No style). To test our initial hypothetical assumption, structural equation modeling and path analysis would be more suitable. The use of such approaches is essential to advance research in this area. Furthermore, our household survey made it difficult to include individual demographics in the analysis.

Author Contributions: Conceptualization, F.L. and F.I.; methodology, D.P.; software, D.S.; validation, D.S.; formal analysis, Y.J. and D.S.; investigation, F.L.; resources, D.S. and Y.J.; data curation, D.P.; writing—original draft preparation, Y.J.; writing—review and editing, D.S. and F.I.; visualization, D.S.; supervision, F.L. and F.I.; project administration, F.I.; funding acquisition, A.Z., D.S., and F.I. All authors have read and agreed to the published version of the manuscript.

Funding: This work was partially supported by the Japan Society for the Promotion of Science KAK-ENHI (grant number JP19K20540). The authors appreciate the financial support for the household survey implementation provided by Inter-Graduate School Doctoral Degree Program on Science for Global Safety, at Tohoku University. Financial support provided by International Research Collaboration of University of Indonesia enabled us to coordinate with local stakeholders (grant number NKB-1943/UN2.R3.1/HKP.05.00/2019). The collaborative research project of the International Research Institute of Disaster Science (IRIDeS), Tohoku University, supported us to complete the manuscript development.

Institutional Review Board Statement: The study was conducted according to the guidelines of the Declaration of Helsinki and approved by the Institutional Review Board of International Research Institute of Disaster Science, at Tohoku University (protocol code 2017–009, 22 January 2018).

Informed Consent Statement: Informed consent was obtained from all subjects involved in the study.

Data Availability Statement: The data presented in this study are available on request from the corresponding author.

Acknowledgments: The authors highly appreciate accommodation provided by the Lebak Gede and Gunung Sugih communities, the Indonesian Red Cross, the Pertamina, the city government offices of Cilegon, the provincial offices of Banten, and the Ministry of Health of Indonesia.

Conflicts of Interest: The authors declare no conflict of interest.

References

1. OECD. Natech Risk Management: 2017–2020 Project Results, OECD Environment, Health and Safety Publications Series on Chemical Accidents. No. 32. 2020. Available online: http://www.oecd.org/officialdocuments/publicdisplaydocumentpdf/?cote=env/jm/mono%282020%294&doclanguage=en (accessed on 17 March 2021).
2. UNDRR-APSTAAG. Asia-Pacific Regional Framework for NATECH (Natural Hazards Triggering Technological Disasters) Risk Management, United Nations Office for Disaster Risk Reduction—Asia-Pacific Science, Technology and Academia Advisory Group. 2020. Available online: https://www.undrr.org/media/48023/download (accessed on 17 March 2021).
3. UNISDR. Words into Action Guidelines: Man-made and Technological Hazards. 2018. Available online: https://www.preventionweb.net/publications/view/54012 (accessed on 17 March 2021).
4. UNISDR. Words into Action Guidelines: Implementation Guide for Man-made and Technological Hazards. 2018. Available online: https://unece.org/environment-policy/publications/words-action-guidelines-implementation-guide-man-made-and (accessed on 17 March 2021).
5. Girgin, S. The Natech events during the 17 August 1999 Kocaeli Earthquake: Aftermath and lessons learned. *Nat. Hazards Earth Syst. Sci.* **2011**, *11*, 1129–1140. [CrossRef]
6. Adiningsih, S.; Lestari, M.; Rahutami, A.I.; Wijaya, A.S. Sustainable Development Impacts of Investment Incentives: A Case Study of the Chemical Industry in Indonesia. 2009. Available online: https://www.iisd.org/publications/sustainable-development-impacts-investment-incentives-case-study-chemical-industry (accessed on 17 March 2021).
7. Moon, S. Justice, geography, and steel: Technology and national identity in Indonesian industrialization. *Osiris* **2009**, *24*, 253–277. [CrossRef]
8. Hudalah, D.; Viantari, D.; Firman, T.; Woltjer, J. Industrial land development and manufacturing deconcentration in greater Jakarta. *Urban Geogr.* **2013**, *34*, 950–971. [CrossRef]
9. Cahyandito, M.F. The effectiveness of community development and environmental protection program in oil and gas industry in Indonesia: Policy, institutional, and implementation review. *J. Manag. Sustain.* **2017**, *7*, 115–126. [CrossRef]
10. Lestari, F.; Pelupessy, D.; Jibiki, Y.; Putri, F.A.; Yurianto, A.; Widyaputra, G.; Maulana, S.; Maharani, C.F.; Imamura, F. Analysis of complexities in Natech disaster risk reduction and management: A case study of Cilegon, Indonesia. *J. Disaster Res.* **2018**, *13*, 1298–1308. [CrossRef]
11. Lestari, F.; Jibiki, Y.; Pelupessy, D.; Imamura, F.; Zulys, A.; Kadir, A.; Paramitasari, D. Exploratory study for strengthening education sectors for responding to complexities due to NATECH (Natural-Hazard Triggered Technological disasters) disasters. In *IOP Conference Series: Earth and Environmental Science*; IOP Publishing: Bristol, UK, 2021; Volume 630, No. 1; p. 012022.
12. Rahayu, H.P. Integrated Logic Model of Effective Tsunami Early Warning System. Doctoral Thesis, Kochi University of Technology, Kochi, Japan, 2012.
13. Wiryadinata, R.; Pratama, A.; Fahrizal, R.; Firmansyah, T.; Widyani, R. Design of linked sirens for tsunami early warning system using telecontrol system (case study at PUSDALOPS PB BPBD of Cilegon city). In *IOP Conference Series: Materials Science and Engineering*; IOP Publishing: Bristol, UK, 2019; Volume 673, No. 1; p. 012057.
14. AHA Centre (ASEAN Coordinating Centre for Humanitarian Assistance on Disaster Management). Press Release: Indonesia Hosts the 7th ASEAN Regional Disaster Emergency Response Simulation Exercise. Available online: https://ahacentre.org/press-release/press-releasethe-aha-centre-launches-the-first-asean-risk-monitor-and-disaster-management-reviewindonesia-hosts-the-7th-asean-regional-disaster-emergency-response-simulation-exercise/2018 (accessed on 24 August 2020).
15. Jibiki, Y.; Pelupessy, D.; Susilowati, I.H.; Putri, F.A.; Lestari, F.; Imamura, F. Exploring Community Preparedness for Complex Disaster: A Case Study in Cilegon (Banten Province in Indonesia). In Proceedings of the International Conference of Occupational Health and Safety ICOHS-2017, Bali, Indonesia, 1–2 November; KnE Life Sciences: Dubai, UAE, 2018; pp. 237–249. [CrossRef]
16. Pelupessy, D.; Jibiki, Y.; Lestari, F.; Zulys, A.; Imamura, F. Different vantage points amongst different stakeholders in NATECH (NAtural hazard-triggered TECHnological) disasters: A case from the 2018 Mt. Anak Krakatau eruption and tsunami. In *IOP Conference Series: Earth and Environmental Science*; IOP Publishing: Bristol, UK, 2021; Volume 630, No. 1; p. 012024.
17. BMKG. Ulasan Guncangan Tanah Akibat Gempabumi Kabupaten Lebak Banten 23 Januari 2018 (Review of Ground Shaking Due to the Lebak Banten Earthquake, 23 January 2018). 2018. Available online: https://www.bmkg.go.id/berita/?p=ulasan-guncangan-tanah-akibat-gempabumi-kabupaten-lebak-banten-23-januari-2018&lang=ID&tag=gempabumi (accessed on 17 March 2021).
18. Kabar Banten. Dilanda Gempa, Karyawan PT. Chandra Asri Berhamburan Keluar (Earthquake hit Chandra Asri Company Rushed into Exit). 2018. Available online: https://www.kabar-banten.com/dilanda-gempa-karyawan-pt-chandra-asri-berhamburan-keluar/ (accessed on 23 May 2019).
19. Yu, J.; Cruz, A.M.; Hokugo, A. Households' risk perception and behavioral responses to natech accidents. *Int. J. Disaster Risk Sci.* **2017**, *8*, 1–15. [CrossRef]
20. Steinberg, L.J.; Basolo, V.; Burby, R.; Levine, J.N.; Maria Cruz, A. Joint seismic and technological disasters: Possible impacts and community preparedness in an urban setting. *Nat. Hazards Rev.* **2004**, *5*, 159–169. [CrossRef]
21. Mitchell, J.T.; Edmonds, A.S.; Cutter, S.L.; Schmidtlein, M.C.; McCarn, R.L.; Hodgson, M.E.; Duhe, S. *Evacuation Behavior in Response to the Graniteville, South Carolina, Chlorine Spill*; Quick Response Research Report 178; University of South Carolina: South Carolina, Columbia, USA, 2005.

22. Baker, E.J. Hurricane evacuation behavior. *Int. J. Mass Emerg. Disasters* **1991**, *9*, 287–310.
23. Lindell, M.K.; Perry, R.W. Household adjustment to earthquake hazard a review of research. *Environ. Behav.* **2000**, *32*, 461–501. [CrossRef]
24. Lindell, M.K.; Prater, C.S.; Gregg, C.E.; Apatu, E.J.; Huang, S.-K.; Wu, H.C. Households' immediate responses to the 2009 American Samoa Earthquake and Tsunami. *Int. J. Disaster Risk Reduct.* **2015**, *12*, 328–340. [CrossRef]
25. Lindell, M.K.; Prater, C.S.; Wu, H.C.; Huang, S.K.; Johnston, D.M.; Becker, J.S.; Shiroshita, H. Immediate behavioural responses to earthquakes in Christchurch, New Zealand, and Hitachi, Japan. *Disasters* **2016**, *40*, 85–111. [CrossRef]
26. Quarantelli, E.L. The nature and conditions of panic. *Am. J. Sociol.* **1954**, *60*, 267–275. [CrossRef]
27. Aguirre, B.E. Emergency evacuations, panic, and social psychology. *Psychiatry* **2005**, *68*, 121–129. [CrossRef]
28. Mawson, A.R. Understanding mass panic and other collective responses to threat and disaster. *Psychiatry* **2005**, *68*, 95–113. [CrossRef] [PubMed]
29. Feng, Z.; González, V.A.; Trotter, M.; Spearpoint, M.; Thomas, J.; Ellis, D.; Lovreglio, R. How people make decisions during earthquakes and post-earthquake evacuation: Using verbal protocol analysis in immersive virtual reality. *Saf. Sci.* **2020**, *129*, 104837. [CrossRef]
30. Nilsson, D.; Johansson, A. Social influence during the initial phase of a fire evacuation—Analysis of evacuation experiments in a cinema theatre. *Fire Saf. J.* **2009**, *44*, 71–79. [CrossRef]
31. Donald, I.; Canter, D. Intentionality and fatality during the King's Cross underground fire. *Eur. J. Soc. Psychol.* **1992**, *22*, 203–218. [CrossRef]
32. Gill, K.; Laposata, E.A.; Dalton, C.F.; Aguirre, B.E. *Analysis of Severe Injury and Fatality in the Station Nightclub Fire*; University of Delaware: Newark, NJ, USA, 2011.
33. Johnson, N.R.; Feinberg, W.E.; Johnston, D.M. Microstructure and panic: The impact of social bonds on individual action in collective flight from the Beverly Hills Supper Club fire. In *Disasters, Collective Behavior and Social Organizations*; University of Delaware Press: Newark, Delaware, USA, 1994; pp. 168–189.
34. Nascimento, K.R.D.S.; Alencar, M.H. Management of risks in natural disasters: A systematic review of the literature on NATECH events. *J. Loss Prev. Process Ind.* **2016**, *44*, 347–359. [CrossRef]
35. Suarez-Paba, M.C.; Perreur, M.; Munoz, F.; Cruz, A.M. Systematic literature review and qualitative meta-analysis of Natech research in the past four decades. *Saf. Sci.* **2019**, *116*, 58–77. [CrossRef]
36. Yu, J.; Hokugo, A. Understanding household mobilization time during Natech accident evacuation. *J. Disaster Res.* **2015**, *10*, 973–980. [CrossRef]
37. BPS—Statistics Indonesia (2021). Statistical Yearbook of Indonesia 2021. Available online: https://www.bps.go.id/publication/2021/02/26/938316574c78772f27e9b477/statistik-indonesia-2021.html (accessed on 17 March 2021).
38. BPS—Statistics Indonesia (2019). Statistical Yearbook of Indonesia 2019. Available online: https://www.bps.go.id/publication/2019/07/04/daac1ba18cae1e90706ee58a/statistik-indonesia-2019.html (accessed on 17 March 2021).
39. Nakaya, N.; Nemoto, H.; Yi, C.; Sato, A.; Shingu, K.; Shoji, T.; Sato, S.; Tsuchiya, N.; Nakamura, T.; Narita, A.; et al. Effect of tsunami drill experience on evacuation behavior after the onset of the Great East Japan Earthquake. *Int. J. Disaster Risk Reduct.* **2018**, *28*, 206–213. [CrossRef]
40. Paton, D. Disaster preparedness: A social-cognitive perspective. *Disaster Prev. Manag. Int. J.* **2003**, *12*, 210–216. [CrossRef]
41. Paton, D.; Smith, L.; Johnston, D. When good intentions turn bad: Promoting natural hazard preparedness. *Aust. J. Emerg. Manag.* **2005**, *20*, 25–30.
42. Lindell, M.K.; Perry, R.W. The protective action decision model: Theoretical modifications and additional evidence. *Risk Anal.* **2012**, *32*, 616–632. [CrossRef]
43. Gizzi, F.T.; Kam, J.; Porrini, D. Time windows of opportunities to fight earthquake under-insurance: Evidence from Google Trends. *Humanit. Social Sci. Commun.* **2020**, *7*, 1–11. [CrossRef]
44. Tan, Y.J.; Maharjan, R. What googling trends tell us about public interest in earthquakes. *Seismol. Res. Lett.* **2018**, *89*, 653–657. [CrossRef]
45. Zannettou, S.; Sirivianos, M.; Blackburn, J.; Kourtellis, N. The web of false information: Rumors, fake news, hoaxes, clickbait, and various other shenanigans. *J. Data Inf. Qual.* **2019**, *11*, 1–37. [CrossRef]
46. Kwanda, F.A.; Lin, T.T. Fake news practices in Indonesian newsrooms during and after the Palu earthquake: A hierarchy-of-influences approach. *Inf. Commun. Soc.* **2020**, *23*, 849–866. [CrossRef]

International Journal of
Environmental Research and Public Health

MDPI

Article

People's Perception of Well-Being during the COVID-19 Pandemic: A Case Study in Japan

Daisuke Sasaki [1,*], Anawat Suppasri [1], Haruka Tsukuda [2], David N. Nguyen [1], Yasuaki Onoda [2] and Fumihiko Imamura [1]

[1] International Research Institute of Disaster Science (IRIDeS), Tohoku University, Sendai 980-8572, Japan
[2] Department of Architecture and Building Science, School of Engineering, Tohoku University, Sendai 980-8579, Japan
* Correspondence: dsasaki@irides.tohoku.ac.jp

Abstract: This study aims to examine people's perception of well-being during the COVID-19 pandemic in Japan and quantitatively clarify key factors towards realizing evidence-based policymaking. In March 2022, 400 participants responded to a survey conducted through Rakuten Insight. The authors applied an ordinal logistic regression (OLR), followed by principal component analysis (PCA), to create a new compound indicator (CI) to represent people's perception of well-being during the pandemic in addition to ordinary least squares (OLS) regression with a forward-backward stepwise selection method, where the dependent variable is the principal component score of the first principal component (PC1), while the independent variables are the same as the abovementioned OLR. Consequently, while analyzing OLR, some independent variables showed statistical significance, while the CI provided an option to grasp people's perception of well-being. Furthermore, family structure was statistically significant in all cases of OLR and OLS. Moreover, in terms of the standardized coefficients (beta) of OLS, the family structure had the greatest impact on the CI. Based on the study results, the authors advocate that the Japanese government should pay more attention to single-person households affected by the COVID-19 pandemic.

Keywords: COVID-19; disaster science; evidence-based policymaking; ordinal logistic regression; principal component analysis; compound indicator; single-person households; Japan

Citation: Sasaki, D.; Suppasri, A.; Tsukuda, H.; Nguyen, D.N.; Onoda, Y.; Imamura, F. People's Perception of Well-Being during the COVID-19 Pandemic: A Case Study in Japan. *Int. J. Environ. Res. Public Health* **2022**, *19*, 12146. https://doi.org/10.3390/ijerph191912146

Academic Editor: Paul B. Tchounwou

Received: 13 August 2022
Accepted: 17 September 2022
Published: 25 September 2022

Publisher's Note: MDPI stays neutral with regard to jurisdictional claims in published maps and institutional affiliations.

1. Introduction

Coronavirus disease (COVID-19), an infectious disease caused by the SARS-CoV-2 virus, was first detected in late 2019 and rapidly spread to the rest of the world in 2020, leading the World Health Organization (WHO) to declare the outbreak as a global pandemic in March 2020 [1]. Researchers have explored the psychological, social, and neuroscientific effects of COVID-19 and presented longer-term strategies for mental health science research [2]. Along with the approach of mental health science research, some researchers have investigated COVID-19 and its implications from a resilience point of view and made some recommendations for the sake of disaster risk reduction [3–5]. In terms of the well-being of people during the COVID-19 pandemic, although we can find abundant literature focusing on how well-being has been affected by COVID-19 mainly in Europe, the US, and China, there is a lack of literature conducting a case study in Japan [6].

The existing literature regarding well-being is as follows. Saladino et al. [7] highlighted the impact on the psychological well-being of the groups most exposed to COVID-19, such as children, college students, and health workers. Sibley et al. [8] focused on the effects of the COVID-19 pandemic and nationwide lockdown on trust, attitudes toward government, and well-being. Brodeur et al. [9] utilized Google Trends data to examine whether COVID-19 and the associated lockdowns implemented in Europe and the United States led to changes in search terms related to the topic of well-being. Lesser and Nienhuis [10] assessed how preemptive measures, such as social distancing and closure of municipal and provincial recreation facilities, impacted physical activity behavior and the well-being of Canadians; they suggested that health-promoting measures directed towards inactive individuals may be essential to improving well-being. Nienhuis and Lesser [11] have also assessed whether sex differences exist in physical activity and well-being since COVID-19 and explored how barriers or facilitators to physical activity may explain these differences. Dahlen et al. [12] demonstrated a positive and robust association between changes in daily activity levels and corresponding changes in psychological well-being. Feitelson et al. [13] assessed the well-being effects of COVID-19 in Israel by analyzing the pandemic's impact on several well-being indicators. Other studies have highlighted the psychological well-being of parents and their children. For example, Gassman-Pines et al. [14] examined the hypothesis that the crisis had worsened the psychological well-being of parents and children using daily survey data collected before and after the crisis started. Patrick et al. [15] investigated how the pandemic and mitigation efforts affected the physical and emotional well-being of parents and children in the United States in early June 2020. Huebener et al. [16] suggest that public policy measures taken to contain COVID-19 can have large effects on family well-being, based on a novel representative survey of parental well-being collected between May and June 2020 in Germany.

Furthermore, previous studies have examined the factors that affect well-being during the pandemic era. O'Connor et al. [17] indicate that the mental health and well-being of the UK adult population appear to have been affected during the initial phase of the COVID-19 pandemic; they state that the increasing rates of suicidal thoughts across waves, especially among young adults, are concerning. Coppola et al. [18] suggest that family is a protective factor with respect to mental health because the perceived mental health of those who did not live alone, and especially those who had to take care of small children, appear to be higher due to a seemingly greater ability to activate coping resources. Özmen et al. [19] stated that the scores of the participants in the survey conducted in Turkey in April 2020, regarding the fear of COVID-19, showed statistically significant differences according to the following variables: age, gender, education level, working status, the presence of pre-existing chronic diseases, regular drug use, and income level. Tomaz et al. [20] advocated that a larger social network, more social contact, and better perceived social support seemed to protect against loneliness and poor well-being; thus, addressing loneliness and social support in older adults is of significance. Fingerman et al. [21] also suggest that older adults who live alone may be more reactive to social contact during the COVID-19 outbreak than those who reside with others. In addition to the abovementioned previous studies, we can also find useful findings for evidence-based policymaking (EBPM) in the field of disaster science [22–27].

Under such circumstances, Suppasri et al. [28] conducted a survey between 5 and 9 November 2020, with a total of 600 respondents in Japan, based on a survey conducted by the European Commission's Joint Research Centre (EU-JRC) and University College London (UCL) to facilitate future international comparisons [29]. One of the preliminary results obtained from a simple tabulation showed that the respondents believed that support for basic needs, such as goods and other utilities, should be prioritized, followed by support for low-income persons and support for persons who own their own businesses [28]. However, the influence of the COVID-19 pandemic on people's perceptions of their well-being in Japan remains an open question. This study is follow-up research, in which the authors developed and conducted a second set of surveys based on Suppasri et al. [28]. Thus, this study aims to examine people's perception of well-being during the COVID-19 pandemic in Japan and clarify key factors in a quantitative manner toward the realization of EBPM in the future. In this regard, this study could contribute to the literature on the well-being of people in Japan during the COVID-19 pandemic, in that it provides lessons learned from Japanese case.

2. Methodology

The questionnaire survey was conducted through Rakuten Insight in March 2022 with a total of 400 respondents. The targeted areas that were selected as follows: Hokkaido (n = 72, 18.0%), Iwate Prefecture (n = 37, 9.3%), Miyagi Prefecture (n = 31, 7.8%), Saitama Prefecture (n = 20, 5.0%), Chiba Prefecture (n = 9, 2.3%), Tokyo Metropolis (n = 38, 9.5%), Kanagawa Prefecture (n = 25, 6.3%), Kyoto Prefecture (n = 12, 3.0%), Osaka Prefecture (n = 43, 10.8%), Hyogo Prefecture (n = 33, 8.3%), Fukuoka Prefecture (n = 38, 9.5%), Saga Prefecture (n = 4, 1.0%), Nagasaki Prefecture (n = 10, 2.5%), Kumamoto Prefecture (n = 15, 3.8%), Oita Prefecture (n = 5, 1.3%), Miyazaki Prefecture (n = 3, 0.8%), and Kagoshima Prefecture (n = 5, 1.3%). Thus, the targeted areas were Hokkaido, the Tohoku region, the capital area, the Kansai region, and the Kyushu region. There were 218 (54.5%) male and 182 (45.5%) female respondents. The average age was 49.1 years, ranging from 25 to 69 years. In terms of the respondents' educational level and employment status, 193 (48.3%) graduated from university, 160 (40.0%) graduated from high school, 256 (64.0%) were employed, and 55 (13.8%) were homemakers. All the questions were presented in Japanese. The survey had 59 questions, which included demographic questions, such as gender, age, annual household income, and family structure. The software package used for statistical analysis in this study was SPSS Statistics 28.

First, the authors apply ordinal logistic regression (OLR). OLR is a regression method for ordinal dependent variables that have been used in social data analysis in the existing literature, such as DeMaris [30]. The dependent variables in this study are the four proxies of people's perception of well-being during the COVID-19 pandemic: (i) change in job satisfaction, (ii) change in satisfaction with family, (iii) change in psychological well-being, and (iv) change in economic well-being. In this study, the authors have chosen the above four variables as the dependent variables while referring to the investigation regarding well-being by the Cabinet Office, Government of Japan [31]. These dependent variables used a 5-point Likert scale ranging from Heavily deteriorated, Deteriorated, Unchanged, Improved, and Heavily improved. The independent variables are the following 22 variables: change in daily food, water, electricity, and heat consumption, change in the use of public transportation, change in use of private transportation, change in use of medical and hospital services, change in use of banking and financial services, change in use of telephone and Internet services, concerns about the lack of economic recovery measures, concerns about the risk of a new wave of COVID-19 infection spreading, concerns about possible disruption of essential and basic services, concerns about the possibility of simultaneous occurrence of natural hazards, concerns about the risk of simultaneous acts of terrorism, cyber-attacks, riots, age, number of households, gender, education level, family structure, length of residency, existence of dependents, existence of pets, employment, annual household income, and residency in the Greater Tokyo Area, which consists of the Tokyo Metropolis, Kanagawa, Chiba, and Saitama Prefectures. The first six variables and the subsequent five variables also utilized 5-point Likert scales ranging from -2 (Heavily decrease) to 2 (Heavily increase) and from 0 (None at all) to 4 (Quite a lot), respectively. Age and the number of household variables are set on a ratio scale. The last nine variables are used as dummy variables. Gender takes a value of 1 (Female) or 0 (Male). Similarly, the education level is 1 (university graduate or above) or 0 (otherwise). The family structure is 1 (single-person household) or 0 (otherwise). The length of residence is 1 (10 years or more) or 0 (otherwise). The existence of dependents is 1 (yes) or 0 (no). The existence of pets is 1 (yes) or 0 (no). Employment was scored as 1 (employed) or 0 (otherwise). The annual household income is 1 (less than five million yen) or 0 (otherwise). Residency in the Greater Tokyo Area is either 1 or 0 (otherwise).

Subsequently, we conducted a principal component analysis (PCA) to create a new compound indicator (CI). The origin of PCA dates back to early 20th-century literature, such as Hotelling [32]. Jolliffe and Cadima [33] explain that PCA is a methodology for reducing the dimensionality of a dataset, which minimizes information loss while increasing interpretability by reducing dimensionality. In particular, PCA creates new uncorrelated variables while maximizing variance by solving an eigenvalue/eigenvector problem. In this study, we adopt the first principal component (PC1) of the four variables of people's perception of well-being during the COVID-19 pandemic as a new CI; then, we calculate the principal component score (PCS) of PC1. Furthermore, we apply ordinary least squares (OLS) regression with a forward-backward stepwise selection method, where the dependent variable is the PCS of the PC1, while the independent variables are the same as the abovementioned OLR, assuming that the residuals follow a normal distribution.

3. Results

3.1. Descriptive Statistics of the Dependent Variables

The frequency distributions of the dependent variables are listed in Table 1. With regard to changes in job satisfaction, three-fourths of the respondents answered, "Unchanged", while almost one-fifth answered, "Deteriorated/Heavily deteriorated". Regarding the change in satisfaction with family, about four-fifths answered, "Unchanged", while almost one-tenth answered, "Deteriorated/Heavily deteriorated" or "Improved/Heavily Improved", respectively. Meanwhile, regarding the change in psychological well-being, about half of the total respondents answered, "Unchanged", while almost two-fifths answered, "Deteriorated/Heavily deteriorated". Furthermore, regarding the change in economic well-being, almost two-thirds answered, "Unchanged", while almost everyone else answered, "Deteriorated/Heavily deteriorated".

Table 1. Frequency distribution of the dependent variables.

	Total n (%)	Heavily Deteriorated n (%)	Deteriorated n (%)	Unchanged n (%)	Improved n (%)	Heavily Improved n (%)
Change in job satisfaction	400 (100.0)	19 (4.8)	63 (15.8)	300 (75.0)	17 (4.3)	1 (0.3)
Change in satisfaction with family	400 (100.0)	4 (1.0)	40 (10.0)	314 (78.5)	36 (9.0)	6 (1.5)
Change in psychological well-being	400 (100.0)	31 (7.8)	128 (32.0)	216 (54.0)	22 (5.5)	3 (0.8)
Change in economic well-being	400 (100.0)	32 (8.0)	101 (25.3)	256 (64.0)	11 (2.8)	0 (0.0)

Based on these results, it appears that both psychological and economic well-being has deteriorated more than job satisfaction and satisfaction with family. In addition, the proportion of Improved/Heavily Improved for change in satisfaction with family is almost the same as that of Deteriorated/Heavily deteriorated; thus, it seems to imply that COVID-19 may influence satisfaction with family, both positively and negatively.

3.2. OLR

3.2.1. Change in Job Satisfaction

The results of the OLR, whose dependent variable is change in job satisfaction, are shown in Table 2. For model fitting, the chi-square test for -2 log-likelihood (-2LL) values of the intercept-only model and the final model indicates statistical significance ($p = 0.003$) at the 5% level, which means that the final model has significant improvement over the intercept-only model. All thresholds are statistically significant, while two independent variables, namely, change in daily food, water, electricity, and heat consumption and family structure, show statistical significance. Notably, positive coefficients lead to a decrease in cumulative logit and vice versa in SPSS OLR, and we find that the more daily food, water, electricity, and heat consumption increase, the more job satisfaction deteriorates. Similarly, the job satisfaction of people who do not belong to a single-person household appears to have improved. This seems to be because remote work has become more popular owing to the COVID-19 pandemic. It costs more in daily food, water, electricity, and heat consumption to perform remote work, which may lead to a decrease in job satisfaction. Meanwhile, remote working can provide more time to stay with families. Therefore, it is possible that people who do not have a single-person household are satisfied with their jobs because of the introduction of remote work.

Table 2. Parameter estimates for change in job satisfaction.

		Estimate	SE	Wald	df	Sig.	95% CI	
							Lower	Upper
Threshold	Heavily deteriorated	−4.578	0.952	23.112	1	<0.001	−6.445	−2.712
	Deteriorated	−2.771	0.924	8.995	1	0.003	−4.582	−0.960
	Unchanged	2.052	0.921	4.962	1	0.026	0.246	3.857
	Improved	4.975	1.329	14.017	1	<0.001	2.371	7.580
Location	Change in daily food, water, electricity and heat consumption	−0.428	0.213	4.060	1	0.044	−0.845	−0.012
	Change in use of public transportation	0.093	0.231	0.162	1	0.688	−0.360	0.545
	Change in use of private transportation	0.101	0.242	0.175	1	0.675	−0.372	0.575
	Change in use of medical and hospital services	−0.052	0.213	0.060	1	0.807	−0.469	0.365
	Change in use of banking and financial services	0.075	0.264	0.080	1	0.777	−0.443	0.592
	Change in use of telephone and internet services	−0.035	0.202	0.029	1	0.864	−0.430	0.361
	Concerns about the lack of economic recovery measures	−0.056	0.122	0.213	1	0.644	−0.296	0.183
	Concerns about the risk of a new wave of COVID-19 infection spreading	−0.199	0.130	2.351	1	0.125	−0.453	0.055
	Concerns about the possible disruption of essential and basic services	−0.013	0.161	0.006	1	0.936	−0.328	0.302
	Concerns about the possibility of simultaneous occurrence of natural hazards	−0.226	0.161	1.975	1	0.160	−0.541	0.089
	Concerns about the risk of simultaneous acts of terrorism, cyber-attacks, riots	0.028	0.151	0.035	1	0.851	−0.267	0.323
	Age	−0.018	0.014	1.685	1	0.194	−0.044	0.009
	Number of households	−0.008	0.024	0.107	1	0.744	−0.054	0.039
	[Gender = 0]	−0.327	0.273	1.434	1	0.231	−0.862	0.208
	[Education level = 0]	−0.194	0.247	0.615	1	0.433	−0.677	0.290
	[Family structure = 0]	**0.967**	**0.318**	**9.252**	**1**	**0.002**	**0.344**	**1.590**
	[Length of residency = 0]	−0.061	0.262	0.053	1	0.817	−0.574	0.453
	[Existence of dependents = 0]	0.297	0.275	1.164	1	0.281	−0.243	0.837
	[Existence of pets = 0]	0.105	0.297	0.125	1	0.723	−0.477	0.687
	[Employment = 0]	0.276	0.278	0.987	1	0.321	−0.268	0.820
	[Annual household income = 0]	0.178	0.258	0.475	1	0.491	−0.328	0.684
	[Residency in the Greater Tokyo area = 0]	−0.202	0.291	0.482	1	0.488	−0.772	0.368
	Pseudo R-square							
	Cox and Snell	0.105						
	Nagelkerke	0.131						
	McFadden	0.069						

Bold font indicates statistical significance at the 5% level.

3.2.2. Change in Satisfaction with Family

The results of the OLR, whose dependent variable is change in satisfaction with family, are shown in Table 3. The chi-square test for -2LL values indicates statistical significance ($p = 0.019$) at the 5% level. Thresholds other than Deteriorated are statistically significant, while four independent variables, namely, concerns about the possibility of simultaneous occurrence of natural hazards, family structure, length of residency, and the existence of dependents, show statistical significance at the 5% level. This may imply that satisfaction with a family of people who are concerned about the possibility of a simultaneous occurrence of disasters caused by natural hazards, tends to deteriorate during the COVID-19 pandemic. We also found that satisfaction with a family of people, who do not belong to a single-person household, seems to have improved as well as job satisfaction, while the short length of residency (less than 10 years) and the inexistence of dependents appear to have a positive impact on change in satisfaction with family during the COVID-19 pandemic era. Notably, a latent variable may exist behind these independent variables, and further study is needed to better understand the results.

Table 3. Parameter estimates for change in satisfaction with family.

		Estimate	SE	Wald	df	Sig.	95% CI	
							Lower	Upper
Threshold	Heavily deteriorated	−3.412	1.051	10.534	1	0.001	−5.473	−1.352
	Deteriorated	−0.864	0.942	0.841	1	0.359	−2.710	0.983
	Unchanged	3.815	0.974	15.345	1	<0.001	1.906	5.723
	Improved	5.941	1.050	32.016	1	<0.001	3.883	7.999
Location	Change in daily food, water, electricity and heat consumption	0.205	0.229	0.807	1	0.369	−0.243	0.654
	Change in use of public transportation	−0.142	0.237	0.359	1	0.549	−0.607	0.323
	Change in use of private transportation	0.149	0.252	0.352	1	0.553	−0.344	0.643
	Change in use of medical and hospital services	−0.172	0.230	0.559	1	0.455	−0.624	0.279
	Change in use of banking and financial services	−0.066	0.291	0.051	1	0.821	−0.636	0.505
	Change in use of telephone and internet services	0.191	0.216	0.783	1	0.376	−0.233	0.616
	Concerns about the lack of economic recovery measures	0.015	0.129	0.013	1	0.909	−0.238	0.268
	Concerns about the risk of a new wave of COVID-19 infection spreading	0.084	0.133	0.397	1	0.528	−0.177	0.344
	Concerns about the possible disruption of essential and basic services	0.145	0.171	0.723	1	0.395	−0.190	0.481
	Concerns about the possibility of simultaneous occurrence of natural hazards	**−0.356**	**0.170**	**4.403**	**1**	**0.036**	**−0.688**	**−0.023**
	Concerns about the risk of simultaneous acts of terrorism, cyber-attacks, riots	0.047	0.160	0.087	1	0.767	−0.265	0.360
	Age	−0.007	0.014	0.212	1	0.645	−0.034	0.021
	Number of households	0.035	0.026	1.801	1	0.180	−0.016	0.087
	[Gender = 0]	0.395	0.289	1.876	1	0.171	−0.170	0.961
	[Education level = 0]	−0.272	0.262	1.075	1	0.300	−0.786	0.242
	[Family structure = 0]	**1.148**	**0.351**	**10.706**	**1**	**0.001**	**0.460**	**1.835**
	[Length of residency = 0]	**0.558**	**0.276**	**4.099**	**1**	**0.043**	**0.018**	**1.098**
	[Existence of dependents = 0]	**0.574**	**0.289**	**3.942**	**1**	**0.047**	**0.007**	**1.141**
	[Existence of pets = 0]	0.357	0.314	1.297	1	0.255	−0.258	0.972
	[Employment = 0]	−0.243	0.292	0.693	1	0.405	−0.816	0.329
	[Annual household income = 0]	0.090	0.274	0.107	1	0.743	−0.448	0.627
	[Residency in the Greater Tokyo area = 0]	−0.442	0.307	2.078	1	0.149	−1.043	0.159
Pseudo R-square								
	Cox and Snell	0.090						
	Nagelkerke	0.117						
	McFadden	0.063						

Bold font indicates statistical significance at the 5% level.

3.2.3. Change in Psychological Well-Being

The results of the OLR, whose dependent variable is change in psychological well-being, are shown in Table 4. The chi-square test for -2LL values indicates statistical significance ($p < 0.001$) at the 5% level. Thresholds other than Deteriorated are statistically significant, while six independent variables, namely, change in use of private transportation, change in the use of telephone and Internet services, education level, family structure, the existence of dependents, and residency in the Greater Tokyo Area, show statistical significance at the 5% level. It seems that an increase in the use of private transportation has a positive impact on psychological well-being, while an increase in the use of telephone and Internet services, which may be caused by remote work due to the COVID-19 pandemic, had an adverse impact. It also appears that the psychological well-being of people, whose education level is not at university graduation or above, has deteriorated and that of people, who do not belong to a single-person household, has a tendency to improve. Furthermore, the existence of dependents appears to have a positive impact on change in psychological well-being, while residency in a place other than the Greater Tokyo area seems to have a negative impact.

Table 4. Parameter estimates for change in psychological well-being.

		Estimate	SE	Wald	df	Sig.	95% CI	
							Lower	Upper
Threshold	**Heavily deteriorated**	**−3.640**	**0.793**	**21.054**	**1**	**<0.001**	**−5.194**	**−2.085**
	Deteriorated	−1.283	0.770	2.777	1	0.096	−2.791	0.226
	Unchanged	**2.204**	**0.781**	**7.971**	**1**	**0.005**	**0.674**	**3.734**
	Improved	**4.398**	**0.946**	**21.602**	**1**	**<0.001**	**2.543**	**6.252**
Location	Change in daily food, water, electricity and heat consumption	−0.247	0.184	1.796	1	0.180	−0.608	0.114
	Change in use of public transportation	−0.302	0.199	2.292	1	0.130	−0.693	0.089
	Change in use of private transportation	**0.445**	**0.210**	**4.520**	**1**	**0.034**	**0.035**	**0.856**
	Change in use of medical and hospital services	0.328	0.185	3.144	1	0.076	−0.035	0.690
	Change in use of banking and financial services	0.129	0.234	0.307	1	0.580	−0.328	0.587
	Change in use of telephone and internet services	**−0.346**	**0.174**	**3.961**	**1**	**0.047**	**−0.687**	**−0.005**
	Concerns about the lack of economic recovery measures	−0.072	0.104	0.477	1	0.490	−0.277	0.132
	Concerns about the risk of a new wave of COVID-19 infection spreading	−0.182	0.109	2.760	1	0.097	−0.396	0.033
	Concerns about the possible disruption of essential and basic services	−0.065	0.140	0.215	1	0.643	−0.338	0.209
	Concerns about the possibility of simultaneous occurrence of natural hazards	0.036	0.138	0.067	1	0.796	−0.236	0.307
	Concerns about the risk of simultaneous acts of terrorism, cyber-attacks, riots	−0.031	0.131	0.057	1	0.811	−0.289	0.226
	Age	−0.012	0.012	1.017	1	0.313	−0.034	0.011
	Number of households	0.035	0.023	2.291	1	0.130	−0.010	0.081
	[Gender = 0]	0.008	0.232	0.001	1	0.974	−0.447	0.463
	[Education level = 0]	**−0.529**	**0.211**	**6.271**	**1**	**0.012**	**−0.943**	**−0.115**
	[Family structure = 0]	**1.005**	**0.280**	**12.842**	**1**	**<0.001**	**0.455**	**1.554**
	[Length of residency = 0]	0.354	0.223	2.518	1	0.113	−0.083	0.792
	[Existence of dependents = 0]	**0.556**	**0.233**	**5.700**	**1**	**0.017**	**0.100**	**1.013**
	[Existence of pets = 0]	−0.138	0.254	0.297	1	0.586	−0.636	0.359
	[Employment = 0]	−0.182	0.234	0.604	1	0.437	−0.642	0.277
	[Annual household income = 0]	0.233	0.220	1.120	1	0.290	−0.198	0.664
	[Residency in the Greater Tokyo area = 0]	**−0.563**	**0.253**	**4.971**	**1**	**0.026**	**−1.058**	**−0.068**
	Pseudo R-square							
	Cox and Snell	0.165						
	Nagelkerke	0.186						
	McFadden	0.083						

Bold font indicates statistical significance at the 5% level.

3.2.4. Change in Economic Well-Being

The results of the OLR, whose dependent variable is change in economic well-being, are shown in Table 5. The chi-square test for -2LL values indicates statistical significance ($p < 0.001$) at the 5% level. All thresholds are statistically significant, while five independent variables, namely, change in daily food, water, electricity, and heat consumption, concerns about the lack of economic recovery measures, concerns about possible disruption of essential and basic services, education level, and family structure, show statistical significance at the 5% level. It seems plausible that increases in daily food, water, electricity, and heat consumption that may be caused by remote work, as well as concerns about the lack of economic recovery measures and concerns about possible disruption of essential and basic services, have a negative impact on economic well-being because these appear to have a straightforward relationship. In addition, the economic well-being, as well as psychological well-being, of people, whose education level is not at university graduation or above, seems to have deteriorated and that of people who do not belong to a single-person household appeared to have a tendency to improve, as observed for the other three dependent variables.

Table 5. Parameter estimates for change in economic well-being.

		Estimate	SE	Wald	df	Sig.	95% CI	
							Lower	Upper
Threshold	**Heavily deteriorated**	**−4.633**	**0.871**	**28.281**	**1**	**<0.001**	**−6.341**	**−2.926**
	Deteriorated	**−2.578**	**0.845**	**9.304**	**1**	**0.002**	**−4.234**	**−0.921**
	Unchanged	**2.218**	**0.862**	**6.625**	**1**	**0.010**	**0.529**	**3.906**
Location	**Change in daily food, water, electricity and heat consumption**	**−0.492**	**0.197**	**6.238**	**1**	**0.013**	**−0.878**	**−0.106**
	Change in use of public transportation	−0.002	0.216	0.000	1	0.993	−0.425	0.421
	Change in use of private transportation	0.241	0.223	1.171	1	0.279	−0.196	0.678
	Change in use of medical and hospital services	0.192	0.195	0.966	1	0.326	−0.191	0.575
	Change in use of banking and financial services	−0.143	0.248	0.331	1	0.565	−0.629	0.344
	Change in use of telephone and internet services	0.019	0.185	0.010	1	0.920	−0.344	0.382
	Concerns about the lack of economic recovery measures	**−0.222**	**0.112**	**3.921**	**1**	**0.048**	**−0.441**	**−0.002**
	Concerns about the risk of a new wave of COVID-19 infection spreading	−0.174	0.117	2.203	1	0.138	−0.405	0.056
	Concerns about the possible disruption of essential and basic services	**−0.316**	**0.149**	**4.510**	**1**	**0.034**	**−0.607**	**−0.024**
	Concerns about the possibility of simultaneous occurrence of natural hazards	0.290	0.151	3.705	1	0.054	−0.005	0.586
	Concerns about the risk of simultaneous acts of terrorism, cyber-attacks, riots	−0.171	0.139	1.515	1	0.218	−0.443	0.101
	Age	−0.018	0.012	2.170	1	0.141	−0.043	0.006
	Number of households	0.025	0.025	0.972	1	0.324	−0.025	0.075
	[Gender = 0]	−0.106	0.250	0.180	1	0.671	−0.596	0.384
	[Education level = 0]	**−0.617**	**0.228**	**7.297**	**1**	**0.007**	**−1.065**	**−0.169**
	[Family structure = 0]	**0.599**	**0.297**	**4.072**	**1**	**0.044**	**0.017**	**1.181**
	[Length of residency = 0]	0.130	0.240	0.295	1	0.587	−0.340	0.601
	[Existence of dependents = 0]	0.204	0.250	0.664	1	0.415	−0.286	0.694
	[Existence of pets = 0]	0.310	0.267	1.356	1	0.244	−0.212	0.833
	[Employment = 0]	−0.391	0.250	2.432	1	0.119	−0.881	0.100
	[Annual household income = 0]	0.440	0.235	3.500	1	0.061	−0.021	0.901
	[Residency in the Greater Tokyo area = 0]	−0.378	0.274	1.904	1	0.168	−0.915	0.159
Pseudo R-square								
Cox and Snell		0.183						
Nagelkerke		0.216						
McFadden		0.108						

Bold font indicates statistical significance at the 5% level.

3.3. Creation of a New CI

3.3.1. PCA

As mentioned in the Methodology section, the four variables of people's perception of well-being under the COVID-19 pandemic, namely, change in job satisfaction, change in satisfaction with family, change in psychological well-being, and change in economic well-being, are input; furthermore, we assume that they range from −2 (Heavily deteriorated) to 2 (Heavily improved). The total explained variance is presented in Table 6. Based on this, we can find that only one principal component was to be extracted by the Kaiser–Guttman criterion; that is, components whose eigenvalues exceed 1 should be extracted, and 51% of the total variance is explained by PC1. The component matrix, as shown in Table 7, implies that changes in psychological well-being and changes in economic well-being may be slightly more correlated with PC1 than changes in job satisfaction and change in satisfaction with family.

Subsequently, the authors calculated the PCS of PC1, namely a new CI, while adjusting its values so that the CI became zero (Unchanged) when all four original input variables have the value of zero (Unchanged). The descriptive statistics and histograms of the CI are shown in Table 8 and Figure 1. We find that the distribution of the CI is skewed toward the negative side, although many are distributed near zero (unchanged).

Table 6. Total variance explained of PCA.

Component	Initial Eigenvalues			Extraction Sums of Squared Loadings		
	Total	% of Variance	Cumulative %	Total	% of Variance	Cumulative %
1	2.057	51.422	51.422	2.057	51.422	51.422
2	0.883	22.066	73.488			
3	0.598	14.942	88.430			
4	0.463	11.570	100.000			
	Extraction Method: Principal Component Analysis					

Table 7. Component matrix of PCA.

	Component 1
change in psychological well-being	0.815
change in economic well-being	0.753
change in job satisfaction	0.678
change in satisfaction with family	0.605
Extraction Method: Principal Component Analysis	

Table 8. Descriptive statistics of the CI.

	CI
Mean	−0.5382
Median	−0.0047
Std. Deviation	1.00000
Variance	1.000
Skewness	−0.577
Std. Error of Skewness	0.122
Kurtosis	1.407
Std. Error of Kurtosis	0.243
Minimum	−3.80
Maximum	2.72

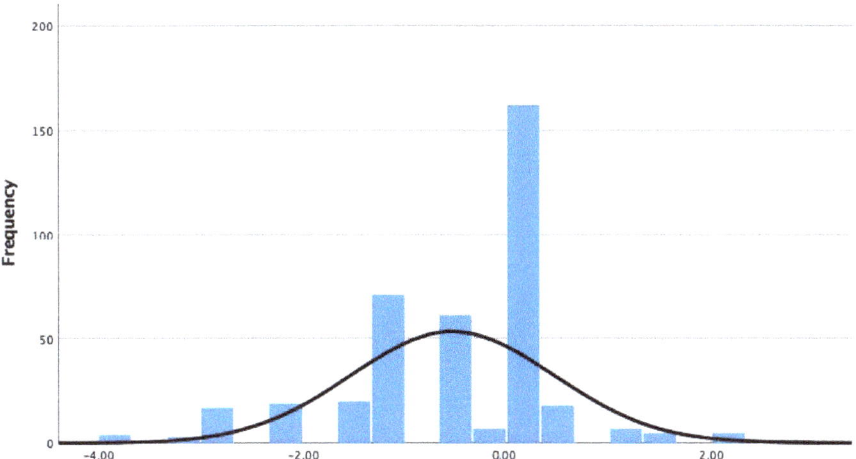

Figure 1. Histogram of the CI.

3.3.2. OLS Regression

We conducted an OLS regression with a forward-backward stepwise selection method, whose dependent variable was the CI, while the independent variables were the same as in the abovementioned OLR. The coefficients of the final selected model and the histogram of standardized residuals are shown in Table 9 and Figure 2. The result of the analysis of variance (ANOVA) is significant ($p < 0.001$), and the adjusted R-square is 0.143. All variance inflation factors (VIF) are less than 10.0, which implies no multicollinearity. The Durbin–Watson ratio is 1.842. It is generally acceptable to assume that the residuals are normally distributed.

Table 9. Coefficients of the finally selected model.

	Unstandardized Coefficients		Standardized Coefficients	t	Sig.	95% Confidence Interval for B	
	B	Std. Error	Beta			Lower	Upper
(Constant)	0.559	0.264		2.118	0.035	0.040	1.078
Change in daily food, water, electricity and heat consumption	−0.155	0.076	−0.100	−2.044	0.042	−0.305	−0.006
Concerns about the risk of a new wave of COVID-19 infection spreading	−0.098	0.046	−0.111	−2.117	0.035	−0.189	−0.007
Concerns about the possible disruption of essential and basic services	−0.112	0.049	−0.119	−2.276	0.023	−0.210	−0.015
Education level	0.274	0.094	0.137	2.902	0.004	0.088	0.460
Age	−0.011	0.005	−0.111	−2.348	0.019	−0.020	−0.002
Family structure	−0.534	0.124	−0.221	−4.301	<0.001	−0.779	−0.290
Existence of dependents	−0.226	0.104	−0.107	−2.163	0.031	−0.431	−0.021
Annual household income	−0.210	0.100	−0.103	−2.097	0.037	−0.406	−0.013

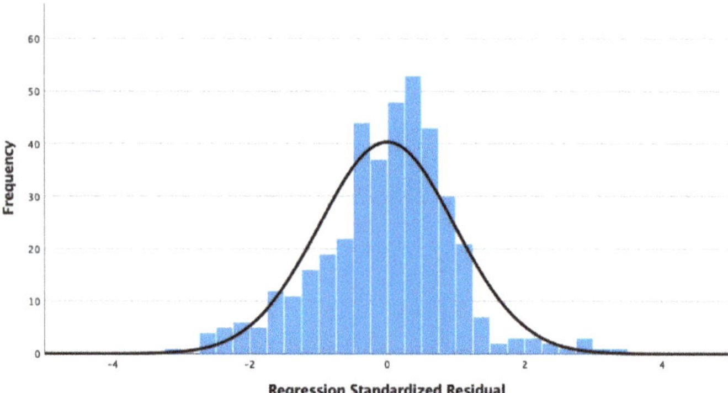

Figure 2. Histogram of the regression standardized residual.

4. Discussion

Based on the results of OLS, eight independent variables, namely changes in daily food, water, electricity, and heat consumption, concerns about the risk of a new wave of COVID-19 infection spreading, concerns about possible disruption of essential and basic services, education level, age, family structure, the existence of dependents, and annual household income, show statistical significance at the 5% level. Three out of eight independent variables, namely concerns about the risk of a new wave of COVID-19 infection spreading, age, and annual household income, do not show statistical significance in either case of OLR. Meanwhile, the family structure shows statistical significance in all cases of OLR. In terms of standardized coefficients (Beta), it seems that family structure has the greatest impact on CI, which is assumed to represent people's perception of well-being under the COVID-19 pandemic in general terms, followed by education level.

It should also be noted that there seems to be a non-significant difference between genders in neither case of OLR/OLS. Meanwhile, Nienhuis and Lesser [11] stated that the analysis based on the data provided by 1098 Canadians, 215 men and 871 women, showed sex differences in physical activity and well-being. Considering that approaches to family for males and females in Japan are different, this result has implications.

In general, it is imperative to prioritize policy targets due to time and budget constraints. As a result of this study, we can assert with evidence that policies for single-person households would improve their well-being effectively and efficiently. This argument seems to be unfamiliar in Japan at the moment, and thus, it is worth reconsidering how the government should allocate limited policy resources to address the ongoing pandemic.

5. Conclusions

In this study, we quantitatively examined people's perceptions of well-being during the COVID-19 pandemic in Japan. In the OLR analysis, some independent variables, which were not common but specific for each dependent variable, demonstrated statistical significance. Meanwhile, the CI created by utilizing PCA in this study provides an option to grasp people's perceptions of well-being. As discussed above, eight independent variables, namely, change in daily food, water, electricity, and heat consumption, concerns about the risk of a new wave of COVID-19 infection spreading, concerns about the possible disruption of essential and basic services, education level, age, family structure, the existence of dependents, and annual household income, are statistically significant at the 5% level in the OLS analysis, whose dependent variable is the CI. Furthermore, we found that family structure had the greatest impact on CI, which was consistent with the results of the OLR analysis. Therefore, we can identify the family structure as a key factor in the realization of EBPM in the future.

Based on the results of this study, the authors advocate that the Japanese government should pay more attention to single-person households affected by the COVID-19 pandemic. Some policies regarding COVID-19 in Japan seemingly tend to be implemented for households consisting of more than one person, such as households with children. The literature review in the Introduction section also indicates that it may be of great significance to address loneliness in the COVID-19 era. We hope that our study can also contribute to the provision of evidence for future policymaking for single-person households in Japan.

The future research focus should be two-fold: (i) to expand to research areas outside Japan so that we can compare results in a cross-sectional manner and verify the validity of the CI created in this study and (ii) to acquire time series data in Japan to assess Japanese policies regarding the COVID-19 pandemic. This cross-sectional and time series analysis could establish a comprehensive and exhaustive framework for evaluating people's perception of well-being during the COVID-19 pandemic and assess relevant policies in a quantitative manner, thus, contributing to the literature on EBPM in the field of disaster science.

Author Contributions: Conceptualization, D.S. and A.S.; methodology, D.S.; software, D.S.; validation, D.S., A.S., H.T., D.N.N. and Y.O.; formal analysis, D.S.; investigation, D.S.; resources, D.S. and A.S.; data curation, D.S.; writing—original draft preparation, D.S.; writing—review and editing, A.S, H.T., D.N.N., Y.O. and F.I.; visualization, D.S.; supervision, F.I.; project administration, A.S.; funding acquisition, A.S. and F.I. All authors have read and agreed to the published version of the manuscript.

Funding: This research was financially supported by UCL-Tohoku University Strategic Partner Funds and the COVID-19-related Research Support Project (IRIDeS).

Institutional Review Board Statement: This study was approved by the Ethics Committee of the International Research Institute of Disaster Science (IRIDeS), Tohoku University (protocol code 2021-047 and approved on 21 January 2022).

Informed Consent Statement: Informed consent was obtained from all subjects involved in the study.

Data Availability Statement: The data presented in this study are available upon request.

Acknowledgments: The authors would like to express their sincere appreciation to the editor, anonymous reviewers, and IRIDeS' ethics committee for their valuable comments on the questionnaire and the manuscript. The authors thank David Alexander and Gianluca Pescaroli at the Institute for Risk and Disaster Reduction (IRDR), UCL. We would also like to express our sincere gratitude to Monica Cardarelli, Luca Galbusera, and Georgios Giannopoulos from the European Commission's Joint Research Centre (JRC), Ispra, Italy, who agreed to use some questions from this survey and provided valuable insights into framing the questionnaire.

Conflicts of Interest: The authors declare no conflict of interest. The funders had no role in the design of the study; in the collection, analyses, or interpretation of data; in the writing of the manuscript; or in the decision to publish the results.

References

1. Saengtabtim, K.; Leelawat, N.; Tang, J.; Suppasri, A.; Imamura, F. Consequences of COVID-19 on health, economy, and tourism in Asia: A systematic review. *Sustainability* **2022**, *14*, 4624. [CrossRef]
2. Holmes, E.; O'Connor, R.; Perry, V.; Tracey, I.; Wessely, S.; Arseneault, L.; Ballard, C.; Christensen, H.; Silver, R.C.; Everall, I.; et al. Multidisciplinary research priorities for the COVID-19 pandemic: A call for action for mental health science. *Lancet Psychiatry* **2020**, *7*, 547–560. [CrossRef]
3. Chatterjee, R.; Bajwa, S.; Dwivedi, D.; Kanji, R.; Ahammed, M.; Shaw, R. COVID-19 Risk Assessment Tool: Dual application of risk communication and risk governance. *Prog. Disaster Sci.* **2020**, *7*, 100109. [CrossRef] [PubMed]
4. Djalante, R.; Shaw, R.; DeWit, A. Building resilience against biological hazards and pandemics: COVID-19 and its implications for the Sendai Framework. *Prog. Disaster Sci.* **2020**, *6*, 100080. [CrossRef]
5. Shaw, R.; Kim, Y.; Hua, J. Governance, technology and citizen behavior in pandemic: Lessons from COVID-19 in East Asia. *Prog. Disaster Sci.* **2020**, *6*, 100090. [CrossRef] [PubMed]
6. Aknin, L.; De Neve, J.; Dunn, E.; Fancourt, D.; Goldberg, E.; Helliwell, J.F.; Jones, S.P.; Karam, E.; Layard, R.; Lyubomirsky, S.; et al. Mental Health During the First Year of the COVID-19 Pandemic: A Review and Recommendations for Moving Forward. *Perspect. Psychol. Sci.* **2022**, *17*, 915–936. [CrossRef] [PubMed]
7. Saladino, V.; Algeri, D.; Auriemma, V. The psychological and social impact of Covid-19: New perspectives of well-being. *Front. Psychol.* **2020**, *11*, 577684. [CrossRef]
8. Sibley, C.G.; Greaves, L.M.; Satherley, N.; Wilson, M.S.; Overall, N.C.; Lee, C.H.J.; Milojev, P.; Bulbulia, J.; Osborne, D.; Milfont, T.L.; et al. Effects of the COVID-19 pandemic and nationwide lockdown on trust, attitudes toward government, and well-being. *Am. Psychol.* **2020**, *75*, 618–630. [CrossRef] [PubMed]
9. Brodeur, A.; Clark, A.E.; Fleche, S.; Powdthavee, N. COVID-19, lockdowns and well-being: Evidence from Google Trends. *J. Public Econ.* **2021**, *193*, 104346. [CrossRef]
10. Lesser, I.A.; Nienhuis, C.P. The impact of COVID-19 on physical activity behavior and well-being of Canadians. *Int. J. Environ. Res. Public Health* **2020**, *17*, 3899. [CrossRef]
11. Nienhuis, C.P.; Lesser, I.A. The impact of COVID-19 on women's physical activity behavior and mental well-being. *Int J Env Res Public Health* **2020**, *17*, 9036. [CrossRef] [PubMed]
12. Dahlen, M.; Thorbjørnsen, H.; Sjåstad, H.; von Heideken Wågert, P.; Hellström, C.; Kerstis, B.; Lindberg, D.; Stier, J.; Elvén, M. Changes in physical activity are associated with corresponding changes in psychological well-being: A pandemic case study. *Int. J. Environ. Res. Public Health* **2021**, *18*, 10680. [CrossRef] [PubMed]
13. Feitelson, E.; Plaut, P.; Salzberger, E.; Shmueli, D.; Altshuler, A.; Ben-Gal, M.; Israel, F.; Rein-Sapir, Y.; Zaychik, D. The effects of COVID-19 on wellbeing: Evidence from Israel. *Sustainability* **2022**, *14*, 3750. [CrossRef]
14. Gassman-Pines, A.; Ananat, E.O.; Fitz-Henley, J. COVID-19 and parent-child psychological well-being. *Pediatrics* **2020**, *146*. [CrossRef] [PubMed]
15. Patrick, S.W.; Henkhaus, L.E.; Zickafoose, J.S.; Lovell, K.; Halvorson, A.; Loch, S.; Letterie, M.; Davis, M.M. Well-being of parents and children during the COVID-19 pandemic: A national survey. *Pediatrics* **2020**, *146*, e2020016824. [CrossRef] [PubMed]
16. Huebener, M.; Waights, S.; Spiess, C.K.; Siegel, N.A.; Wagner, G.G. Parental well-being in times of Covid-19 in Germany. *Rev. Econ. Househ.* **2021**, *19*, 91–122. [CrossRef] [PubMed]
17. O'Connor, R.; Wetherall, K.; Cleare, S.; McClelland, H.; Melson, A.; Niedzwiedz, C.L.; O'Carroll, R.E.; O'Connor, D.B.; Platt, S.; Scowcroft, E.; et al. Mental health and well-being during the COVID-19 pandemic: Longitudinal analyses of adults in the UK COVID-19 Mental Health & wellbeing study. *Br. J. Psychiatry* **2020**, *218*, 326–333. [CrossRef]
18. Coppola, I.; Rania, N.; Parisi, R.; Lagomarsino, F. Spiritual well-being and mental health during the COVID-19 pandemic in Italy. *Front. Psychiatry* **2021**, *12*, 626944. [CrossRef]
19. Özmen, S.; Özkan, O.; Özer, Ö.; Yanardağ, M.Z. Investigation of COVID-19 fear, well-being and life satisfaction in Turkish society. *Soc. Work Public Health* **2021**, *36*, 164–177. [CrossRef]
20. Tomaz, S.A.; Coffee, P.; Ryde, G.C.; Swales, B.; Neely, K.C.; Connelly, J.; Kirkland, A.; McCabe, L.; Watchman, K.; Andreis, F.; et al. Loneliness, wellbeing, and social activity in Scottish older adults resulting from social distancing during the COVID-19 pandemic. *Int. J. Environ. Res. Public Health* **2021**, *18*, 4517. [CrossRef]

21. Fingerman, K.; Ng, Y.; Zhang, S.; Britt, K.; Colera, G.; Birditt, K.; Charles, S. Living alone during COVID-19: Social contact and emotional well-being among older adults. *J. Gerontol. Ser. B* **2020**, *76*, e116–e121. [CrossRef] [PubMed]
22. Egawa, S.; Jibiki, Y.; Sasaki, D.; Ono, Y.; Nakamura, Y.; Suda, T.; Sasaki, H. The correlation between life expectancy and disaster risk. *J. Disaster. Res.* **2018**, *13*, 1049–1061. [CrossRef]
23. Moriyama, K.; Sasaki, D.; Ono, Y. Comparison of global databases for disaster loss and damage data. *J. Disaster. Res.* **2018**, *13*, 1007–1014. [CrossRef]
24. Sasaki, D.; Moriyama, K.; Ono, Y. Hidden common factors in disaster loss statistics: A case study analyzing the data of Nepal. *J. Disaster. Res.* **2018**, *13*, 1032–1038. [CrossRef]
25. Sasaki, D.; Ono, Y. Overview of the special issue on the development of disaster statistics. *J. Disaster. Res.* **2018**, *13*, 1002–1006. [CrossRef]
26. Sakamoto, M.; Sasaki, D.; Ono, Y.; Makino, Y.; Kodama, E.N. Implementation of evacuation measures during natural disasters under conditions of the novel coronavirus (COVID-19) pandemic based on a review of previous responses to complex disasters in Japan. *Prog. Disaster. Sci.* **2020**, *8*, 100127. [CrossRef]
27. Sasaki, D.; Moriyama, K.; Ono, Y. Main features of the existing literature concerning disaster statistics. *Int. J. Disaster. Risk Reduc.* **2020**, *43*, 101382. [CrossRef]
28. Suppasri, A.; Kitamura, M.; Tsukuda, H.; Boret, S.P.; Pescaroli, G.; Onoda, Y.; Imamura, F.; Alexander, D.; Leelawat, N.; Syamsidik. Perceptions of the COVID-19 pandemic in Japan with respect to cultural, information, disaster and social issues. *Prog. Disaster. Sci.* **2021**, *10*, 100158. [CrossRef]
29. Pescaroli, G.; Galbusera, L.; Cardarilli, M.; Giannopoulos, G.; Alexander, D. Linking healthcare and societal resilience during the Covid-19 pandemic. *Saf. Sci.* **2021**, *140*, 105291. [CrossRef]
30. DeMaris, A. A tutorial in logistic regression. *J. Marriage Fam.* **1995**, *57*, 956. [CrossRef]
31. Cabinet Office (CAO). *Activities Regarding Well-Being*; CAO, 2022. Available online: https://www5.cao.go.jp/keizai2/wellbeing/index.html (accessed on 4 September 2022).
32. Hotelling, H. Analysis of a complex of statistical variables into principal components. *J. Educ. Psychol.* **1933**, *24*, 417–441. [CrossRef]
33. Jolliffe, I.T.; Cadima, J. Principal component analysis: A review and recent developments. *Philos. Trans. A Math. Phys. Eng. Sci.* **2016**, *374*, 20150202. [CrossRef] [PubMed]

International Journal of
Environmental Research
and Public Health

MDPI

Article

Exploring People's Perception of COVID-19 Risk: A Case Study of Greater Jakarta, Indonesia

Dicky C. Pelupessy [1], Yasuhito Jibiki [2,3,*] and Daisuke Sasaki [2]

[1] Faculty of Psychology, Universitas Indonesia, Kampus UI Depok, Depok City 16424, West Java, Indonesia
[2] International Research Institute of Disaster Science, Tohoku University, 468-1 Aoba Aramaki, Aoba, Sendai 980-8572, Japan
[3] Center for Integrated Disaster Information Research, The University of Tokyo, 7-3-1 Hongo, Bunkyo-ku, Tokyo 113-0033, Japan
* Correspondence: yasuhito.jibiki.e3@tohoku.ac.jp

Abstract: This study aims to understand people's perceptions of COVID-19 risk in Greater Jakarta, Indonesia. In response to the COVID-19 pandemic, the Indonesian government enacted a health protocol campaign and highlighted the community as an important unit of protocol compliance. We hypothesized that people's perception of the likelihood of being infected with COVID-19 is associated with health protocol compliance at the community level and their perception of community resilience. As the number of infected persons drastically increased, the "family cluster" also became a significant issue in the pandemic response, especially in Indonesia. In this study, we explored both community and family aspects that influence people's perceptions. We conducted an online survey in March 2021 with 370 respondents residing in the Greater Jakarta area. The respondents were classified into four age groups (20s, 30s, 40s, and 50-and-over), with gender-balanced samples allocated to each group. We used a questionnaire to measure the perception of COVID-19 risk along with the Conjoint Community Resiliency Assessment Measure (CCRAM). Multiple regression analysis revealed that family factors have a much larger influence on the individual perception of the likelihood of contracting COVID-19 than community factors. The results suggest that the link between family-level efforts against COVID-19 and individual-level perceptions cannot be separated in response to the pandemic.

Keywords: perception; COVID-19; family; community; Jakarta; Indonesia

Citation: Pelupessy, D.C.; Jibiki, Y.; Sasaki, D. Exploring People's Perception of COVID-19 Risk: A Case Study of Greater Jakarta, Indonesia. *Int. J. Environ. Res. Public Health* **2023**, *20*, 336. https://doi.org/10.3390/ijerph20010336

Academic Editor: Paul B. Tchounwou

Received: 22 November 2022
Revised: 19 December 2022
Accepted: 20 December 2022
Published: 26 December 2022

1. Introduction

In the wake of the COVID-19 pandemic, certain countries have opted to focus their responses on the community level [1]. In Indonesia, the national government enacted the Large-Scale Social Restriction Policy (*Pembatasan Sosial Berskala Besar*: PSBB) in April 2020, followed by the Enforcement of Restrictions on Community Activities (*Pemberlakuan Pembatasan Kegiatan Masyarakat*: PPKM) in January 2021 [2]. These policies highlight "communities" as important units for health protocol implementation in Indonesia. As the COVID-19 situation was updated, the government enacted the Emergency PPKM (PPKM Darurat), Micro-Based PPKM (known as PPKM Mikro), and PPKM levels. These new regulations do not substantially alter the importance of communities in health protocol compliance, but rely heavily on the capacity of communities in response to the pandemic. Such a social background reminds us of the need to consider the concept of "community resilience." As communities consist of individuals, communities and individuals cannot be understood separately. We assume that individuals perceive COVID-19 through their communities and resiliency.

In addition to the importance of the community-level perspective, during the second wave of COVID-19 infection in Indonesia (May–September 2021), "self-isolation" in each household was a critical issue in health protocol implementation in Indonesia. The number of infected individuals increased drastically during this period. Each family faced great difficulty in managing health protocols in the household in the event that one of the family members tested positive because the hospitals and other medical facilities were fully occupied.

In the Indonesian context, community-level efforts and family-level responses may influence individual perceptions, which, however, need to be verified. Furthermore, recent research has reported that some types of individual demographics may be related to perceptions of COVID-19. Extant literature, such as a study from Peru and China [3] and another from Italy [4], noted that risk perception is the basis for examining the decision-making and behavior of each individual. Bavel et al. [5] indicated that it is important to better understand the risk perception of COVID-19 under pandemic conditions that nobody had experienced before.

Thus, using statistical analysis, we aimed to examine whether these three factors (demographic, family, and community) are relevant to people's perceptions of COVID-19. As the COVID-19 situation has evolved, the issue of family cluster or household transmission has emerged in many countries (e.g., in Italy [6], India [7], the UK [8], Madagascar [9], Switzerland [10]). However, these studies indicate that the effects caused by family clusters and household transmission have not yet been well investigated, and further research is required. Moreover, due to the boundaries of academic disciplines, findings of community-related studies and those of family-related studies have not been integrated. In Greater Jakarta, people face both family/household issues and communities at the same time. As the demographic variables are the basis of individual characteristics and a large amount of research, it is reasonable to include them to examine family/household and community factors. Therefore, our aim to study these three factors is justified. For that reason, we targeted people who lived in Greater Jakarta, Indonesia, which has been considered the epicenter of COVID-19 in Indonesia since the pandemic started in 2020 [11].

2. Literature Review

During the COVID-19 pandemic, many researchers have examined how people perceive this phenomenon as unprecedented in their lifetime [3,12]. While some studies have focused on psychological factors at the individual level, others have considered the family and community perspectives. Although individual demographics also have an effect on people's perceptions, generalized and consistent tendencies have not yet been observed. Furthermore, the findings were based on results from different countries using different methodologies. We note such limitations but review earlier research that investigated people's perceptions to build our hypotheses based on the synthesized empirical findings.

2.1. Individual Risk Perception and COVID-19

Researchers have used a variety of words, such as worry, anxiety, fear, stress, attitude, threat, and perception. A simple definition, such as "Risk perceptions are interpretations of the world" [13] (p. 3), would be replicable and provide a basis for our study.

According to one literature review [3], many studies have demonstrated a link between risk perception and COVID-19 prevention [14–16]. However, Monge-Rodríguez et al. [3] and Yıldırım et al. [16] showed moderate effects of risk perception on protective behavior compared with the results of de Bruin and Bennett [14]. Rubaltelli et al. [17] reported a weak positive correlation between anxiety and risk perception. Existing literature has noted that risk perception is the basis for examining the decision-making and behavior of each individual [3,4,18–22]. Lohiniva et al. [19] and Shahin and Hussien [21] argued that understanding public risk perception is critical for risk perception. As Bavel et al. [5] indicated, it is important to better understand risk perception during the pandemic that nobody had experienced before.

2.2. The "Family Cluster" in Indonesia

To the best of our knowledge, family-level infections have not been studied extensively in some countries. However, in Indonesia, the so-called "family cluster" problem was considered an urgent issue during the pandemic. *The Jakarta Post*, one of the most trustworthy media outlets in Indonesia, reported that the Jakarta Special Province Governor stated that the family cluster was one of the major causes of the surge in COVID-19 infection after long holidays in October and November 2020 [23]. Supriyati et al. also found that COVID-19 transmission mostly occurs in families in Indonesia [24]. Furthermore, Nasrudin et al. noted that individual anxiety caused by COVID-19 influences family-level compliance with health protocols [25].

2.3. Role of the Community

Some studies have examined collectivistic (rather than individualistic) aspects of COVID-19 responses. There is a theoretical assumption that risk is perceived through the lens of group membership [26]. According to Stevenson et al. [27], the "group process" [28,29] and "group-level nature" [30,31] are crucial perspectives for understanding the current pandemic situation, and community identity plays a pivotal role in effective behavioral responses to COVID-19 [27]. Moreover, one study [32] found that social support eased the negative mental health impact of COVID-19. These findings suggest that local settings, including the communities to which people belong, are important for examining how people perceive COVID-19.

2.4. Individual Risk Perception and Demographic Factors
2.4.1. Age

Some studies indicate that age affects the relationship between individual risk perceptions and age. Chan et al. demonstrated that the elderly are less likely to worry about COVID-19 [33]. Similarly, Megatsari et al. found that older individuals experienced less anxiety [34]. Savadori and Lauriola found that younger participants were less worried about getting infected with the coronavirus [4], whereas Adiyoso and Wilopo found that younger individuals showed a stronger relationship with risk perception [35]. Harapan et al. demonstrated that age was a significant predictor of the risk of infection and the study participants aged between 21 and 30 years had the highest perceived risk [36]. Bernabe-Valero et al. showed that younger people exhibit higher stress levels [12]. Based on our review, it seems difficult to determine whether older or younger individuals show a consistent tendency, even though age itself matters. However, other studies have not clearly indicated that age is relevant to individual risk perception [3,37,38].

2.4.2. Gender

Only a few studies indicate that gender is unrelated to risk perception. Two studies analyzed the perception of the risk of contracting coronavirus, but their analysis suggested that gender was not a significant predictor of risk perception [36,37]. In contrast, most studies have reported that women show more negative responses to risk perception [3,4,12,15,16,34,39–43].

2.4.3. Education

Three studies [12,34,37] demonstrated a similar tendency in the effect of educational level on risk perception, such that those with lower educational levels showed higher levels of negative feelings about COVID-19. However, two studies indicated that education was not significant for risk perception [4,36].

2.4.4. Marital Status

Harapan et al. revealed that the perceived risk of infection was lower among married than unmarried respondents [36]. Similarly, Bernabe-Valero et al. showed that individuals who were single exhibited higher stress levels [12].

3. Hypotheses

Based on the empirical findings summarized in the literature reviewed, demographic, family, and community-level factors are relevant to people's perceptions of COVID-19. In this study, we define people's perception of the likelihood of contracting COVID-19 at the individual level as the dependent variable (see Table 1 and Figure 1).

Table 1. Types of the dependent and independent variables.

Variable Name	Types of Variables
Dependent variable	
Perception of the likelihood of contracting COVID-19 at the individual level	Five-point Likert scale, 1 = very unlikely, 5 = very likely
Independent variable	
Perception of the likelihood that a family member would contract COVID-19	Five-point Likert scale, 1 = very unlikely, 5 = very likely
Perception of community compliance with health protocols	Five-point Likert scale, 1 = very unlikely, 5 = very likely
CCRAM [*1]	Continuous variable
Age	Continuous variable
Gender	Binary, 1 = male, 0 = female
Marital status	Binary, 1 = married, 0 = not married
Education [*2]	Quantitative variable, 1 = the lowest (Elementary), 6 = the highest (Post-graduate)
Household income	Quantitative variable, 1 = the lowest (< IDR 1,500,000), 7 = the highest (>IDR 10,000,000)

[*1] CCRAM: Conjoint Community Resiliency Assessment Measure. [*2] We interpreted the education variable as the degree of educational level (relatively higher to lower), and the level was dealt with as a quantitative variable. In Table 2, the elementary and the junior high school were merged due to very small-sized samples. Five (1.4%) and eleven (3.0%) respondents had elementary and junior high education, respectively.

Table 2. Demographic statistics summary ($n = 370$).

Variable	Percent	Variable	Percent
Gender		**Marital status**	
Male ($n = 185$)	50.0	Married ($n = 239$)	64.6
Female ($n = 185$)	50.0	Unmarried ($n = 131$)	35.4
Age			
20–29 ($n = 100$)	27.0		
30–39 ($n = 100$)	27.0	**Household Income**	
40–49 ($n = 100$)	27.0	<IDR 1,500,000 ($n = 20$)	5.4
>50 ($n = 70$)	18.9	IDR 1,500,000–2,499,999 ($n = 21$)	5.7
Education		IDR 2,500,000–3,499,999 ($n = 28$)	7.6
Below High School level ($n = 16$)	4.4	IDR 3,500,000–4,999,999 ($n = 42$)	11.4
Senior High ($n = 101$)	27.3	IDR 5,000,000–7,499,999 ($n = 74$)	20.0
Diploma 1–4 ($n = 52$)	14.1	IDR 7,500,000–9,999,999 ($n = 60$)	16.2
University (Bachelor's) ($n = 176$)	47.6	>IDR 10,000,000 ($n = 125$)	33.8
Post-graduate (Master's and Ph.D.) ($n = 25$)	6.7		

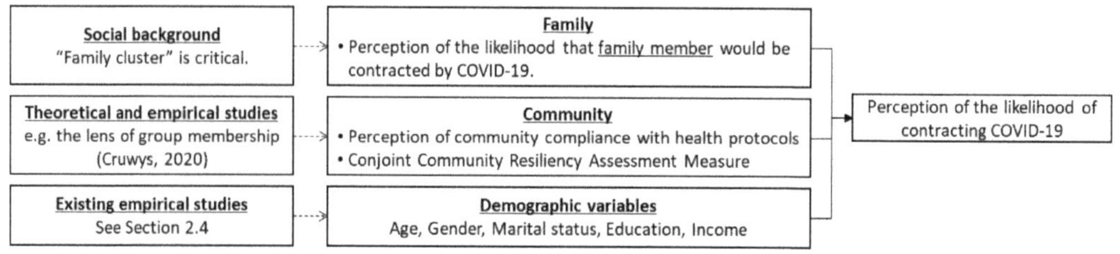

Figure 1. Hypothetical framework of our analysis.

The first independent variable was the family's role. We used the variable of perception of the likelihood that a family member would contract COVID-19. We assume that people perceive their own likelihood of contracting COVID-19 to be greater when they perceive that of their family members to be greater.

The second independent variable was the community. We used two types of variables: perception of community compliance with health protocols and joint community resilience assessment measure (CCRAM). While the former directly reflects how people perceive a health response, the latter assesses more generalized crisis contexts. The details of the CCRAM are explained in the next section.

We set some types of demographics as the first independent variables. As noted earlier, it is difficult to determine whether older or younger individuals show a consistent tendency, even though age itself matters. Female respondents reported a more negative response to perceived risk. As the effects of education and marital status have not yet been determined, we sought to verify whether they were significant.

We used multiple regression analysis to control for the effects of demographic variables.

To clarify our terminology, we define communities as *kelurahan*s (towns) that could be treated as a unit of locality that is smaller than the city but larger than the neighborhood. Hence, in some ways, it can be considered a town.

4. Methods

Although the survey implementation, data, and CCRAM measurements were the same as those of Pelupessy et al. [44], the research purpose in this article is original, and the hypothetical idea has no duplication with Pelupessy et al. [44].

4.1. COVID-19 in Greater Jakarta

Greater Jakarta is one of the largest urban areas in the world [45]. As the COVID-19 pandemic continues, Greater Jakarta has faced an increasing number of infections [11].

Communities are among the key actors in Greater Jakarta's response to the pandemic. Pangaribuan and Munandar argued that governmental organizations in Greater Jakarta faced significant difficulty in responding to the community's non-compliance with health protocols under PSBB implementation [46]. Yakhamid and Zaqi indicated that the Jakarta Special Province government achieved PPKM Darurat due to the synergy between community compliance in implementing health protocols and the achievements of the vaccination program [47].

Family clustering in Greater Jakarta is also a critical issue. Handayani et al. found that the stricter health protocol in Greater Jakarta was rarely applied, especially by those who lived in the same house, resulting in family members infected with the rest of their family from workplace cluster cases [48].

Both community and family aspects are observed in Greater Jakarta; therefore, we think that it is reasonable to test our hypothesis using Greater Jakarta as a case study.

4.2. On-Line Survey Implementation and Our Respondents

We conducted an online survey in March 2021 with 370 respondents residing in the Greater Jakarta area. The period in which we implemented the online survey was immediately after the Indonesian people experienced the first wave of COVID-19 (around February 2021). As the survey was conducted during the period between the first wave of COVID-19 in Indonesia and the second wave (May–September 2021), the social situation around Greater Jakarta was relatively calm, and the respondents were able to participate in the survey.

Participants (n = 370) were adults aged 18–59 years (M = 37.7, SD = 10.5 years). They were recruited using quota sampling while maintaining an equal proportion of male and female participants to account for gendered perceptions. Respondents were classified into four age groups (20s, 30s, 40s, and 50-and-over). In the 50-and-over age group,

35 respondents were allocated to the male and female groups. All participants completed online and web-based questionnaires.

To the best of our knowledge, the surveys of the existing literature were largely conducted online, except for [33,37]. Considering the current pandemic situation, online surveys are considered the most feasible. It should be noted that the samples could be biased, and such a methodological challenge should be a major concern in future research.

To identify the sample size, we adopted the following steps: Based on the fact that the population size of Greater Jakarta is sufficiently large, we considered the required sample size in the interval estimation of the population proportion. We set the margin of error to 5%, the confidence level to 95%, and the population proportion to 0.5. We identified 385 samples as target values. Considering that the older generation seems less familiar with using online survey platforms, we assumed that fewer survey participants were in the older generation than the younger generation. Practically, we secured 70 samples in the age bracket of over 50s, while we received 100 samples in their 20s, 30s, and 40s. Finally, we collected 370 samples in total using quota sampling.

A summary of the demographic information is presented in Table 2. Regarding the primary occupation of the respondents, 21.6% were in the retail sector, the largest sector in the sample. Housewives and those working in the manufacturing sector accounted for 10.8% of the respondents. Others were employed in construction, transportation, education, government, and other sectors.

4.3. Measures

To analyze perceptions of COVID-19, we administered the CCRAM. As reviewed in Section 2.4, some studies suggest the importance of collectivistic (rather than individualistic) thinking in the case of COVID-19. Although there are several instruments to assess how people perceive matters through a collectivistic lens, we used the CCRAM in this study. The COVID-19 pandemic, caused by community transmission, highlights the significance of the community. The CCRAM is not only a community resilience measurement tool but also one fit for use in both daily and emergency/crisis situations in terms of community-related parameters [49]. Several recent studies have measured community resilience related to and during the COVID-19 pandemic using the CCRAM [50–52].

The CCRAM is a measure of community resilience. Leykin et al. [53] developed it to establish an integrated multidimensional instrument to assess community resilience and later [54] tested whether their measurement was robust as a psychological indicator showing peoples' evaluations of their communities' capacity to respond to emergencies in diverse contexts. It assesses the strength of five important dimensions of community function (leadership, collective efficacy, preparedness, place attachment, and social trust) that can be used to profile and predict community resilience [53–55].

The CCRAM assesses leadership factors in community resilience through six items representing general faith in decision-makers, specific faith in local leaders, perception of fairness in the way local authority provides services, and functioning of the community [53]. Collective efficacy was evaluated using five items representing collective efficacy, support, involvement in the community, and mutual assistance [53]. The collective efficacy items echo arguments that they comprise a composite of mutual trust and a shared willingness to work for the common good of a neighborhood [56,57]. The preparedness factor comprises four items representing family and community acquaintances with emergency situations and a view of the town's preparedness for emergency situations [53]. The place attachment factor is composed of four items representing emotional attachment to the community, sense of belonging, pride in the community, and ideological identification with the community [53]. This perspective regarding place attachment aligns well with the assertions of Manzo and Perkins, who, based on a cross-disciplinary literature review, pointed out that individuals' feelings toward their place are connected to community-level perceptions [58]. Finally, the social trust factor is composed of two items representing trust and the quality of relationships between members in the community [53].

In this study, we used a 21-item self-report CCRAM questionnaire. Each item was measured on a five-point Likert scale, and the answers provided scores. The total CCRAM score was calculated as the sum of all the scores for each item. These five dimensions were generated in accordance with Leykin et al.'s categorization [53]. For example, "leadership" consists of six items, and its value is calculated as the sum of the scores of these six items. Details of the 21 items of the CCRAM, the 5 dimensions, and their basic statistics are presented in Appendices A and B.

4.4. Models

Multiple regression analysis was conducted to verify our hypotheses. The five ways (or written "models" hereinafter) of the multiple regression analysis are presented in Table 3. In the multiple regression analysis, the dependent variable "Perception of the likelihood of contracting COVID-19" was consistently used in all analyses.

The independent variables "Demographic" (age, gender, marital status, education, and household income), "Family" (perception of the likelihood that family member would contract COVID-19), and "Community" (perception of community compliance with health protocols) were used in all analyses.

Only CCRAM was exceptional: In Models 1 and 2, we used the total CCRAM score. Models 3 and 4 tested the five dimensions of the CCRAM (leadership, collective efficacy, preparedness, place attachment, and social trust). In Model 5, we input all 21 CCRAM items.

For multiple regression analysis, the forced imputation method was adopted in Models 1 and 3. A stepwise method was used for Models 2, 4, and 5.

While our analysis provides baseline data based on a survey carried out in March 2021, continuing research is necessary to understand people's perceptions more deeply, considering the evolving COVID-19 situation.

Table 3. Results of the multiple regression analyses. Note: Dependent variable for all models was "perception of the likelihood of contracting COVID-19".

Model 1. Forced imputation method				
	Standardized Coefficient (β)	t-value	p-value	VIF
Constant		0.846	0.398	
Age	-0.051	-1.705	0.089	1.117
Gender	0.026	0.913	0.362	1.019
Marital Status	0.069	2.203	0.028	1.205
Education	0.050	1.528	0.127	1.310
Household Income	-0.036	-1.049	0.295	1.483
Family would be infected	0.838	28.404	0.000	1.068
Perception of health protocol compliance at the community level.	-0.029	-0.933	0.351	1.170
CCRAM Total Score	0.038	1.233	0.219	1.194
R	0.840			
adj R^2	0.699			
n	370			

Model 2. Stepwise method				
	Standardized Coefficient (β)	t-value	p-value	VIF
Constant		4.307	0.000	
Family would be infected.	0.835	29.069	0.000	1.000
R	0.835			
adj R^2	0.696			
n	370			

Table 3. *Cont.*

Model 3. Forced imputation method				
	Standardized Coefficient (β)	*t*-value	*p*-value	VIF
Constant		0.764	0.446	
Age	−0.054	−1.808	0.072	1.121
Gender	0.019	0.642	0.521	1.035
Marital Status	0.068	2.176	0.030	1.208
Education	0.044	1.329	0.185	1.335
Household Income	−0.029	−0.831	0.406	1.495
Family would be infected.	0.838	28.491	0.000	1.075
Perception of health protocol compliance at the community level	−0.024	−0.791	0.429	1.191
CCRAM_1. Leadership	−0.103	−1.729	0.085	4.426
CCRAM_2. Collective Efficacy	0.125	2.043	0.042	4.687
CCRAM_3. Preparedness	0.055	1.001	0.318	3.795
CCRAM_4. Place Attachment	0.026	0.615	0.539	2.236
CCRAM_5. Social Trust	−0.066	−1.390	0.165	2.801
R	0.844			
adj R^2	0.703			
n	370			

Model 4. Stepwise method				
	Standardized Coefficient (β)	*t*-value	*p*-value	VIF
Constant		0.524	0.601	
Family would be infected.	0.839	29.248	0.000	1.006
Collective Efficacy	0.057	1.972	0.049	1.006
R	0.837			
adj R^2	0.698			
n	370			

Model 5. Stepwise method				
	Standardized Coefficient (β)	*t*-value	*p*-value	VIF
Constant		1.849	0.065	
Family would be infected.	0.830	29.353	0.000	1.005
I can depend on people in my town to come to my assistance in a crisis. [Collective Efficacy]	0.119	3.867	0.000	1.195
The relations between the various groups in my town are good. [Social Trust] [Social Trust]	−0.065	−2.116	0.035	1.199
R	0.842			
adj R^2	0.706			
n	370			

5. Results

Multiple regression analysis was conducted to verify whether the three hypothetical factors (demographic, family, and community levels) were relevant to people's perceptions of COVID-19 (see Table 3).

Among the five models, Model 5 was identified as the best fitting- model with reference to its highest value of the adjusted R^2, which was 0.706. As the value of the adjusted R^2 in all models was approximately 0.7, our models reasonably explained the dependent variable. Regarding the effects of demographic variables, only "marital status" was significant for the dependent variable, such that married individuals perceived themselves as more likely to contract COVID-19. This result is inconsistent with earlier findings [12,36]. However, "marital status" was significant in Models 1 and 3, but not in Model 5, the best-fitting model. Although some researchers have reported that age, sex, education, and income may be relevant to risk perception, our analysis did not reveal any significant associations.

The significance of the family level aspect was prominent in multiple regression analysis. In all models, the value of the standardized coefficient (β) for the perception of the likelihood that family members would contract COVID-19 was the highest. These results indicate that those who think that their family members will be infected with COVID-19 perceive that they are also likely to be infected, corroborating Nasrudin et al. [25]. As indicated in the previous paragraph, "marital status" was significant. This result indicates that married individuals are more likely to contract COVID-19, reflecting that the family aspect is an important factor.

For the community-level factor, the independent variable "perception of health protocol compliance at the community level" did not demonstrate significance in any model. By contrast, the collective efficacy dimension of the CCRAM and an item of the social trust dimension were significant. In Models 3, 4, and 5, the collective efficacy dimension was significant. However, the value of the standardized coefficient (β) was positive, indicating that those who maintained higher collective efficacy tended to think that they were more likely to contract COVID-19. However, these results are contradictory. In our interpretation, due to the seriousness of the pandemic, even though someone maintains collective efficacy, such a degree of perception cannot overcome its influence. The social trust dimension of CCRAM was significant only in Model 5, indicating that those who put greater trust in various groups perceived that they were less likely to be infected by COVID-19. This result implies that higher social trust can reduce negative perceptions of COVID-19.

6. Discussion and Conclusions

This study aimed to examine people's perceptions of COVID-19. Based on our multiple regression analysis, the "family" factor seems highly relevant to people's perceptions. The community-level factor showed significance, but compared with the family factor, it did not contribute much to people's perceptions. However, as noted earlier, the findings should be verified using case studies with different data for greater generalizability.

In future research, we will highlight the definition of communities and characteristics of the data. First, in the present study, we defined "communities" as "*kelurahan*" (towns). However, if we adopt a different definition of communities, such as neighborhood associations (locally known as *Rukun Tetangga* and *Rukun Warga*) or traditional units *kampung*, we might obtain different results. Second, data characteristics should be carefully considered. The period of survey implementation may have affected the data and results. If we carried out another survey in a relatively calm situation or if the capacities of communities were improved gradually, the family factor would not be much stronger. In terms of survey implementation, our data were collected through an online survey; however, this method has limitations. As older people may be less familiar with the internet, they may hesitate to participate online. Even though we controlled for age in the multiple regression analysis, both the original data and the potential respondents for online surveys were biased to some extent. Future research needs to consider the possibility of different sampling techniques and explore whether qualitative data analysis can be a useful method to supplement the limitations of quantitative methods.

Based on our analysis, we adduce two practical implications: the identification of "communities" and "family" affairs. For more effective community-level interventions, it is important to consider which types of "communities" should be targeted. As described above, town and neighborhood associations can be distinguished. In the best-fitting model of our multiple regression analysis, the independent variable "The relations between the various groups in my town are good" was significant. This result implies that there are subgroups within a town, and, thus, it would appear meaningful to specifically identify how these subgroups contribute to the response to COVID-19. We hardly found that infected families were isolated without assistance, even though they faced difficulties. Communities and their subgroups may have filled this gap to achieve seamless assistance. Kuno [59], based on his observation in the central area of Jakarta, reported a "neighborhood lockdown" guided by neighborhood associations. Hosobuchi [60], also based on her study in Jakarta during the COVID-19 pandemic, introduced activities at neighborhood associations for juvenile delinquency prevention and provided religious education programs. This empirical evidence suggests that communities and their subgroups may have proactively contributed to the management of local needs, and more detailed studies are necessary in future research.

The results of our multiple regression analysis indicate that "I (person A) would be more likely to contract COVID-19 if my family member (person B) contracted COVID-19". Person B's perception of the likelihood of contracting COVID-19 was highly dependent on the health condition of person A as B's family member. This interpretation suggests a chain reaction in each family and highlights the importance of mutuality among the family members.

As broader implications of the findings for similar contexts of other pandemics, not just the COVID-19 pandemic, we consider two aspects: the necessity of long-term studies and more standardized surveys. First, the necessity of longer-term studies reflects the longer and evolving process of COVID-19 expansion. Approximately three years have passed since the COVID-19 pandemic began; we have been experiencing some changes in its expansion stages. In the early stage, social lockdown and public health protocol compliance at the community level were our main concerns. Then, vaccination became a critical issue. However, not many people were better aware of family cluster/household transmission in the beginning. Such dynamics of the current pandemic, noting that the virus mutation continues, may have complicated effects on people's perceptions. Long-term studies are necessary to better understand the evolving nature of the disease. Second, more standardized surveys should be conducted in the future. Facing unprecedented social deficiencies, many surveys, including ours, have been conducted worldwide using different methodologies without common definitions of critical terminologies. Although each study has the right to determine its own methods and concepts, standardized surveys aiming at more harmonized efforts to monitor people's perceptions should be a challenge in future research.

Author Contributions: Conceptualization, D.C.P.; methodology, D.C.P.; validation, D.C.P., Y.J. and D.S.; formal analysis, Y.J.; investigation, D.C.P.; resources, D.S.; data curation, Y.J.; writing—original draft preparation, Y.J.; writing—review and editing, D.C.P. and D.S.; visualization, Y.J.; supervision, D.C.P.; project administration, D.S.; funding acquisition, D.S. All authors have read and agreed to the published version of the manuscript.

Funding: This work was partially supported by the Collaborative Research Project of the International Research Institute of Disaster Science (IRIDeS), Tohoku University, and partially supported by JSPS KAKENHI Grant Number 21H03680.

Institutional Review Board Statement: The study was conducted according to the guidelines of the Declaration of Helsinki and approved by the Institutional Review Board of International Research Institute of Disaster Science, Tohoku University (protocol code 2020–051, 18 February 2021).

Informed Consent Statement: Informed consent was obtained from all subjects involved in the study.

Data Availability Statement: The data presented in this study are available on request from the corresponding author.

Conflicts of Interest: The authors declare no conflict of interest.

Appendix A. Basic Statistics of the Five Dimensions of CCRAM (*n* = 370)

	Min	Max	Mean	*SD*
CCRAM Total Sum	23	105	74.797	12.511
CCRAM: Leadership	6	30	21.51	4.17
CCRAM: Collective Efficacy	5	25	18.18	3.09
CCRAM: Preparedness	4	20	13.96	2.72
CCRAM: Place Attachment	4	20	13.61	2.64
CCRAM: Social Trust	2	10	7.54	1.39

Appendix B. Basic Statistics of the 21 Items of CCRAM (*n* = 370)

	Mean	*SD*
1. The municipal authority (regional council) of my town functions well [1: Leadership]	3.66	0.83
2. There is mutual assistance and concern for others in my town [2: Collective Efficacy]	3.85	0.74
3. My town is organized for emergency situations [3: Preparedness]	3.40	0.85
4. I am proud to tell others where I live [4: Place Attachment]	3.61	0.87
5. The relations between the various groups in my town are good [5: Social Trust]	3.78	0.81
6. I have faith in the decision-makers in the municipal authority (regional council) [1: Leadership]	3.62	0.83
7. I can depend on people in my town to come to my assistance in a crisis [2: Collective Efficacy]	3.32	0.89
8. The residents of my town are acquainted with their role in an emergency situation [3: Preparedness]	3.67	0.78
9. I feel a sense of belonging to my town [4: Place Attachment]	3.61	0.79
10. There is trust among the residents of my town [5: Social Trust]	3.76	0.73
11. In my town, appropriate attention is given to the needs of children [1: Leadership]	3.50	0.87
12. There are people in my town who can assist in coping with an emergency [2: Collective Efficacy]	3.63	0.82
13. In my town, there are sufficient public protection facilities (such as shelters) [3: Preparedness]	3.34	0.89
14. I remain in this town for ideological reasons [4: Place Attachment]	3.11	0.92
15. I have faith in the ability of the elected/nominated head of my town to lead the transit from routine to emergency management of the town [1: Leadership]	3.58	0.82
16. I believe in the ability of my community to overcome an emergency situation [2: Collective Efficacy]	3.72	0.74
17. My family and I are acquainted with the emergency system of my town (to be activated in times of emergency) [3: Preparedness]	3.55	0.84
18. I would be sorry to leave the town where I live [4: Place Attachment]	3.29	0.91
19. The municipal authority (regional council) provides its services in fairness [1: Leadership]	3.52	0.90
20. The residents of my town are greatly involved in what is happening in the community [2: Collective Efficacy]	3.65	0.80
21. The residents of my town will continue to receive municipal services during an emergency situation [1: Leadership]	3.62	0.78

Note: "[]" at the end of each item refers to the names of the five dimensions. For example, "leadership" consists of six items, one of which is Item 1.

References

1. World Health Organization. Risk Communication and Community Engagement Readiness and Response to Coronavirus Disease (COVID-19): Interim Guidance. Available online: https://apps.who.int/iris/bitstream/handle/10665/331513/WHO-2019-nCoV-RCCE-2020.2-eng.pdf (accessed on 6 June 2021).
2. Muhyiddin, M.; Nugroho, H. A Year of COVID-19: A long road to recovery and acceleration of Indonesia's development. *J. Perenc. Pembang.* **2021**, *5*, 1–19. [CrossRef]

3. Monge-Rodríguez, F.S.; Jiang, H.; Zhang, L.; Alvarado-Yepez, A.; Cardona-Rivero, A.; Huaman-Chulluncuy, E.; Torres-Mejía, A. Psychological factors affecting risk perception of COVID-19: Evidence from Peru and China. *Int. J. Environ. Res. Public Health* **2021**, *18*, 6513. [CrossRef] [PubMed]

4. Savadori, L.; Lauriola, M. Risk perception and protective behaviors during the rise of the COVID-19 outbreak in Italy. *Front. Psychol.* **2021**, *11*, 577331. [CrossRef] [PubMed]

5. Bavel, J.J.V.; Baicker, K.; Boggio, P.S. Using social and behavioural science to support COVID-19 pandemic response. *Nat. Hum. Behav.* **2020**, *4*, 460–471. [CrossRef]

6. Costenaro, P.; Di Chiara, C.; Boscolo, V.; Barbieri, A.; Tomasello, A.; Cantarutti, A.; Cozzani, S.; Liberati, C.; Oletto, S.; Giaquinto, C.; et al. Perceived Psychological Impact on Children and Parents of Experiencing COVID-19 Infection in One or More Family Members. *Children* **2022**, *9*, 1370. [CrossRef]

7. Ranshing, S.; Lavania, M.; Potdar, V.; Patwardhan, S.; Prayag, P.S.; Jog, S.; Kelkar, D.; Sawant, P.; Shinde, M.; Chavan, N. Transmission of COVID-19 infection within a family cluster in Pune, India. *Indian J. Med. Res.* **2021**, *153*, 555.

8. Hall, J.A.; Harris, R.J.; Zaidi, A.; Woodhall, S.C.; Dabrera, G.; Dunbar, J.K. HOSTED—England's Household Transmission Evaluation Dataset: Preliminary findings from a novel passive surveillance system of COVID-19. *Int. J. Epidemiol.* **2021**, *50*, 743–752. [CrossRef]

9. Ratovoson, R.; Razafimahatratra, R.; Randriamanantsoa, L.; Raberahona, M.; Rabarison, H.J.; Rahaingovahoaka, F.N.; Andriamasy, E.H.; Herindrainy, P.; Razanajatovo, N.; Andriamandimby, S.F.; et al. Household transmission of COVID-19 among the earliest cases in Antananarivo, Madagascar. *Influenza Other Respir. Viruses* **2022**, *16*, 48–55. [CrossRef]

10. Siegrist, M.; Luchsinger, L.; Bearth, A. The impact of trust and risk perception on the acceptance of measures to reduce COVID-19 cases. *Risk Anal.* **2021**, *41*, 787–800. [CrossRef]

11. Jakarta Heath Department. Data Monitoring on COVID-19 Cases. Available online: https://corona.jakarta.go.id/en (accessed on 26 September 2022).

12. Bernabe-Valero, G.; Melero-Fuentes, D.; De Lima Argimon, I.I.; Gerbino, M. Individual differences facing the COVID-19 pandemic: The role of age, gender, personality, and positive psychology. *Front. Psychol.* **2021**, *12*, 644286. [CrossRef]

13. Cori, L.; Bianchi, F.; Cadum, E.; Anthonj, C. Risk perception and COVID-19. *Int. J. Environ. Res. Public Health* **2020**, *17*, 3114. [CrossRef] [PubMed]

14. de Bruin, W.B.; Bennett, D. Relationships between initial COVID-19 risk perceptions and protective health behaviors: A national survey. *Am. J. Prev. Med.* **2020**, *59*, 157–167. [CrossRef] [PubMed]

15. Dryhurst, S.; Schneider, C.R.; Kerr, J.; Freeman, A.L.; Recchia, G.; Van Der Bles, A.M.; Spiegelhalter, D.; Van Der Linden, S. Risk perceptions of COVID-19 around the world. *J. Risk Res.* **2020**, *23*, 994–1006. [CrossRef]

16. Yıldırım, M.; Geçer, E.; Akgül, Ö. The impacts of vulnerability, perceived risk, and fear on preventive behaviours against COVID-19. *Psychol. Health Med.* **2021**, *26*, 35–43. [CrossRef]

17. Rubaltelli, E.; Tedaldi, E.; Orabona, N.; Scrimin, S. Environmental and psychological variables influencing reactions to the COVID-19 outbreak. *Br. J. Health Psychol.* **2020**, *25*, 1020–1038. [CrossRef] [PubMed]

18. Chen, Y.; Feng, J.; Chen, A.; Lee, J.E.; An, L. Risk perception of COVID-19: A comparative analysis of China and South Korea. *Int. J. Disaster Risk Reduct.* **2021**, *61*, 102373. [CrossRef]

19. Lohiniva, A.L.; Sane, J.; Sibenberg, K.; Puumalainen, T.; Salminen, M. Understanding coronavirus disease (COVID-19) risk perceptions among the public to enhance risk communication efforts: A practical approach for outbreaks, Finland, February 2020. *Eurosurveillance* **2020**, *25*, 2000317. [CrossRef]

20. Motta Zanin, G.; Gentile, E.; Parisi, A.; Spasiano, D. A preliminary evaluation of the public risk perception related to the COVID-19 health emergency in Italy. *Int. J. Environ. Res. Public Health* **2020**, *17*, 3024. [CrossRef]

21. Shahin MA, H.; Hussien, R.M. Risk perception regarding the COVID-19 outbreak among the general population: A comparative Middle East survey. *Middle East Curr. Psychiatry* **2020**, *27*, 1–19. [CrossRef]

22. Wang, J.; Guo, C.; Wu, X.; Li, P. Influencing factors for public risk perception of COVID-19—Perspective of the pandemic whole life cycle. *Int. J. Disaster Risk Reduct.* **2022**, *67*, 102693. [CrossRef]

23. The Jakarta Post. Jakarta Imposes Stricter Restrictions ahead of Year-End Holidays to Prevent Spike in COVID-19 Cases. Available online: https://www.thejakartapost.com/news/2020/12/17/jakarta-imposes-stricter-restrictions-ahead-of-year-end-holidays-to-prevent-spike-in-covid-19-cases.html (accessed on 7 October 2022).

24. Supriyati, S.; Wahyuni, A.; Wahab, R.A.A.; Halim, K.S.; Nugroho, E.A.; Soddiq, M.S. Family social capital on public respond to COVID-19 in Indonesia. *J. Commun. Empower. Health* **2021**, *4*, 196–202. [CrossRef]

25. Nasrudin, N.; Urifah, S.; Prihaninuk, D. Family compliance in implementing health protocols: Factor analysis of knowledge, values and beliefs infecting COVID-19. *Int. J. Nurs. Midwifery Sci.* **2022**, *6*, 88–95.

26. Cruwys, T. Risk perception. In *Together Apart: The Psychology of COVID-19*; Jetten, J., Reicher, S.D., Haslam, S.A., Cruwys, T., Eds.; SAGE Publications Limited: London, UK, 2020; pp. 68–72.

27. Stevenson, C.; Wakefield, J.R.; Felsner, I.; Drury, J.; Costa, S. Collectively coping with coronavirus: Local community identification predicts giving support and lockdown adherence during the COVID-19 pandemic. *Br. J. Soc. Psychol.* **2021**, *60*, 1403–1418. [CrossRef] [PubMed]

28. Biddlestone, M.; Green, R.; Douglas, K.M. Cultural orientation, power, belief in conspiracy theories, and intentions to reduce the spread of COVID-19. *Br. J. Soc. Psychol.* **2020**, *59*, 663–673. [CrossRef]

29. Goldberg, M.; Maibach, E.W.; Van der Linden, S.; Kotcher, J. Social norms motivate COVID-19 preventive behaviors. *PsyArXiv* **2020**. [CrossRef]

30. Drury, J.; Reicher, S.; Stott, C. COVID-19 in context: Why do people die in emergencies? It's probably not because of collective psychology. *Br. J. Soc. Psychol.* **2020**, *59*, 686–693. [CrossRef]

31. Jetten, J.; Reicher, S.D.; Haslam, S.A.; Cruwys, T. (Eds.) *Together Apart: The Psychology of COVID-19*; SAGE Publications: London, UK, 2020.

32. Bauer, L.L.; Seiffer, B.; Deinhart, C.; Atrott, B.; Sudeck, G.; Hautzinger, M.; Rösel, I.; Wolf, S. Associations of exercise and social support with mental health during quarantine and social distancing measures during the COVID-19 pandemic: A cross-sectional survey in Germany. *MedRxiv* **2020**. [CrossRef]

33. Chan, E.Y.Y.; Huang, Z.; Lo, E.S.K.; Hung, K.K.C.; Wong, E.L.Y.; Wong, S.Y.S. Sociodemographic predictors of health risk perception, attitude and behavior practices associated with health-emergency disaster risk management for biological hazards: The case of COVID-19 pandemic in Hong Kong, SAR China. *Int. J. Environ. Res. Public Health* **2020**, *17*, 3869. [CrossRef]

34. Megatsari, H.; Laksono, A.D.; Ibad, M.; Herwanto, Y.T.; Sarweni, K.P.; Geno, R.A.P.; Nugraheni, E. The community psychosocial burden during the COVID-19 pandemic in Indonesia. *Heliyon* **2020**, *6*, e05136. [CrossRef]

35. Adiyoso, W.; Wilopo. Social distancing intentions to reduce the spread of COVID-19: The extended theory of planned behavior. *BMC Public Health* **2021**, *21*, 1836. [CrossRef]

36. Harapan, H.; Anwar, S.; Nainu, F.; Setiawan, A.M.; Yufika, A.; Winardi, W.; Gan, A.; Sofyan, H.; Mudatsir, M.; Oktari, R.; et al. Perceived risk of being infected with SARS-CoV-2: A perspective from Indonesia. *Disaster Med. Public Health Prep.* **2022**, *16*, 455–459. [CrossRef] [PubMed]

37. Kuang, J.; Ashraf, S.; Das, U.; Bicchieri, C. Awareness, risk perception, and stress during the COVID-19 pandemic in communities of Tamil Nadu, India. *Int. J. Environ. Res. Public Health* **2020**, *17*, 7177. [CrossRef] [PubMed]

38. Linardi, V.; Syakurah, R.A.; Moudy, J. Demography factors influencing Indonesian general knowledge on COVID-19. *Int. J. Public Health Sci.* **2021**, *10*, 113–118. [CrossRef]

39. Broche-Pérez, Y.; Fernández-Fleites, Z.; Jiménez-Puig, E.; Fernández-Castillo, E.; Rodríguez-Martin, B.C. Gender and fear of COVID-19 in a Cuban population sample. *Int. J. Ment. Health Addict.* **2020**, *20*, 83–91. [CrossRef] [PubMed]

40. Dwipayanti, N.M.U.; Lubis, D.S.; Harjana, N.P.A. Public perception and hand hygiene behavior during COVID-19 pandemic in Indonesia. *Front. Public Health* **2021**, *9*, 543. [CrossRef] [PubMed]

41. Rana, I.A.; Bhatti, S.S.; Aslam, A.B.; Jamshed, A.; Ahmad, J.; Shah, A.A. COVID-19 risk perception and coping mechanisms: Does gender make a difference? *Int. J. Disaster Risk Reduct.* **2021**, *55*, 102096. [CrossRef]

42. Reznik, V.; Gritsenko, V.; Konstantinov, N.; Khamenka, R.; Isralowitz, R. COVID-19 fear in Eastern Europe: Validation of the Fear of COVID-19 Scale. *Int. J. Ment. Health Addict.* **2020**, *19*, 1903–1908. [CrossRef]

43. Sengeh, P.; Jalloh, M.B.; Webber, N.; Ngobeh, I.; Samba, T.; Thomas, H.; Nordenstedt, H.; Winters, M. Community knowledge, perceptions and practices around COVID-19 in Sierra Leone: A nationwide, cross-sectional survey. *BMJ Open* **2020**, *10*, e040328. [CrossRef]

44. Pelupessy, D.; Jibiki, Y.; Sasaki, D. Exploring people's Perception of Disaster Risk Reduction Investment for Flood Management: The Case of Jakarta Floods in Indonesia. In *Together Apart: Financing Investment in Disaster Risk Reduction and Climate Change Adaptation*; Ishiwatari, M., Sasaki, D., Eds.; Springer: Singapore, 2022; pp. 51–69.

45. Martinez, R.; Masron, I.N. Jakarta: A city of cities. *Cities* **2020**, *106*, 102868. [CrossRef]

46. Pangaribuan, M.T.; Munandar, A.I. Kebijakan Pemerintah DKI Jakarta Menangani Pandemi COVID-19. *J. Ilmu Pemerintah.* **2021**, *14*, 1–9.

47. Yakhamid, R.Y.; Zaqi, N.A.R. Efektivitas PPKM Darurat Dalam Penanganan Lonjakan Kasus COVID-19. *Semin. Nas. Off. Stat.* **2021**, *2021*, 235–244. [CrossRef]

48. Handayani, W.; Insani, T.D.; Fisher, M.; Gim, T.H.T.; Mardhotillah, S.; Adam, U.E.F. Effects of COVID-19 restriction measures in Indonesia: A comparative spatial and policy analysis of selected urban agglomerations. *Int. J. Disaster Risk Reduct.* **2022**, *76*, 103015. [CrossRef] [PubMed]

49. Shapira, S. Trajectories of community resilience over a multi-crisis period: A repeated cross-sectional study among small rural communities in Southern Israel. *Int. J. Disaster Risk Reduct.* **2022**, *76*, 103006. [CrossRef]

50. Kimhi, S.; Eshel, Y.; Marciano, H.; Adini, B. Distress and resilience in the days of COVID-19: Comparing two ethnicities. *Int. J. Environ. Res. Public Health* **2020**, *17*, 3956. [CrossRef] [PubMed]

51. Xu, J.; Zeng, Z.; Hong, Y.; Xi, Z.; Zhu, X.; Peng, Z. Grassroots Mirroring under COVID-19: Does Community Resilience Affect Residents' Responses? The Case of Shenzhen, China. *Sustainability* **2022**, *14*, 10159. [CrossRef]

52. Zhang, J.; Wang, Y.; Zhou, M.; Ke, J. Community resilience and anxiety among Chinese older adults during COVID-19: The moderating role of trust in local government. *J. Community Appl. Soc. Psychol.* **2022**, *32*, 411–422. [CrossRef]

53. Leykin, D.; Lahad, M.; Cohen, O.; Goldberg, A.; Aharonson-Daniel, L. Conjoint community resiliency assessment measure-28/10 items (CCRAM28 and CCRAM10): A self-report tool for assessing community resilience. *Am. J. Commun. Psychol.* **2013**, *52*, 313–323. [CrossRef]

54. Leykin, D.; Lahad, M.; Cohen, R.; Goldberg, A.; Aharonson-Daniel, L. The dynamics of community resilience between routine and emergency situations. *Int. J. Disaster Risk Reduct.* **2016**, *15*, 125–131. [CrossRef]

55. Cohen, O.; Leykin, D.; Lahad, M.; Goldberg, A.; Aharonson-Daniel, L. The conjoint community resiliency assessment measure as a baseline for profiling and predicting community resilience for emergencies. *Technol. Forecast. Soc. Chang.* **2013**, *80*, 1732–1741. [CrossRef]
56. Norris, F.H.; Stevens, S.P.; Pfefferbaum, B.; Wyche, K.F.; Pfefferbaum, R.L. Community resilience as a metaphor, theory, set of capacities, and strategy for disaster readiness. *Am. J. Commun. Psychol.* **2008**, *41*, 127–150. [CrossRef]
57. Sampson, R.; Raudenbush, S.; Earls, F. Neighborhoods and violent crime: A multilevel study of collective efficacy. *Science* **1997**, *277*, 918–924. [CrossRef] [PubMed]
58. Manzo, L.C.; Perkins, D.D. Finding common ground: The importance of place attachment to community participation and planning. *J. Plan. Lit.* **2006**, *20*, 335–350. [CrossRef]
59. Kuno, G. Neighborhood Lockdown as the New Normal? Jakarta's COVID-19 Experience. 2020. Available online: https://covid-19chronicles.cseas.kyoto-u.ac.jp/en/post-037-html/ (accessed on 21 November 2022).
60. Hosobuchi, M. Social Security in Poor and Low-income Communities in the COVID-19 Disaster: A case study in Jakarta, Indonesia. *Glob. Urban Stud.* **2021**, *14*, 25–44.

International Journal of
Environmental Research and Public Health

Article

The Impact of Climate-Change-Related Disasters on Africa's Economic Growth, Agriculture, and Conflicts: Can Humanitarian Aid and Food Assistance Offset the Damage?

Go Shimada

School of Information and Communication, Meiji University, Chiyoda-ku, Tokyo 101-8301, Japan;
go_shimada@meiji.ac.jp

Abstract: This study analyzed the impact of climate-related natural disasters (droughts, floods, storms/rainstorms) on economic and social variables. As the Africa-specific empirical literature is limited, this study used panel data from 1961–2011 on Africa. The study used a panel data regression model analysis. The results showed that climate change-related natural disasters affected Africa's economic growth, agriculture, and poverty and caused armed conflicts. Among the disasters, droughts are the main cause of negative impact, severely affecting crops such as maize and coffee and resulting in increased urban poverty and armed conflicts. In contrast, international aid has a positive effect but the impact is insignificant compared to the negative consequences of climate-related natural disasters. Cereal food assistance has a negative crowding-out effect on cereal production. International donors should review their interventions to support Africa's adaptative capacity to disasters. Government efficiency has reduced the number of deaths, and this is an area that supports Africa's adaptative efforts.

Keywords: climate change; natural disasters; agricultural production; food aid; official development assistance; conflict; poverty; cereal production; humanitarian aid

Citation: Shimada, G. The Impact of Climate-Change-Related Disasters on Africa's Economic Growth, Agriculture, and Conflicts: Can Humanitarian Aid and Food Assistance Offset the Damage? *Int. J. Environ. Res. Public Health* **2022**, *19*, 467. https://doi.org/10.3390/ijerph19010467

Academic Editors: Mikio Ishiwatari and Daisuke Sasaki

Received: 21 November 2021
Accepted: 30 December 2021
Published: 1 January 2022

Publisher's Note: MDPI stays neutral with regard to jurisdictional claims in published maps and institutional affiliations.

1. Introduction

With the increasing threat of global warming, there is a drastic increase in the number of catastrophic natural disasters. Every year, irregular and extreme weather events are reported worldwide. In 2021, Europe experienced an oppressive heatwave and was hit by devastating floods in July [1]. The magnitude of destruction was severe in Germany and Belgium, causing several deaths. In Asia, tropical cyclones have become stronger in recent years [2]. Additionally, Japan has recorded torrential rainfall, flooding, and landslides more often than before. Global warming is a worldwide phenomenon, but African countries are disproportionately punished, even though they contributed the least toward greenhouse gas emissions compared with developed countries. Therefore, do donor countries contribute sufficient aid to African countries to promote Africa's adaptive capacity? To answer this question, we must understand the extent of the collateral damage of global warming in Africa. This study analyzed the damage caused by climate-related natural disasters, such as floods, droughts, and storms. (The phenomenon of extreme temperature is not included in this study, as the frequency is shallow in Africa.) The remainder of this paper is organized as follows. Section 1 examines the current damage trends caused by climate-related natural disasters. Section 2 presents the study methods and data. Section 3 discusses the analysis results. Based on the results, the discussion is presented in Section 4. Finally, conclusions are drawn from the results and discussion.

1.1. Background: Climate Change and Natural Disasters in Africa

This section reviews climate change and natural disasters in Africa. Subsequently, it reviews the existing literature on the impact of climate change and related natural disasters [3,4]. For instance, using satellite images, Tellman et al. [5] found that the majority of the population was exposed to floods between 2000 and 2015, especially in Asia and sub-Saharan Africa. The authors also argued that their projections for 2030 indicate that a larger number of people will be exposed to flood threats.

1.2. GDP Per Capita

Gross domestic product (GDP) per capita helps measure the economic and social impact of climate-related natural disasters in Africa. There are several studies on the overall impact of natural disasters on economic growth, but not specific to climate-related disasters [6–10]. While there is no consensus regarding its impact on economic growth, some found a significantly negative long-term effect [11–16]. Conversely, some studies show a positive impact on economic growth as disasters promote the "Schumpeterian creative destruction" process [17–21]. According to this view, disasters promote innovation and investment, destroying existing practices, products, or services. Other studies reveal mixed results on economic growth due to natural disasters [22,23]. Therefore, it is necessary to clarify the effect of climate-related disasters in Africa.

1.3. Agricultural Production and Aid

The socioeconomic impact on agriculture and food security have been identified as critical sectors in the age of climate change [24]. Lesk et al. [25] found that droughts have reduced national cereal production by 9–10% in terms of impact on crop production. However, they found statistically no significant effect of floods. Nonetheless, the question arises if this is truly a global phenomenon or specific to Africa.

Africa is particularly vulnerable to climate change, as the region's adaptive capacity has certain constraints [26–30]. Notably, most farmers in Africa are small-holder farmers without adequate education or skills to adapt to warming temperatures and damages caused by natural disasters [31]. As there is limited empirical researched, this study examines the impact of climate-related disasters on African agriculture because this sector is most critical for people's livelihoods compared to other sectors. Therefore, if disasters decrease agricultural production, it increases poverty in both rural and urban areas.

In this regard, the literature does not consider social variables. Agricultural production is not determined by disasters. Other factors, such as farmers' human capital and market demand for agricultural commodities, are essential production factors [32,33]. The prior literature does not control for these social variables and may produce biased results. Therefore, this study estimated the impact of climate-related disasters by controlling for these factors.

A critical control variable is disaster relief provided by international donors, including the United Nations (UN), World Bank, and bilateral governments. Some literature is available on disaster relief based on case studies [34]. In contrast, there are few quantitative studies, especially on how disaster relief mitigates damage or contributes toward recovery [35,36]. One study [35] found that increased foreign aid resulted in higher fatality rates. According to another study [36], international aid increases social strife rather than decreasing it, promoting a new conflict over the distribution of resources.

Some studies have shown that aid increased after a disaster [7,37]. (There are some studies on the impact of domestic government aid to mitigate the damage, but not international aid (see, for example, [38]).) However, none applied an aid disaggregation approach to examine the impact of different aid types. These studies only used aggregated aid data. This is challenging because aid has a distinct impact. For instance, aid for primary education shows different results from assistance provided for infrastructure or humanitarian food aid. If this study uses aggregated aid data, it may lead to erroneous policy recommendations. For instance, assistance for infrastructural growth tends to expand in budgets compared to agricultural development aid. It is essential to distinguish between various aid types. Therefore, there is a research gap in this regard. There is no study on whether humanitarian aid, food aid, and other assistance forms have mitigated the damage caused by natural disasters.

1.4. Impact on Poverty and Armed Conflict

Regarding the impact on poverty, Kahn [6] found that the Gini coefficient is positively correlated with disaster-related deaths. As Barrett [39] discussed, food insecurity is associated with sudden catastrophe-like disasters and chronic poverty. Damage due to disasters is of two types: direct and indirect. Notably, in direct damage, the disaster itself kills people. In indirect damage, there are cases where people die not because of the disaster itself but due to displacement (i.e., losing their jobs after the disaster) and resultant poverty. Therefore, it is essential to measure the impact of such indirect consequences.

There is no consensus in the previous literature on the natural disaster–conflict nexus. Some studies found a link, whereas some others did not. For instance, O'Loughlin et al. [40] studied the link between climate variability and armed conflicts and found that, in general, extremely high temperatures are associated with greater conflict levels. They also found that the link varies depending on the conflict type and different subregions of Africa. Burke et al. [41] found a strong historical association between civil war and temperature in Africa, indicating that by 2030 armed conflict is likely to increase by approximately 54%. However, Buhaug [42] argued against this nexus, reporting that the incidence of armed conflict has declined in Africa since 2002 despite rising temperature levels. Following this counterargument, Burke et al. [43] stated that there are some econometric issues in Buhaug's [42] study. (Burke also admits that the climate–conflict nexus still stands, but the nexus has weakened since 2002.)

These previous studies examined the temperature–conflict nexus. However, this nexus does not have a direct causal relationship. There are several indirect links between the two: natural disasters and vegetation. High temperatures cause crop damage, and damage occurs when the temperature is above 30 °C or when the average temperature is above 25 °C for a prolonged period. Tolerance to high temperatures varies among crops. Therefore, these two consequences would have some social impact, such as reducing income. Consequently, such outcomes potentially lead to conflicts. The past literature did not consider the indirect causal relationship and treats the temperature–conflict link as a black box. The impact of high temperatures on vegetation was beyond the scope of this study; therefore, this research examined the impact of disasters on conflicts.

1.5. Government Effectiveness

Strömberg [7] studied whether government effectiveness is essential for dealing with disasters. The study used the government effectiveness index, an indicator produced by the World Bank, to test its importance of governance effectiveness. Government effectiveness is essential during disasters. At the time of a crisis, the government needed to handle everything quickly and within a limited period. Even in developed countries, handling crises after disasters is challenging and sometimes governments fail to cope with them. However, this is likely more difficult for developing countries. This raises the question: How important is government effectiveness in Africa? Strömberg found that government effectiveness reduces the number of people killed by natural disasters globally and not specifically in Africa. This study examined the importance of government effectiveness to mitigate damage caused by climate-related disasters.

This section reviewed the current trends in climate-related natural disasters in Africa and examined the literature on the impact of disasters on GDP, agriculture, poverty and conflict, and government effectiveness.

First, based on the identified research gaps, the next section examines the impact of climate-related natural disasters and international aid on GDP per capita. Second, the nexus between disaster–agriculture is discussed, focusing on major crops. Third, the impact on poverty and conflict was tested. Finally, factors contributing to decreasing (or increasing) the death toll due to natural disasters were tested. One of the factors examined was international aid extended to African countries.

Based on the research gap identified in this section, the next section discusses the following. First, it overviews the impact of climate-related natural disasters and international aid on GDP per capita. Second, the nexus between the disasters–agriculture is discussed, focusing on major crops. Third, the impacts on poverty and conflicts are investigated. Finally, factors contributing to decreasing (or increasing) death tolls by natural disasters are tested. One of the factors examined is international aid extended to African countries.

2. Methods and Data

This section describes the analytical framework for empirical analysis. This study tested the impact of three elements of climate-related disasters on (1) GDP per capita and agricultural production, (2) impact on poverty and conflict, and (3) factors contributing toward reducing the impact of disasters.

Research gaps were identified based on the literature review in Section 1. There is no consensus on the impact of natural disasters on GDP, and there is no empirical analysis focusing on the impact of climate-related natural disasters on agricultural production in Africa. Even the literature on global agricultural production does not control for other socioeconomic conditions. Agricultural production is a part of economic activities; therefore, there is a need to control for these variables. Otherwise, there is a potential for result bias. Therefore, it is important to understand the impact of climate-related natural disasters on GDP and agricultural production, considering other socio-economic factors.

To estimate the impact on GDP per capita growth, the following formula was used, following the model used by Skidmore and Toya [20].

$$\Delta\left(\frac{Y}{P}\right)_{i,t} = \alpha_{i,t} + \beta_1\Delta\left(\frac{Y}{P}\right)_{i,\,t-1} + \beta_2 Dis_{i,t} + \beta_3 Aid_{i,t} + \beta_4 Gov_{i,t} + \beta_5 X_{i,t} + \varepsilon_{i,t} \tag{1}$$

Y denotes GDP and P represents the population. Therefore, $\Delta\left(\frac{Y}{P}\right)_{i,t}$ is GDP per capita growth, i is a country index to capture country-specific effects, and t is the time (year) index. The lagged GDP per capita growth $(\Delta\left(\frac{Y}{P}\right)_{i,\,t-1})$ is included because the previous year's growth trend greatly affects the current year's economic activities. $Dis_{i,t}$ is a measure of the impact of disasters specific to country i at time t. This study used the number of people affected by disasters for this variable. This is because these data represent the impact of disasters. The number of occurrences does not necessarily equate to the impact. If a disaster occurs in an uninhabited area, then the impact on human activities is limited, as people do not live there. *Aid* denotes international aid. This is global official development assistance (ODA) data and not a specific country's ODA. This study used a different type of aid data, as impact varies depending on the aid type. For instance, the impact of cereal food aid is not the same as a medical aid. Therefore, it is necessary to disaggregate aid data. This study used the following data: aid for agriculture, humanitarian aid, and cereal food aid. $Gov_{i,t}$ denotes government expenditure. $X_{i,t}$ are the other control variables, including the following variables: education, fertility rate, and government effectiveness index. The education variable represents human capital [32–44]. This study used the government effectiveness index developed by the World Bank, which measures the quality of public services, infrastructure, and civil service based on the World Bank's survey [45].

The following analytical framework will be used to estimate the impacts on agricultural production, reformulating Equation (1) to focus on agriculture.

$$Agr_{i,t} = \alpha_{i,t} + \beta_1 Dis_{i,t} + \beta_2\Delta\left(\frac{Y}{P}\right)_{i,t} + \beta_3 Aid_{i,t} + \beta_4 X_{i,t} + \varepsilon_{i,t} \tag{2}$$

$Agr_{i,t}$ denotes the variable for agricultural production. $Dis_{i,t}$, $\Delta\left(\frac{Y}{P}\right)_{i,t}$, and $Aid_{i,t}$ indicate the same as the above equation. $\Delta\left(\frac{Y}{P}\right)_{i,t}$ is included because agricultural production is affected by economic activities. $X_{i,t}$ denotes other control variables, such as inequality in educational attainment and the government effectiveness variable. The former represents human capital.

As this estimation used panel data, three methods were used to assess the results. (This study used Stata/SE 17.0 for the estimation.) They estimate fixed effects (FE), random effects (RE), and pooling. Among the three estimation methods, the most appropriate estimation method is determined by the results of the F-test, Hausman test, and the Breusch–Pagan test. The method determined for each model is reported at the bottom of the result tables.

Table 1 shows the descriptive statistics of the data used in the empirical study. This dataset covers 90 African countries (including both sub-Saharan African and North African countries). This is unbalanced panel data. The period varies depending on the data, and they are annual data with gaps. This dataset was constructed using the following four datasets.

Table 1. Descriptive Statistics.

Variable	Obs.	Mean	Std. Dev.	Min.	Max.	Year	Data Source
Number of People Affected by Climate-related Disasters	3394	171,620.9	906,675.7	0	23,000,000	1900–2021	EM-DAT
Number of People Affected by Drought	3143	0.0140703	0.080764	0	1	1900–2021	EM-DAT
Number of People Affected by Flood	3143	0.0024037	0.0179437	0	0	1927–2021	EM-DAT
Number of People Affected by Storm	3143	0.0012416	0.0250523	0	1	1948–2021	EM-DAT
Number of Deaths by Climate-related Disasters	3394	267.0533	6475.562	0	300,000	1900–2021	EM-DAT
Control of Corruption	1185	−0.542972	0.687297	−1.869	2	1996–2019	V-Dem
Local Government Index	3224	0.4376833	0.3243581	0	1	1900–2020	V-Dem
Educational Inequality, Gini	2283	60.83778	22.18283	11.875	99.804	1927–2010	V-Dem
Net ODA	2286	10.48105	11.82156	−0.251879	147	1960–2011	WDI
Humanitarian ODA	503	66,800,000	171,000,000	1387	1,380,000,000	2002–2011	WDI
ODA for reconstruction relief and rehabilitation	296	5,796,040	13,200,000	−52,185	96,900,000	2002–2011	WDI
ODA for Agriculture	513	29,000,000	40,400,000	7069	387,000,000	2002–2011	WDI
Emergency ODA	493	64,100,000	165,000,000	1387	1,280,000,000	2002–2011	WDI
ODA for Disaster Prevention and Preparedness	249	1,209,090	2,266,356	−75,420	16,700,000	2002–2011	WDI
Cereal Food Aid	1309	67,786.64	165,895	0	1,900,805	1988–2012	WDI
Agriculture Production Index	2583	70.33316	28.66801	13.42	193	1961–2011	WDI
Cereal Production Index	2512	82.68087	77.53881	5.79	1925	1961–2011	WDI
Maize Production (ton)	1950	584,838.4	1,210,479	4	10,500,000	1961–2019	FAOSTAT
Sorghum Production (ton)	1522	245,611.4	506,910.9	0	5,265,580	1961–2019	FAOSTAT
Millet Production (ton)	1323	131,864.3	262,178.4	54	1,878,527	1961–2019	FAOSTAT
Rice Production (ton)	1669	372,630.1	938,768.1	0	7,253,373	1961–2019	FAOSTAT
Wheat Production (ton)	1105	607,994	1,521,939	0	9,607,736	1961–2019	FAOSTAT
Barley Production (ton)	552	449,506.7	749,312.6	100	3,831,130	1961–2019	FAOSTAT
Fonio Production (ton)	381	36,381.73	83,020.09	100	530,227	1961–2020	FAOSTAT
Poverty gap at the urban poverty line (%)	76	11.90395	9.049087	1.8	40	1961–2011	WDI
Poverty gap at the rural poverty line (%)	77	22.07273	9.499953	3.6	53	1961–2012	WDI
Battle-related deaths (number of people)	278	1411.522	4618.351	0	50,293	1961–2013	WDI

First, the Emergency Events Database (EM-DAT) was used for natural disaster-related data [46]. The EM-DAT is an international disasters' database widely used to analyze natural disasters and has been managed by the Center for Research on the Epidemiology of Disasters (CRED) since 1988. For disasters to be recorded as an extreme event, one of the following criteria must be met: (1) 10 or more people reportedly killed, (2) 100 or more people affected, (3) a declaration of a state of emergency, and (4) a call for international assistance.

Second, the WDI (World Development Indicators) dataset is compiled by the World Bank [45]. The Food and Agriculture Organization Corporate Statistical Database (FAO-

STAT) is a dataset on agricultural production provided by the Food and Agriculture Organization (FAO) [47]. Finally, the V-dem is used, produced by the Varieties of Democracy Project on government effectiveness and educational inequality data [48].

Five types of variables are used as measures for disaster impact. The four variables represent the number of people affected by the following: (1) climate-related disasters (aggregate variable), (2) droughts, (3) floods, and (4) storms. The variable of climate-related disasters is the aggregate variable of droughts, floods, and storms. This study also used the number of deaths caused by climate-related disasters. In addition, this study used two other variables: corruption control and the local government index as measures of government effectiveness. If corruption control is inadequate, the government's effectiveness is considered low and would affect the implementation of recovery after a disaster. For international aid, seven different types of data were prepared to examine how aid works to mitigate the impact of natural disasters. The aid for agricultural development works differently from humanitarian aid. For agricultural production, the agricultural production index is the aggregate index. Production data for major cereals, such as maize, sorghum, and millet, were used to examine the impact of crops.

3. Results

This section reviews the current trends of natural disasters in Africa before analyzing the impacts of natural disasters. Figure 1 shows the number of people affected by droughts, floods, and storms caused by climate change. The data used the Emergency Events Database (EM-DAT), mentioned in the previous section [4]. As a measure of natural disasters, Figure 1 uses the number of people affected rather than the number of disasters that occurred because this is a better measure of disaster severity. Each disaster is different in scale and impact. If a disaster occurs in a remote mountainous area, the social impact is less, but the same type of event would have a devastating effect in an urban area. Therefore, Figure 1 shows the number of people affected, rather than the number of occurrences.

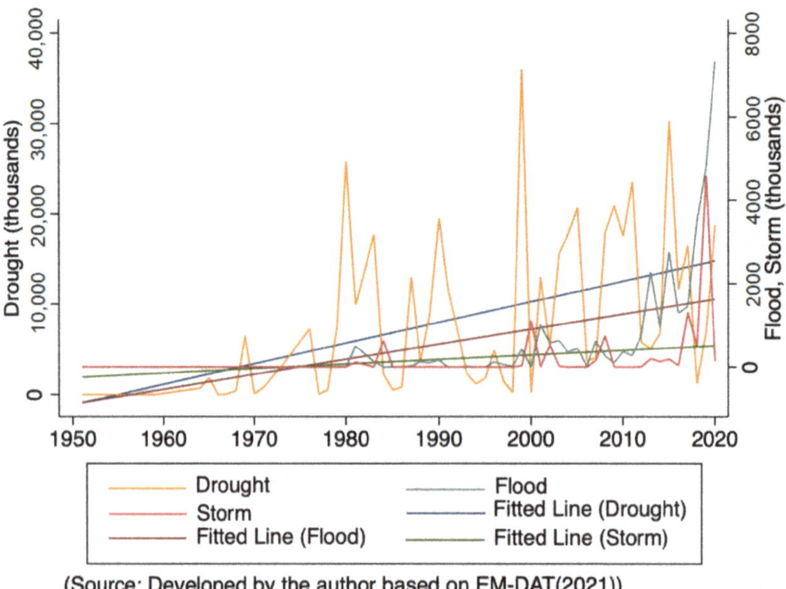

Figure 1. The total number of people affected by droughts, floods, and storms in Africa.

Furthermore, an essential point must be considered when interpreting these numbers. The disaster number, such as the number of people affected by a disaster, may be underreported compared with the actual situation [7]. For instance, some authoritarian African regimes may underreport the damage caused by disasters to avoid being criticized for their response. Given African governments' capacity for data authenticity, there may be misreporting across countries and over time [49]. In such a case, there is an underlying risk of underestimating the significance of the impact of climate-related disasters.

Figure 1 has two axes because the damage caused by droughts is much larger than that caused by floods and storms. The left axis represents the total number of people affected by droughts, and the right axis shows the number of people displaced by floods and storms. The worst drought affected more than 35,000 people in 1999. However, on the other hand, the impact of floods and storms is much lower than the drought. Due to the nature of disasters, all three lines, representing each natural disaster, fluctuate significantly. For instance, there was a massive drought in 1999, but there was no drought before and after the year. The damage caused by droughts was not significant until the late 1960s, and they seem to occur quite often now. A similar pattern can be observed for floods and storms. However, during the late 1990s, there was an unprecedented increase in the number of people affected by floods and storms.

To clearly understand the long-term trend of natural disasters, fitted lines for drought, floods, and storms are drawn in Figure 1. These three fitted lines show a clear upward trend, especially due to droughts, followed by floods and storms. Therefore, as Figure 1 confirms, the damage caused by climate-related disasters has increased rapidly since the late 1960s, primarily during droughts. Therefore, the next question is how disaster-related damage affects Africa's socio-economic development and growth.

Table 2 presents the estimated coefficients in the impact on GDP per capita growth, and the difference among the four models is the difference of data used. Models 1 and 2 used aggregate disaster measures (number of people affected by climate-related disasters) and Models 3 and 4 used disaggregated disaster variables (droughts, floods, and storms). As the dataset is unbalanced panel data with gaps, the N used for estimation differs depending on the variable used. Models 1 and 2 show that climate-related natural disasters significantly lowered GDP per capita growth. However, as Models 3 and 4 show, the impact differs depending on the natural disaster type. Only droughts had a statistically significant negative impact, as opposed to floods and storms. This indicates the importance of any drought policy to sustain GDP per capita growth in Africa.

Models 1 and 3 examined the impact of humanitarian ODA on GDP per capita. How do global efforts mitigate the negative impact of natural disasters? Similar to Models 2 and 4, which examined the emergency aid, all these variables became insignificant. This is reasonable considering that, compared with the size of a country's GDP, the amount of aid in these categories is too small to capture the impact. Therefore, the study next examined aid impact focusing on crops because the agricultural sector is particularly vulnerable to natural disasters, affecting economic growth and damaging crop production.

Table 3 examines the impact on agricultural production. The number of people affected by natural disasters, an aggregate variable, is used to measure disaster impact. This variable is statistically significant, and the negative coefficient is also substantial. This result indicates that climate-related disasters have a significant negative impact on agricultural production. Educational inequality is also negative. Therefore, human capital is important for agricultural production. As the educational Gini coefficient reflects income inequality, the widening rich–poor gap negatively impacts agricultural production. This is consistent with past studies [32,33]. Therefore, human capital is required to cope with disasters caused by climate change, which are likely to increase in the future.

Table 2. Economic Growth Impacts by Climate-related Disasters and Aid.

Dependent Variable	GDP Per Capita Growth (Annual %)			
	(1)	**(2)**	**(3)**	**(4)**
Lagged GDP per capita growth	0.1694751 ***	0.1543027 **	0.16712 ***	0.1519377 **
	(2.80)	(2.49)	(2.75)	(2.44)
Education (+15 years old)	−1.228877	−1.139392	−1.25669	−1.172042
	(−0.56)	(−0.52)	(−0.58)	(−0.53)
Fertility Rate	1.35599	2.05509	1.479018	2.180705
	(0.66)	(0.97)	(0.71)	(1.02)
HDI	−3.463914	−0.953797	−2.788904	−0.2467038
	(−0.17)	(−0.04)	(−0.13)	(−0.01)
Number of People Affected by Climate-related Disasters	−8.057152 **	−8.057298 **		
	(−2.07)	(−2.05)		
Number of People Affected by Drought			−7.632749 *	−7.614573 *
			(−1.87)	(−1.85)
Number of People Affected by Flood			−13.04042	−13.28393
			(−0.99)	(−1.01)
Number of People Affected by Storm			0.00000684	0.00000685
			(1.05)	(1.05)
Government Expenditure	0.1529741	0.1360603	0.1624341	0.1460737
	(1.53)	(1.31)	(1.62)	(1.40)
Humanitarian ODA	−0.0000000007		−0.000000001	
	(−0.14)		(−0.14)	
Emergency ODA		−0.000000001		−0.000000001
		(−0.180)		(−0.18)
Constant	1.043053	−3.935066	0.1118494	−4.888921
	(0.06)	(−0.22)	(0.01)	(−0.28)
N	272	264	272	264
Type of Regression	FE	FE	FE	FE

Note: Numbers in parentheses are *t*-values; ***, **, and * indicate statistical significance at the 1%, 5%, and 10% levels, respectively.

According to Models 1 and 4, humanitarian aid and ODA for disaster prevention and preparedness are not statistically significant. In contrast, according to Models 2 and 3, agricultural aid is positive. These results are consistent with the expected outcome, as humanitarian aid and disaster prevention assistance do not support agricultural production. However, this analysis shows that disasters have a substantial negative impact on agricultural production. Therefore, agricultural aid has become even more important during the age of global warming, specifically to fight against the long-term negative consequences. However, it is necessary to note that the coefficient of agricultural aid is small, indicating that agricultural aid alone is not enough to mitigate the negative impact of climate-related disasters.

Table 3. Impact on Agricultural Production by Climate-related Disasters and Aid.

Dependent Variable	Agriculture Production Index			
	(1)	(2)	(3)	(4)
GDP per capita growth	−0.0003642 (−0.00)	−0.1086933 (−0.95)	−0.115295 (−0.99)	−0.032736 (−0.10)
Number of People Affected by Climate-related Disasters	−24.41415 *** (−2.74)	−23.02236 *** (−2.58)	−21.84345 ** (−2.41)	−35.02979 ** (−2.06)
Educational Inequality, Gini	−4.608486 *** (−19.84)	−4.224185 *** (−15.48)	−4.253704 *** (−15.29)	−6.260741 *** (−6.89)
Humanitarian ODA	−0.0000000052 (−0.47)			
ODA for Agriculture		0.000000075 *** (2.65)	0.000000076 *** (2.66)	
ODA for disaster prevention & preparedness (lagged)				−0.0000008 (−0.95)
Local Government Index			−8.470382 (−1.35)	3.579998 (0.14)
Constant	319.1347 *** (29.24)	297.3906 *** (22.74)	303.5181 *** (21.33)	397.2466 *** (9.78)
N	370	379	370	120
Type of Regression	FE	FE	FE	FE

Note: Numbers in parentheses are *t*-values; ***, and ** indicate statistical significance at the 1%, and 5% levels, respectively.

Table 4. Impact on Agricultural Production by Climate-related Disasters and Aid.

Dependent Variable	Agriculture Production Index		
	(1)	(2)	(3)
GDP per capita growth	−0.0106388 (−0.09)	−0.1173564 (−1.01)	0.3344108 *** (5.03)
Number of People Affected by Climate-related Disasters	−23.4604 *** (−2.58)		
Number of People Affected by Flood		−17.48007 (−0.41)	−9.036425 (−0.30)
Number of People Affected by Drought		−24.51023 *** (−2.61)	−12.51029 * (−1.94)
Number of People Affected by Storm		−0.000009 (−0.93)	−0.000007 (−1.57)
Educational Inequality, Gini	−4.64529 *** (−19.48)	−4.230962 *** (−15.09)	−2.568219 *** (−26.68)
Local Government Index	−8.113644 (−1.30)	−8.025702 (−1.28)	20.9569 *** (6.93)
Emergency ODA	−0.000000003 (−0.775)		
ODA for Agriculture		0.00000008 *** (2.70)	
Cereal Food Aid			−0.0000148 *** (−3.49)
Constant	326.9576 *** (26.54)	302.2468 *** (21.07)	206.6968 *** (37.24)
N	355	370	958
Type of Regression	FE	FE	FE

Note: Numbers in parentheses are *t*-values; *** and * indicate statistical significance at the 1%, and 10% levels, respectively.

Table 4 investigates the impact of climate-related disasters on agriculture. Compared to Table 3, the impact of different types of aid is tested in Table 4. Models 1, 2, and 3 examine emergency aid, agricultural aid, and cereal food aid, respectively. Emergency aid is not significant, similar to humanitarian aid. This is expected because emergency aid does not aim at agricultural development. The impact on agricultural aid and cereal food aid is in the opposite direction, wherein agricultural aid is positive but cereal food aid is negative due to a substitutional effect. The inflow of cereals from foreign countries seems to have a crowding-out effect on domestically produced cereals. For agricultural aid, even if the coefficient is positive, it is very small and unremarkable compared to the coefficient of disasters. In other words, agricultural aid does not compensate for the negative impact of disasters. The results for the impact of disasters and educational inequality are the same as those in Table 3, further confirming the earlier results.

Tables 5 and 6 examine the impact on a crop-by-crop basis. The difference between these two tables is that Table 5 includes ODA for agriculture, whereas Table 6 examines the impact of cereal aid. This analysis shows that the impact varies by crop. For instance, droughts negatively impact maize. Storms reduce rice and fonio production because, at the heading stage, strong winds can topple the panicles of rice and fonio. Floods also reduce fonio production. However, this result is not robust, as these crops are statistically insignificant (Table 6). Therefore, the impact of disasters on crops varies depending on the vegetation type.

Agricultural ODA has a positive impact on maize, sorghum, millet, and rice, but the impact differs by crop. Furthermore, the coefficient is very small. More importantly, it can be observed in Table 6 that the impact of cereal aid is different from agricultural ODA. Overall, cereal aid has a negative impact, but this influence differs depending on the crop. Maize, sorghum, rice, and wheat are all negatively affected, but millet production increases marginally. However, the reason for increased millet production is unclear, but millet may be used as a substitute for cereals such as wheat, whose production decreased. In many cases, cereal aid is provided during humanitarian crises, including natural disasters. However, cereal aid can negatively affect production because cereals provided through food aid have a crowding-out effect on domestic production.

This study also examined non-cereal crops, such as bananas, cassava, tea, and coffee (Appendix A). It can be observed that only coffee is negatively and strongly affected by droughts because water stress affects coffee production due to water availability sensitivity.

As discussed above, natural disasters negatively impact agricultural production. The next question is what are the consequences of such effects. Table 7 examines the impact of climate-related disasters on poverty and conflicts because reduced agricultural production indicates lower income for farmers. This would probably impact poverty and conflicts. As observed in Section 1 (Literature review), previous studies examine the temperature–conflict link without considering the internal mechanism. As discussed, using disaster data, this study examined the internal nexus between climate change and armed conflicts.

Model 1 examined the impact of aggregate climate-related natural disasters in rural areas and shows that poverty does not increase in rural areas. The growth in GDP per capita reduces poverty in rural areas. However, this situation contrasts in urban areas, where climate-related natural disasters increase poverty. This is probably because people migrate from rural to urban areas in search of jobs after natural disasters occur. Therefore, rather than rural areas, poverty increases in urban areas. Model 3 analyzes the impact of different disaster types, focusing only on urban areas. The results show that climate variability that leads to extreme events such as droughts causes a substantial increase in poverty in Africa.

Int. J. Environ. Res. Public Health **2022**, 19, 467

Table 5. Impact on Cereal Production by Climate-related Disasters and Aid.

Dependent Variable	(1) Cereal Production Index	(2) Maize Production	(3) Sorghum Production	(4) Millet Production	(5) Rice Production	(6) Wheat Production	(7) Barley Production	(8) Fonio Production
GDP per capita growth	0.3519966 (1.22)	1591.614 [0.41]	−1304.271 (−0.74)	830.9917 (0.62)	−1334.704 [−0.45]	9952.268 [1.06]	24,650.14 [0.80]	−2459.17 (−1.21)
Educational Inequality, Gini	−7.099192 *** (−10.44)	−32,610.23 *** [−5.19]	−5121.155 (−1.58)	−989.7303 (−0.40)	−8121.831 [−1.37]	−8631.816 [−0.54]	−1084.82 [−0.05]	−12,828.91 *** (−8.28)
Number of People Affected by Flood	−4.82892 (−0.05])	−910,600.5 [−0.46]	27,443.77 (0.03)	−35,073.24 (−0.06)	−223,888.4 [−0.16]	1,576,617 [0.19]	−1,673,180 [−0.07]	−745,509.2 *** (−2.92)
Number of People Affected by Drought	−51.96536 ** (−2.27)	−1,027,551 *** [−3.75]	−39,551.82 (−0.33)	−102,533.6 (−0.94)	−46,956.45 [−0.19]	91,261.4 [0.15]	−10,5139.6 [−0.09]	118,431.8 (0.46)
Number of People Affected by Storm	0.000000779 (0.03)	−0.1388403 [−0.64]	−0.0418895 (−0.44)	−0.1643853 (−0.62)	−0.66323 *** [−2.77]	−0.0056365 [−0.01]	0.0449533 [0.01]	−51.86224 * (−1.92)
ODA for Agriculture	0.000000116 * (1.65)	0.0040394 *** [5.45]	0.001325 *** (3.95)	0.000796 *** (3.19)	0.0034716 *** [5.11]	0.0011092 [0.53]	0.0014411 [0.42]	−0.0003032 (−2.39)
Constant	433.7546 *** (13.19)	2,230,139 *** [5.79]	461,934.9 *** (2.82)	208,877.2 (1.55)	866,423.4 ** [2.43]	1,128,003 [1.32]	400,714.1 [0.36]	1,043,952 *** (8.78)
N	373	247	324	185	332	158	62	45
Type of Regression	FE	FE	RE	FE	RE	RE	RE	FE

Note: Numbers in brackets are z-values, and in parentheses are t-values; ***, **, and * indicate statistical significance at the 1%, 5%, and 10% levels, respectively.

Table 6. Cereal Food Aid and Cereal Production.

Dependent Variable	(1) Cereal Production Index	(2) Maize Production	(3) Sorghum Production	(4) Millet Production	(5) Rice Production	(6) Wheat Production	(7) Barley Production	(8) Fonio Production
GDP per capita growth	0.9851934 *** (4.12)	6623.565 *** [2.74]	2369.321 (2.18)	1372.234 (1.62)	−645.8084 [−0.42]	15,459.38 *** [2.75]	42,489.02 *** [4.30]	−775.6227 (−0.64)
Educational Inequality, Gini	−1.605586 *** (−4.97)	−20,892.1 *** [−10.16]	−4520.574 *** (−4.98)	−4523.103 *** (−5.85)	−11,114.77 *** [−4.99]	−10,901.36 ** [−2.16]	10845.18 [1.36]	−6237.34 *** (−7.90)
Number of People Affected by Flood	54.60008 (0.50)	−55925.41 [−0.07]	−5490.607 (−0.02)	−1908.055 (−0.01)	−518772.2 [−0.64]	−1131083 [−0.57]	3,794,894 [0.32]	−313,625.6 (−1.02)
Number of People Affected by Drought	−31.99213 (−1.39)	−552,229.1 *** [−3.75]	−61,476.59 (−1.04)	−74,580.37 (−1.21)	−1235.48 [−0.01]	48,256.81 [0.15]	64,481.36 [0.13]	86,418.55 (0.77)

Table 6. *Cont.*

Dependent Variable	(1) Cereal Production Index	(2) Maize Production	(3) Sorghum Production	(4) Millet Production	(5) Rice Production	(6) Wheat Production	(7) Barley Production	(8) Fonio Production
Number of People Affected by Storm	−0.000008 (−0.50)	−0.119833 [−1.37]	−0.0115932 (−0.33)	0.000002 (0.00)	0.0225578 [0.25]	0.0035995 [0.02]	−5.263566 [−0.90]	−66.31653 (−1.54)
Cereal Food Aid	−0.000033 *** (−2.14)	−0.94005 *** [−10.41]	−0.073275 ** (−2.00)	0.1050056 * (1.74)	−1.45745 *** [−15.80]	−1.6366 *** [−8.37]	−0.0561999 [−0.30]	0.1900641 (0.71)
Constant	177.215 *** (10.77)	1,782,212 *** [10.82]	458,177.8 *** (9.45)	403,646.2 *** (9.17)	1,150,479 *** [5.83]	1,287,005 *** [3.85]	−120,938.3 [−0.25]	535,114.8 *** (8.68)
N	952	631	540	457	598	401	165	115
Type of Regression	FE	RE	FE	FE	RE	RE	RE	FE

Note: Numbers in brackets are z-values, and in parentheses are t-values; ***, **, and * indicate statistical significance at the 1%, 5%, and 10% levels, respectively.

Table 7. Poverty and Battle.

Dependent Variable	Poverty in Rural Area (1)	Poverty in Urban Area (2)	Poverty in Urban Area (3)	Battle Related Deaths (4)	Battle Related Deaths (5)
GDP per capita growth	−0.5020925 * [−1.84]	−0.0326405 [−0.15]	−0.1644819 (−0.63)	−87.08594 ** (−2.37)	−87.32139 ** (−2.40)
Number of People Affected by Climate-related Disasters	9.285749 [1.12]	15.6348 ** [2.34]		11,854.74 *** (2.93)	
Number of People Affected by Flood			−64.6435 (−1.50)		12,956.34 (0.57)
Number of People Affected by Drought			20.16717 ** (2.57)		13,122.96 *** (3.89)
Number of People Affected by Storm			−79.01091 (−0.88)		−0.0149493 (−0.28)
Constant	23.394 *** [16.03]	11.288 *** [8.01]	12.342 *** (12.38)	1142.126 *** (3.91)	1105.389 *** (3.77)
N	77	76	75	260	260
Type of Regression	RE	RE	FE	FE	FE

Note: Numbers in parentheses are t-values; ***, **, and * indicate statistical significance at the 1%, 5%, and 10% levels, respectively.

Amid armed conflicts, another possible outcome is reduced agricultural production. Models 4 and 5 studied the impact of natural disasters on the number of battle-related deaths. Using aggregate natural disaster data, Model 4 shows that while growth in GDP per capita reduced battle-related deaths, natural disasters significantly increased the death toll. Model 5 also confirmed this aspect. Among the types of natural disasters, droughts exacerbated battle-related deaths in Africa.

As these results indicate, natural disasters reduce agricultural production and trigger poverty and armed conflicts.

Finally, Table 8 examines the various factors that contributed toward reducing the number of deaths caused by climate-related natural disasters. This analysis aimed to explore how African countries can respond and mitigate the adverse effects of climate warming.

Table 8. Factors Contributed to Reduce the Number of Deaths by Disasters.

Dependent Variable	Total Number of Deaths by Climate-Related Disasters			
	(1)	(2)	(3)	(4)
HDI (Human Development Index)	−504.342 (−1.31)	−621.4799 (−1.50)	−349.3779 * (−1.68)	−354.2997 * (−1.68)
Gov. Effectiveness	−131.8227 ** (−2.37)	−141.7638 ** (−2.48)	−54.9951 * (−1.77)	−55.76964 * (−1.76)
Control of Corruption	36.64476 (0.60)	20.4265 (0.32)	31.70747 (0.96)	31.43229 (0.93)
Regulatory Quality		49.01479 (0.76)	14.14797 (0.48)	13.79879 (0.46)
ODA for Disaster Prevention and Preparedness	−0.000008 *** (−2.78)	−0.0000073 *** (−2.65)		
Humanitarian ODA			−0.0000002 *** (−4.04)	
Emergency ODA				−0.0000002 *** (−4.16)
Constant	187.7321 (1.05)	254.3052 (1.28)	181.7999 * (1.74)	181.3178 * (1.73)
N	234	234	343	334
Type of Regression	FE	FE	FE	FE

Note: Numbers in brackets are z-values, and in parentheses are *t*-values; ***, **, and * indicate statistical significance at the 1%, 5%, and 10% levels, respectively.

Model 1 examined HDI (Human Development Index), government effectiveness, and corruption control. Corruption control is a proxy for government transparency. Government effectiveness is strongly significant in reducing the number of disaster-related deaths. The ODA for disaster prevention and preparedness has also become positive. However, the coefficient is not necessarily large enough to mitigate the impact on deaths. Therefore, the government's capacity building is critical for mitigating the impact of natural disasters. Model 2 includes the regulatory quality of government policies, but the results do not change.

Model 4 tests humanitarian aid, and Model 5 includes emergency aid. Both are statistically significant in reducing the number of deaths caused by natural disasters. This is an excellent outcome for international donors, but it is necessary to note that the coefficients are very small. In other words, to mitigate damage caused by increasing climate-related disasters, these aids are inadequate for coping with the damage. From this analysis, it can be concluded that government effectiveness is key to reducing the number of deaths caused by natural disasters.

4. Conclusions

Unlike previous studies, this study controlled for social variables and examined the crop-by-crop impact. Using panel data from African countries, this study found the following four aspects.

1. Climate-related natural disasters negatively impact per capita GDP growth and agricultural production. The impact is severe on cereal production, especially droughts (maize) and storms (rice and fonio).
2. While ODA for agriculture has a slightly positive impact, cereal aid food negatively impacts cereal production (maize, sorghum, rice, wheat).
3. Climate-related disasters, primarily droughts, increase poverty in urban areas and increase battle-related deaths. This result supports Burke et al.'s [41] argument.
4. Finally, government effectiveness is key to determining the number of deaths caused by climate-related disasters.

There are several important policy implications. First, these findings show that climate change has severe consequences not only for the development of African countries but also armed conflicts. As mentioned earlier, Africa is the least responsible for the increase in greenhouse emissions. Therefore, donor countries must initiate quick action to assist African countries and help them cope with climate change, especially climate-related natural disasters. Among natural disasters that are closely associated with human life, droughts have severe consequences on the following socio-economic activities: GDP per capita growth, agricultural production, poverty, and armed conflicts. International donors must focus on developing measures to prevent droughts and assist African countries' adaptive strategies to combat global warming.

Second, government effectiveness is key to coping with climate-related disasters. International aid must be provided to improve effectiveness. This is clear, as the coefficient of the impact to reduce the number of deaths was small on ODA for disaster prevention and preparedness, humanitarian ODA, and emergency ODA.

Third, there is a need to review if cereal aid is beneficial for African countries as it reduces cereal production, possibly due to a crowding-out effect on domestic production.

As discussed in Section 2, disaster data may be underreported. Most authoritarian governments do not overemphasize the damage caused by natural disasters. These governments have a strong incentive to underreport damages. Thus, our assessment likely underestimates the true damage level. Therefore, it is necessary to plan policies and measures that consider this possibility.

Funding: This work was supported by JSPS KAKENHI Grant Number 18H03606 and a grant from the Institute of Social Sciences, Meiji University.

Institutional Review Board Statement: Not applicable.

Informed Consent Statement: Not applicable.

Data Availability Statement: The data that support the study findings are available from Go Shimada. Restrictions apply to the availability of these data, as they are used under license for this study.

Acknowledgments: I sincerely appreciate Shinya Konaka and Toru Yanagihara for their comments on an earlier version of this study. The author is solely responsible for any errors or inaccuracies reported.

Conflicts of Interest: The author declares that he has no known competing financial interest or personal relationships that could have influenced the work reported in this study.

Appendix A

Table A1. Non-Cereal Crops.

Dependent Variable	(1) Banana Production	(2) Cassava Production	(3) Tea Production	(4) Coffee Production
GDP per capita growth	−3964.216 * (−1.87)	−13,413.36 [−1.51]	493.2195 [0.98]	87.8662 [0.26]
Educational Inequality, Gini	−35,370.73 *** (−7.14)	−123,926.1 *** [−5.87]	−1405.38 [−1.61]	623.709 [1.15]
Number of People Affected by Flood	−172,524.4 (−0.14)	4,010,153 [0.69]	−157,908.5 [−0.60]	53,130.28 [0.26]
Number of People Affected by Drought	117,739.9 (0.59)	−1,717,949 [−1.59]	33,912.43 [1.46]	−81,737.55 ** [−2.08]
Number of People Affected by Storm	0.0329793 (0.25)	−1.00887 [−1.63]	−0.0004288 [−0.04]	0.0114696 [0.55]
ODA for Agriculture	0.0002185 (0.45)	0.0047701 ** [2.15]	0.0001 [1.54]	−0.0000291 [−0.35]
Constant	2,052,232 *** (8.55)	9,049,522 *** [6.25]	106,277.7 ** [1.89]	2775.313 [0.10]
N	197	188	80	144
Type of Regression	FE	RE	RE	FE

Note: Numbers in brackets are z-values, and in parentheses are t-values; ***, **, and * indicate statistical significance at the 1%, 5%, and 10% levels, respectively.

References

1. Centre for Research on the Epidemiology of Disasters. Extreme Weather Events in Europe. Cred Crunch 2021. Available online: https://www.cred.be/publications (accessed on 1 September 2021).
2. Peduzzi, P.; Chatenoux, B.; Dao, H.; De Bono, A.; Herold, C.; Kossin, J.; Mouton, F.; Nordbeck, O. Global trends in tropical cyclone risk. *Nat. Clim. Chang.* **2012**, *2*, 289–294. [CrossRef]
3. Field, C.B.; Barros, V.R. *Climate Change 2014—Impacts, Adaptation and Vulnerability: Regional Aspects*; Cambridge University Press: Cambridge, UK, 2014.
4. Bernard, B.; Vincent, K.; Frank, M.; Anthony, E. Comparison of extreme weather events and streamflow from drought indices and a hydrological model in River Malaba, Eastern Uganda. *Int. J. Environ. Stud.* **2013**, *70*, 940–951. [CrossRef]
5. Tellman, B.; Sullivan, J.A.; Kuhn, C.; Kettner, A.J.; Doyle, C.S.; Brakenridge, G.R.; Erickson, T.A.; Slayback, D.A. Satellite imaging reveals increased proportion of population exposed to floods. *Nature* **2021**, *596*, 80–86. [CrossRef]
6. Kahn, M.E. The death toll from natural disasters: The role of income, geography, and institutions. *Rev. Econ. Stat.* **2005**, *87*, 271–284. [CrossRef]
7. Strömberg, D. Natural Disasters, Economic Development, and Humanitarian Aid. *J. Econ. Perspect.* **2007**, *21*, 199–222. [CrossRef]
8. Noy, I.; Vu, T.B. The economics of natural disasters in a developing country: The case of Vietnam. *J. Asian Econ.* **2010**, *21*, 345–354. [CrossRef]
9. Cavallo, E.A.; Noy, I. The Economics of Natural Disasters: A Survey. 2009. Available online: https://publications.iadb.org/en/publication/economics-natural-disasters-survey (accessed on 4 October 2021).
10. Cavallo, E.; Galiani, S.; Noy, I.; Pantano, J. Catastrophic natural disasters and economic growth. *Rev. Econ. Stat.* **2013**, *95*, 1549–1561. [CrossRef]
11. Shimada, G. The macroeconomic impacts of natural disasters: A case study of Japan. *GSAPs J. Grad. Sch. Asia Pac. Stud.* **2012**, *24*, 121–137.
12. Noy, I. The macroeconomic consequences of disasters. *J. Dev. Econ.* **2009**, *88*, 221–231. [CrossRef]
13. Shimada, G. The role of social capital after disasters: An empirical study of Japan based on Time-Series-Cross-Section (TSCS) data from 1981 to 2012. *Int. J. Disaster Risk Reduct.* **2015**, *14*, 388–394. [CrossRef]
14. Benson, C.; Clay, E.J. Disasters, vulnerability and the global economy: Implications for less-developed countries and poor populations. In *Developmental Entrepreneurship: Adversity, Risk, and Isolation*; Emerald Group Publishing Limited: Bingley, UK, 2006.
15. Shimada, G. A quantitative study of social capital in the tertiary sector of Kobe–Has social capital promoted economic reconstruction since the Great Hanshin Awaji Earthquake? *Int. J. Disaster Risk Reduct.* **2017**, *22*, 494–502. [CrossRef]

16. Rasmussen, T. Macroeconomic implications of natural disasters in the Caribbean WP/04.224. *IMF Work. Pap.* **2004**, *2004*, 1–25.
17. Albala-Bertrand, J. *Political Economy of Large Natural Disasters: With Special Reference to Developing Countries*; Oxford University Press: Oxford, UK, 1993; ISBN 0198287658.
18. Dacy, D.C.; Kunreuther, H. *Economics of Natural Disasters*; Implications for Federal Policy; Free Press: New York, NY, USA, 1969.
19. Toll, R.; Leek, F. Economic Analysis of Natural Disasters. In *Climate Change and Risk*; Downing, T., Olsthoorn, A., Tol, R.S., Eds.; Routledge: London, UK, 1999; pp. 308–327.
20. Skidmore, M.; Toya, H. Do natural disasters promote long-run growth? *Econ. Inq.* **2002**, *40*, 664–687. [CrossRef]
21. Sawada, Y.; Bhattcharyay, R.; Kotera, T. *Aggregate Impacts of Natural and Man-Made Disasters: A Quantitative Comparison*; Research Institute of Economy, Trade and Industry: Tokyo, Japan, 2011.
22. Charvériat, C. Natural disasters in Latin America and the Caribbean: An overview of risk. *Inter-Am. Dev. Bank (IDB) Work. Pap.* **2000**, *434*, 1–104. [CrossRef]
23. Hochrainer, S. *Assessing the Macroeconomic Impacts of Natural Disasters: Are There Any*; World Bank Policy Research Working Paper Series; The World Bank: Washington, DC, USA, 2009.
24. Vermeulen, S.J.; Aggarwal, P.K.; Ainslie, A.; Angelone, C.; Campbell, B.M.; Challinor, A.J.; Hansen, J.W.; Ingram, J.S.I.; Jarvis, A.; Kristjanson, P.; et al. Options for support to agriculture and food security under climate change. *Environ. Sci. Politics* **2012**, *15*, 136–144. [CrossRef]
25. Lesk, C.; Rowhani, P.; Ramankutty, N. Influence of extreme weather disasters on global crop production. *Nature* **2016**, *529*, 84–87. [CrossRef] [PubMed]
26. Connolly-Boutin, L.; Smit, B. Climate change, food security, and livelihoods in sub-Saharan Africa. *Reg. Environ. Chang.* **2016**, *16*, 385–399. [CrossRef]
27. Shimada, G.; Motomura, M. Building Resilience through Social Capital as a Counter-Measure to Natural Disasters in Africa: A Case Study from a Project in Pastoralist and Agro-Pastoralist Communities in Borena, in the Oromia Region of Ethiopia. *Afr. Study Monogr.* **2017**, *53*, 35–51.
28. Higuchi, Y.; Shimada, G. Industrial Policy, Industrial Development, and Structural Transformation in Asia and Africa. In *Paths to the Emerging State in Asia and Africa*; Otsuka, K., Sugihara, K., Eds.; Springer: Singapore, 2019; pp. 195–218.
29. Hosono, A.; Page, J.; Shimada, G. (Eds.) *Workers, Managers, Productivity-Kaizen in Developing Countries*; Palgrave Macmillan: Singapore, 2020.
30. Shimada, G.; Sonobe, T. Impacts of management training on workers: Evidence from Central America and the Caribbean region. *Rev. Dev. Econ.* **2021**, *25*, 1492–1514. [CrossRef]
31. Phiiri, G.K.; Egeru, A.; Ekwamu, A. Climate change and agriculture nexus in Sub-Saharan Africa: The agonizing reality for smallholder farmers. *Int. J. Curr. Res. Rev.* **2016**, *8*, 57.
32. Schultz, T.W. Transforming traditional agriculture. In *Transforming Traditional Agriculture*; University of Chicago Press: Chicago, IL, USA, 1964.
33. Schultz, T.W. Investment in human capital. *Am. Econ. Rev.* **1961**, *51*, 1–17.
34. Cuny, F.C. *Disasters and Development*; Intertect Press: Dallas, TX, USA, 1994.
35. Raschky, P.; Schwindt, M. *Aid, Natural Disasters and the Samaritan's Dilemma*; World Bank Policy Research Working Paper Series; The World Bank: Washington, DC, USA, 2009. [CrossRef]
36. De Juan, A.; Pierskalla, J.; Schwarz, E. Natural disasters, aid distribution, and social conflict–Micro-level evidence from the 2015 earthquake in Nepal. *World Dev.* **2020**, *126*, 104715. [CrossRef]
37. Becerra, O.; Cavallo, E.; Noy, I. Foreign aid in the aftermath of large natural disasters. *Rev. Dev. Econ.* **2014**, *18*, 445–460. [CrossRef]
38. Andor, M.A.; Osberghaus, D.; Simora, M. Natural disasters and governmental aid: Is there a charity hazard? *Ecol. Econ.* **2020**, *169*, 106534. [CrossRef]
39. Barrett, C.B. Measuring food insecurity. *Science* **2010**, *327*, 825–828. [CrossRef]
40. O'Loughlin, J.; Linke, A.M.; Witmer, F.D. Effects of temperature and precipitation variability on the risk of violence in sub-Saharan Africa, 1980–2012. *Proc. Natl. Acad. Sci. USA* **2014**, *111*, 16712–16717. [CrossRef]
41. Burke, M.B.; Miguel, E.; Satyanath, S.; Dykema, J.A.; Lobell, D.B. Warming increases the risk of civil war in Africa. *Proc. Natl. Acad. Sci. USA* **2009**, *106*, 20670–20674. [CrossRef]
42. Buhaug, H. Climate not to blame for African civil wars. *Proc. Natl. Acad. Sci. USA* **2010**, *107*, 16477–16482. [CrossRef]
43. Burke, M.B.; Miguel, E.; Satyanath, S.; Dykema, J.A.; Lobell, D.B. Climate robustly linked to African civil war. *Proc. Natl. Acad. Sci. USA* **2010**, *107*, E185. [CrossRef]
44. Schultz, T.W. *On Investing in Specialized Human Capital to Attain Increasing Returns*; Oxford Blackwell: Oxford, UK, 1988.
45. The World Bank. *World Development Indicators*; The World Bank: Washington, DC, USA, 2021.
46. EM-DAT. *The Emergency Events Database Université Catholique de Louvain (UCL)-CRED, D. Guha-Sapir*; EM-DAT: Brussels, Belgium, 2021; Available online: http://www.emdat.be (accessed on 1 September 2021).
47. Food Agriculture Organization of the United Nations. *FAOSTAT Statistical Database*; Food Agriculture Organization of the United Nations: Rome, Italy, 2021.
48. Coppedge, M.; Gerring, J.; Knutsen, C.H.; Lindberg, S.I.; Teorell, J.; Altman, D.; Bernhard, M.; Fish, M.S.; Glynn, A.; Hicken, A.; et al. *V-Dem Codebook v9*; The V-Dem Institute: Gothenburg, Sweden, 2019.
49. Jerven, M. *Poor Numbers*; Cornell University Press: Ithaca, NY, USA, 2013.

International Journal of
Environmental Research and Public Health

Article

Disaster Risk Reduction Funding: Investment Cycle for Flood Protection in Japan

Mikio Ishiwatari [1,*] and Daisuke Sasaki [2]

1   Graduate School of Frontier Sciences, The University of Tokyo, Kashiwa 277-8563, Japan
2   International Research Institute of Disaster Science, Tohoku University, Sendai 980-8577, Japan;
    dsasaki@irides.tohoku.ac.jp
*   Correspondence: ishiwatari.mikio@jica.go.jp

Abstract: Background: Investment in disaster risk reduction is crucial in order to mitigate disaster damage. However, for many countries, particularly developing ones, financing investment in disaster risk reduction is challenging. This study aims to examine the factors that affect investments in flood protection and the approaches to securing investments by analyzing investment trends in Japan. Methods: This study examines 150 years of flood protection and investment cycles that helped reduce damages in Japan. The dataset of flood protection budgets, flood damage, and national income since 1878 was created from public statistics. Documents and reports concerned with disaster management, river management, and finance were examined. Results: The study found five investment cycles of flood protection from the late 19th century to the present. The country established financing mechanisms, such as legislation and long-term plans, following major flood disasters. However, external shocks such as war, economic recession, disaster, and tightened national finance had a major impact on these investments. The fluctuations in the budget created an investment cycle. The country had increased its budget to 0.9% of its national income in the 1990s. It often experienced flood damage accounting for over 1% of the national income until 1961, but succeeded in decreasing the damage to less than 1%, and currently it is limited to less than 0.4%. Conclusions: The financial mechanisms established from the long-term perspective could support an increase in budgets for flood protection, leading to a decrease in damage. However, established financing mechanisms may weaken the financial flexibility of the country.

Keywords: financing mechanism; flood protection; investment cycle; investment in DRR; Japan; long-term plan; lost decades; Sendai Framework for Disaster Risk Reduction

Citation: Ishiwatari, M.; Sasaki, D. Disaster Risk Reduction Funding: Investment Cycle for Flood Protection in Japan. *Int. J. Environ. Res. Public Health* **2022**, *19*, 3346. https://doi.org/10.3390/ijerph19063346

Academic Editor: Timothy W. Collins

Received: 8 February 2022
Accepted: 8 March 2022
Published: 11 March 2022

Publisher's Note: MDPI stays neutral with regard to jurisdictional claims in published maps and institutional affiliations.

1. Introduction

Investment in disaster risk reduction (DRR) is crucial for mitigating disaster damage, which is increasing in most parts of the world due to socio-economic and climatic changes [1,2]. Since disasters hinder growth and sustainable development, reducing disaster risks could promote the achievement of the Sustainable Development Goals (SDGs) [3]. The Japanese government identified DRR as one of the priority areas for the promotion of SDGs, in particular making cities resilient, relating to SDG 11, and adapting to disaster risks increased by climate change, relating to SDG 13 [4]. The Sendai Framework for DRR (SFDRR), which UN member states agreed to in 2015, shows the paradigm of DRR and emphasizes investing in DRR as one of the four priority actions [5]. However, continuously increasing investments in DRR is a challenge [6–8]. Governments are often forced to reduce flood protection investments or divert parts of this to other sectors because of external factors such as economic recessions and wars. This study examines investment trends and factors that affect investments in flood protection in Japan. It also aims to identify various approaches to securing investment for DRR.

It is well known that capital investment creates an economic cycle with a period ranging from 7–11 years. This cycle, first identified by the French economist Clement Juglar

197

and called the business cycle, is affected by the timing of capital investment corresponding to the durability of the company's equipment [9]. Investment in DRR can also create a cycle affected by fluctuating budgets for flood protection and changing flood damage.

Ishiwatari and Sasaki [10], were able to define an investment cycle for flood protection by examining investment trends in Asia's major flood-prone economies. These economies increased government budgets for flood protection following major flood disasters, but were then unable to sustain these budgets. The share-to-gross domestic product for the budgets of flood protection and flood damage forms cycles. This study analyzes similar investment cycles in Japan's almost 150 years of flood protection funding.

Evolving Mechanisms of Flood Protection in Japan

Recent studies have examined the trends in flood protection in Japan from a wide range of perspectives, such as technology, institutions, environments, and local communities [11–18]. However, only a limited number of studies have examined investment issues.

Historically, the Japanese have always fought against flooding. The earliest reference to disaster management of floods can be traced to the 4th century when the emperor instructed the construction of the first river dikes with irrigation facilities along the Yodogawa River in Osaka Prefecture [19]. Local communities were historically responsible for protecting their own houses and agricultural lands since, during the 16th century, only Bakufu, the military government headed by shoguns, and federal lords protected the major towns and castles because of limited technical and financial resources [11]. Takahasi [12] examined the changes in the relationship between local communities and the government since the federal period and found that local communities became less involved in flood protection, as river works required higher levels of engineering in the modern period.

Following the Meiji revolution in the late 19th century, the country started modernizing its socio-economy using Western technology and introduced flood protection technology from the Netherlands and other Western countries. Takei's [13] review of the evolving trends of technology and institutions in flood protection, since the Meiji era found that governments promoted flood protection works following large-scale flood disasters, but were unable to implement these continuously due to changing political and financial situations. Takahasi and Uitto [14] traced the evolution of river management policies from the Meiji era and stressed that the country was forced to change its policy to include environmental conservation in river management during high growth in the 1970s and 1980s. Kajiwara [15] examined changes in river administration since the Meiji era from the perspectives of legislation, institution, and technology. He found that the Japanese government focused on technological issues led by engineers, integration of flood protection with water resource development through construction of multipurpose dams, and promoting projects via long-term planning. Nakamura and Oki [16] examined the increasing trends in the safety levels of flood protection programs in major rivers since 1910 and argued that Japan's, flood risk management has experienced a paradigm shift. where technology has become a priority due to social, economic, and climate (flooding) evolution. Nakamura [17] reviewed the modern history of flood protection by analyzing approaches to determining targeted flood volumes regarding the safety levels of floods. The government, because of its financial and social constraints, initially used recorded maximum floods to decide targeted volumes. In the high growth period since the 1960s, the government applied probability analysis to formulate river basin plans and increased the safety levels of flooding. For example, the government is currently implementing projects to protect the Tokyo and Osaka metropolitan areas against once-in-200-year floods in the major rivers. The government developed methods for examining investment efficiency in the 1960s and formulated manuals for the economic analysis of the projects for flood protection. These manuals include the methods developed for estimating benefits by analyzing asset values in flood-affected areas [18].

2. Methods

Ishiwatari and Sasaki [10] conceptualized the investment cycle for flood protection by examining investment trends in major flood-prone economies in Asia (Figure 1a). Usually, governments start to increase budgets for flood protection following severe flood disasters and these investments are usually for managing flood damage. Ideally, governments try to maintain the budget at a specific scale and continue to mitigate the damage caused by floods (Figure 1b). In reality, because of socio-economic or political changes, such as wars and economic recession, governments may reduce this budget over time. Again, when severe floods occur, the limited budget is inadequate for managing the disaster damage, and this leads to the start of a new investment cycle.

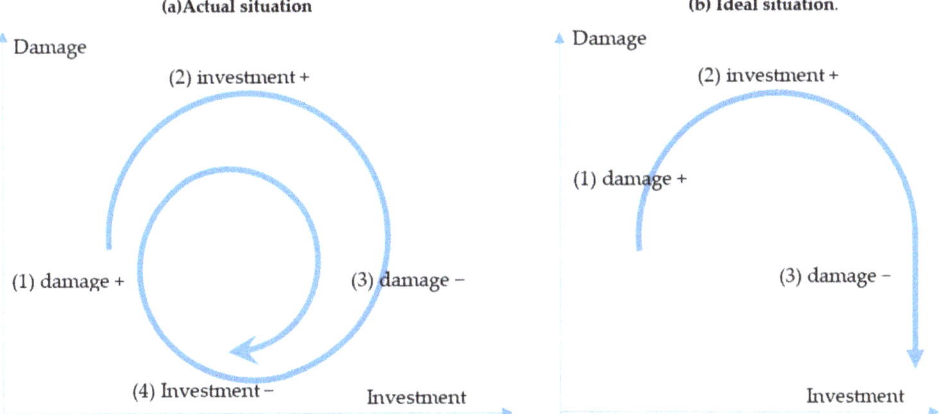

Figure 1. Concept of investment cycle. Source: Authors' elaboration.

The dataset for flood protection budgets, flood damage, and national income (NI) since 1878 used in this study was generated from public statistics. We use NI because government data on gross domestic product, before 1955, were unavailable. The documents and reports of government organizations concerned with disaster management, river management, and finance were examined to study the financial and socioeconomic situations affecting flood protection investment.

This study examines annual damage and budgets in the share-of-NI to identify investment cycles. The investment cycle for flood protection is determined as follows:

1. Large disasters that triggered an increase in the budget for flood protection were identified, and periods of increasing damage were determined.
2. The periods of increasing budgets following large disasters were determined.
3. The periods of decreasing budgets and damage before the next large disaster were determined.

3. Results and Discussion

3.1. Five Investment Cycles for Flood Protection in Japan

In this study, the authors found the following five cycles of investment:

I 1878–1906 Establishing the modernized mechanism of flood protection.
II 1906–1931 Constructing structural frameworks, such as channels and dikes, for major rivers.
III 1931–1945 National land devastation during the wartime regime.
IV 1945–1958 Responding to a series of flood disasters.
V 1958–2014 Implementing flood protection during high growth and recession.

The factors forming these five cycles are summarized in Table 1.

Table 1. Outline of investment cycles.

	(1) Damage Increase	(2) Budget Increase	(3) Damage Decrease	(4) Budget Decrease	Background
	Disasters of Trigger (Share-of-NI)	Instruments to Increase (Share-of-NI)		Causes of Decrease	
		I Establishing a modernized mechanism of flood protection			
1878–1906	1878–1896 1896 flood (11)	1896–1899 River Law (0.9)	1899–1901	1901–1906 Russo-JPN War	Modernization & Industrialization
		II Constructing structural framework in major rivers			
1906–1931	1906–1910 1907 & 1910 floods (4)	1910–1914 Long-term plan Special account (0.6)	1914–1923	1923–1931 Kanto Earthquake, Great Depression	Economic growth
		III National land devastation in wartime regime			
1931–1945	1931–1935 1934 Muroto Typhoon 1935 floods (4)	NA	NA	1935–1945 Sino-JPN War, WWII	Wartime regime
		IV Responding to a series of flood disasters			
1945–1958	1945–1947 1947, Kathleen typhoon (10)	1947–1953 Responding to major disasters (0.6)	1953–1955	1955–1958 Shift to transport & other sectors	a series of severe floods
		V Implementing flood protection during high growth and recession			
1958–2014	1958–1959 1959 Isewan Typhoon (5)	1959–1982 Long-term plan Revising river law (0.9)	1982–2000	2000–2014 Economic Recession & Tight national budget	High growth & lost decades

Source: Authors' elaboration.

3.1.1. Establishing a Modernized Mechanism for Flood Protection (1878–1906)

Flood damage increased as the country modernized. Japan established mechanisms for implementing national flood protection projects based on the enacted river law in 1896 (Figure 2).

Increasing Damage (1878–1896)

Japan modernized and developed its economy by introducing Western technology and systems following the Meiji Revolution in the late 19th century. The country developed its light industry, particularly the spinning and yarn-making industries, and built large-scale factories for this.

Flood damage began to increase because of economic development and modernization. The assets of factories and urban facilities were accumulated in flood prone areas. Forestry in mountainous and hilly regions had diminished during the revolution, leading to an increase in flood volume in rivers.

Damage and human losses from flooding more than tripled in the 1890s compared to the 1880s [13]. Flood damage in 1885, 1889 and 1893 reached over 4% of NI. The 1885 flood submerged most of Osaka city and affected 270,000 people. In 1889 and 1893, typhoons damaged Wakayama and Okayama prefectures. Flood and tsunami disasters in 1896 caused damage throughout the country, the annual economic damage reaching 11% of NI.

While the damage caused by floods increased, the national government's involvement in implementing flood protection measures was limited. The national government conducted river works mainly for navigation to support economic development, whereas prefecture governments and local communities were responsible for flood protection and conducted low-cost works for limited areas with traditional technology. However, prefecture governments and communities were unable to respond to the increasing flood disasters because of their limited technical and financial capacities [20].

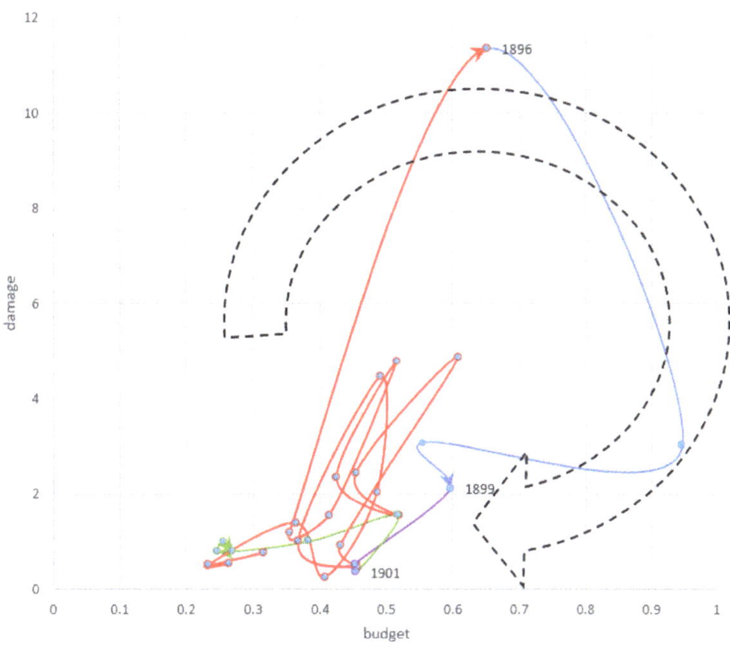

Figure 2. First investment cycle 1878–1906 (Share of the National Income %). Source: Authors' elaboration.

Increasing Budget (1896–1899)

The government launched national flood protection projects following a series of flood disasters in the 1880s and 1890s, and the budget for flood protection reached 0.9% of NI in 1897. The River Law, enacted in 1896, established a scheme for promoting the national government's flood protection projects. These measures, which prefecture governments and local communities were unable to implement, covered multiple prefectures, needed modern technology, and incurred enormous costs [21]. The landowners of farmlands requested that the Imperial Diet take measures to mitigate the increasing flood damage. Flood disasters reduced the rental income of landowners from tenant farmers, and the government's income from property taxes decreased. As major voters, the landowners were able to influence the Diet, since the Diet was established in 1890 under restricted voting rights based on property tax payments [13]. Thereafter, flood protection became the second most important infrastructure policy after railroads [14].

The national government became directly involved in measures taken to prevent floods across ten major rivers [22]. For example, the government promoted the construction of a diversion channel in the Yodogawa River from 1896 to 1910 to protect the downtown area of Osaka from flooding, as well as river improvement works in the Chikugogawa River in Kyusyu in the western Japan region from 1896. This period saw the introduction of Western technology for flood protection. Japanese engineers studying civil engineering in Europe led these projects, while the government invited Dutch engineers as advisors. The concept of flood protection involved collecting as much rainfall as possible in the river and discharging it as quickly as possible through the river channels between the dikes. The river channels were widened and excavated, and continuous dikes were built from the mountains to the sea. These systems prevented floods in urban areas and low-lying farmlands and promoted land use for development activities [23].

Decreasing Damage (1899–1901)

The flood damage decreased to 0.4% of NI. The government started national projects in six major rivers before 1900 [24].

Decreasing Budget (1901–1906)

The government shifted its financing to the Russo-Japanese War in 1904 and 1905 and decreased the budget for other sectors, including flood protection. The budget reduced to less than 0.3% of the NI. The government managed a temporary special account for the military during the war from 1904 to 1906. This account, separated from general revenues and expenditures, was established to manage the expenses necessary for military operations until the end of the war [25].

3.1.2. Constructing a Structural Framework in Major Rivers (1906–1931)

Following major disasters in 1907 and 1910, the government established mechanisms for long-term planning and special accounts to secure multi-year commitment for investment in flood protection. However, the budget for flood protection declined because of the reconstruction after the Great Kanto Earthquake in 1923 and in response to the 1929 Great Depression (Figure 3).

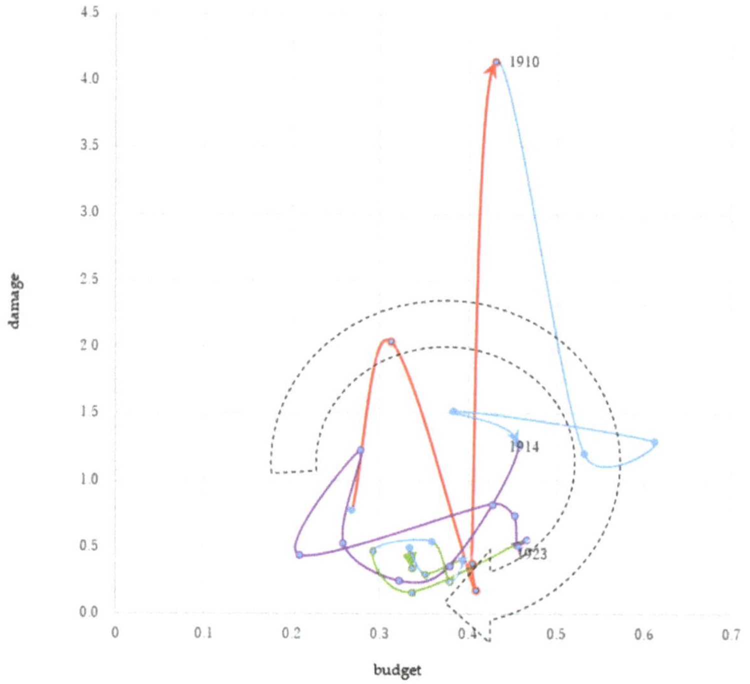

Figure 3. Second investment cycle 1906–1931 (Share of the National Income %). Source: Authors' elaboration.

Increasing Damage (1906–1910)

Two typhoons hit Yamanashi and Kyoto Prefectures simultaneously in 1907. The economic damage reached 2% of NI that year. The flood disasters in 1910 left around 2500 people dead or missing in the Eastern Japan Region, including Tokyo. Economic damage reached over 4% of NI.

Increasing Budget (1910–1914)

Following the 1907 and 1910 disasters, the government formulated its first long-term flood protection plan in 1911. This long-term plan covered 65 rivers and envisaged completing work in 20 major rivers over 18 years as the first phase and 30 rivers as the second phase [26]. For example, a diversion channel project was started in the Arakawa River in 1911 to protect downtown Tokyo, where urbanization was progressing. This plan required 1.7% of the national budget annually. The government enacted the Law for the Special Account of Flood Protection and created a specific account to manage flood protection, separate from the general national account [27]. This account was financed by the national budget, cost-sharing of local governments, and loans from postal savings.

The long-term plan was implemented as planned in 1911 and 1912. In 1913, the government reduced its budgets for flood protection because of the consolidation of state finances.

Decreasing Damage (1914–1923)

In 1915, the government abolished the special account for flood protection and suspended borrowing funds from postal savings to improve the nation's financial situation. Postal savings did not increase during this period. The government could not continue the long-term plan because of the delayed progress of work and the financial conditions affected by inflation after World War I. While the first long-term plan envisaged completing work in all 10 rivers by 1921, in reality the work was completed for only two rivers [26]. The government formulated the second long-term plan in 1921, which covered 73 major rivers [13].

Decreasing Budget (1923–1931)

The budget for flood protection declined to 0.3% of NI, since the government had to allocate funds for rehabilitation from the Great Kanto Earthquake in 1923 and the 1929 Great Depression. The earthquake killed over 100,000 people in Tokyo and its neighboring areas.

Despite this, the government was able to complete continuous high-dike systems for major rivers, because of the flood protection investments made over almost 30 years. These systems formed the structural framework of the major rivers and substantially decreased flood damage by protecting areas previously inundated in alluvial plains. Examples of this are the diversion channel in Tokyo in the Arakawa River and the Okouchi channel in Niigata in the Shinanogawa River, which became functional in 1924 and 1931, respectively.

3.1.3. Wartime Regime (1931–1945)

In the years 1931 to 1945, no investment cycle for flood protection was formed because the national government funds had to be diverted to finance military expenditures. Even though severe floods had occurred in 1934 and 1935, the government decreased the budget for flood protection, reaching its lowest level in history. This period forms an anti-clockwise, reverse investment cycle (Figure 4).

Increasing Damage (1931–1935)

In 1934, the Muroto Typhoon hit Osaka and caused a high tide disaster, leaving nearly 3000 dead or missing. This year's damage reached 3.9% of NI. In 1935, typhoons and heavy rainfall caused flood disasters in Kyoto, Tohoku, and Kyushu. This year's damage reached 2.1% of NI.

The government increased public expenditure, including the budget for flood protection, in 1932 and 1933 to overcome deflation [28], but did not increase it following the 1934 flood. The flood protection budgets exceeded 0.5% of NI in 1933 and 1934 and then decreased to approximately 0.3% until the 1940s.

The government could not start projects in 41 out of the 73 rivers targeted by the second long-term plan of 1921 as the priorities had to change because of flood disasters [13]. The government revised the existing long-term plan and formulated a third long-term plan in 1933. This plan envisaged completing projects in 24 rivers in need of urgent repair for

15 years, and included subsidies for the repair of small- and medium-sized rivers and for erosion control managed by prefectures [26].

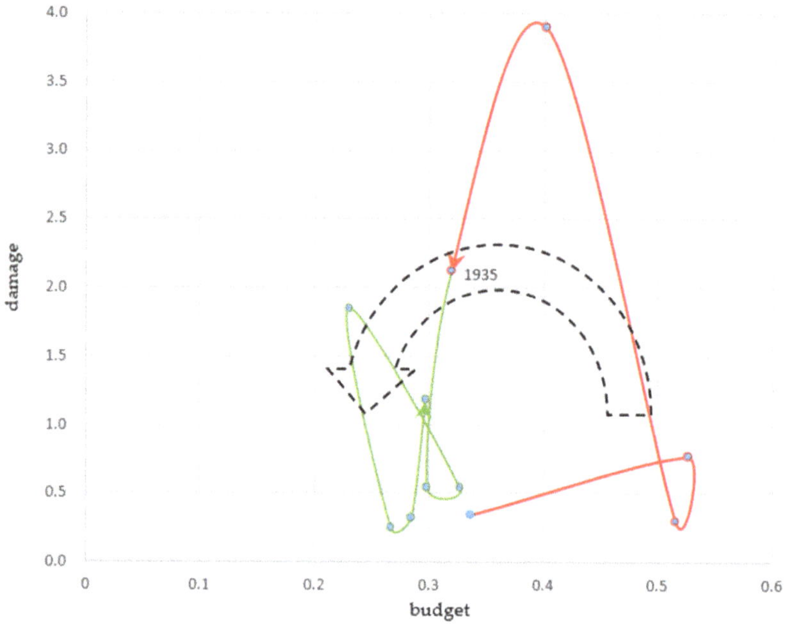

Figure 4. Third investment cycle 1931–1945 (Share of the National Income %). Source: Authors' elaboration.

Decreasing Budget (1935–1945)

The government could not sufficiently start working on its third long-term plan, because Japan had shifted to a wartime regime with the outbreak of the Second Sino-Japanese War in 1937. The government established and managed a temporary military special account until 1946. The war was long, and the government had to drastically cut general expenditures, including the flood protection budget, and increase war expenditures [25]. The budget for flood protection decreased to 0.2% of NI in 1943 and 1944, the lowest since the Meiji era.

3.1.4. A Series of Flood Disasters (1945–1958)

Japan had to recover from World War II and respond to a series of severe floods that occurred in the 1940s and the 1950s (Figure 5).

Increasing Damage (1945–1947)

In 1947, the Kathleen typhoon devastated the Kanto and Tohoku regions. The dikes of the Tonegawa River were broken, and a wide area of the Greater Tokyo was submerged. Some 1000 people died, and this year's damage reached over 10% of NI.

From 1946 to 1954, Japan's economic damage exceeded over 2% of NI each year. The average number of dead and missing individuals during this period was over 1300 per year. This damage was unusual, considering that in the 1910s, 1920s and 1930s, that is, before World War II, there were only two years, 1934 and 1935, when the damage exceeded 2%.

Several factors have been reported to have caused such large damage. The government could not allocate sufficient funds for flood protection during the wartime regime. The budget was less than 0.3% of the NI in 1935. Second, the large-scale destruction of mountains removed a natural barrier against floods. During this period, the amount of timber logged almost doubled, mostly consumed by the military (from the 1930s) and for reconstruction following the war. Since major cities were damaged by the war, and food and goods were in short supply, large amounts of lumber were needed for reconstruction [29]. Takahasi [12] pointed out that flood protection measures, since the Meiji era, caused higher peak flood flows in the middle and lower reaches of rivers, increasing flood damage. Continuous high dikes had been constructed to allow floodwaters to drain quickly, but these also retained floodwaters, which previously would overflow upstream into the river channels.

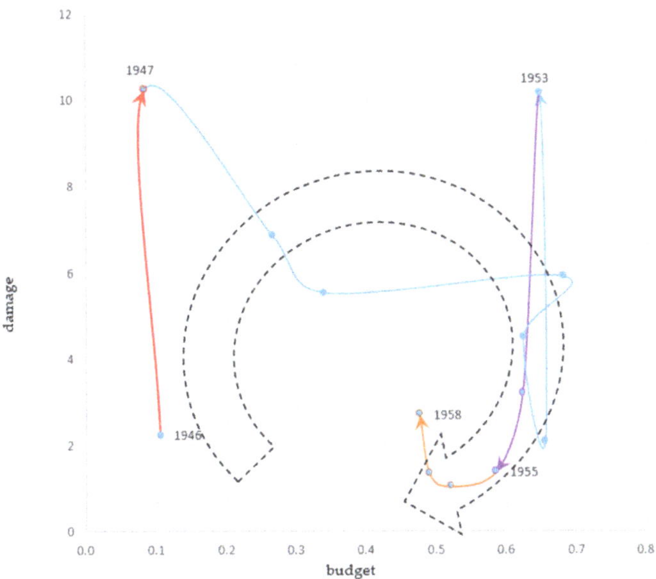

Figure 5. Fourth investment cycle 1945–1958 (Share of the National Income %). Source: Authors' elaboration.

Increasing Budget (1947–1953)

The government formulated a national budget focusing on flood protection in 1950 [25]. The budget for flood protection reached 0.68% of NI in 1950.

While the budget increased, the damage did not decrease substantially and remained over 2% of NI. In 1953, three major flood disasters caused by typhoons and heavy rainfall damaged the western Japan, south Kansai, and Tokai regions. This year's damage reached over 10% of NI.

Decreasing Damage (1953–1955)

The cabinet established the Council of Measures for Forest Protection and Flood Protection in 1953 to promote flood management, considering the severe disasters following World War II [30]. The council consists of ministers of finance, construction, forestry, and academic experts. The council formulated basic guidelines for forest protection and flood protection, with costs accounting for over 40 years of budget. Because of the scale, the concerned ministries could not agree to implement this guideline. From 1954, the damage began to decline and stayed at less than 4% of NI.

Decreasing Budget (1955–1958)

The budget for flood protection decreased from 0.59% to 0.48% of NI. The government allocated more financing to areas directly related to developing industrial infrastructure, such as roads and ports. The Japanese economy recovered to pre-war levels, and the country planned to expand it further [31]. Additionally, the government adopted a tighter fiscal policy to curb demand from 1954 to 1958.

3.1.5. Implementing Flood Protection during High Growth and Recession (1958–2014)

The government developed financing mechanisms for long-term planning and special accounting with legislation following the 1959 Isewan Typhoon disaster. The country could increase the budget for flood protection for two decades during high economic growth and kept it at the highest level of approximately 0.8% of the NI with damage equivalent to less than 0.4% of the NI for the next two decades (Figure 6).

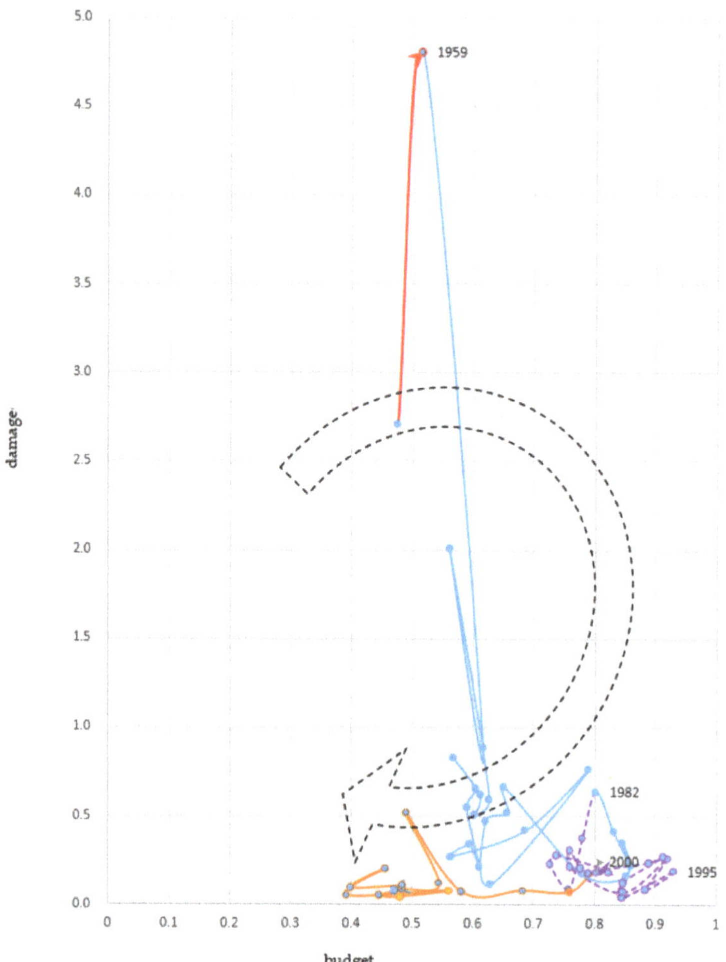

Figure 6. Fifth investment cycle 1958–2014 (Share of the National Income %). Source: Authors' elaboration.

Increasing Damage (1958–1959)

The Isewan Typhoon left 5098 people dead or missing in 1959, the highest number ever recorded for a typhoon disaster. This year's damage reached 4.8% of NI. The storm surge caused by the typhoon damaged Nagoya City, the Aichi Prefecture, and the Mie Prefecture. The bay area around Nagoya City has been created by land reclamation since the 16th century, and most of the land is located in low-lying lands below sea level. This area was urbanized because of the economic boom after World War I, the military boom during World War II, and economic recovery and expansion after the Korean War. However, disaster-prevention measures for this region had not yet been developed.

Increasing Budget (1959–1982)

The government established mechanisms for securing stable budgets to implement flood protection with a long-term perspective following the Isewan Typhoon disaster. The budget for flood protection increased from 0.5% of NI in 1959 to over 0.8% in 1978, and remained at this level until 1982. Areas protected from floods increased to 24% in 1970 and 32% in 1980 [32]. Flood damage decreased to less than 1% of NI during this period, except in 1961. The Second Muroto Typhoon and Baiu rain front damaged 2% of the NI in that year.

The Emergency Measures for Forest Protection and Flood Protection and Special Account for Flood Protection were enacted in 1960. The Cabinet decided on a 10-year plan for flood protection with committed budget plans, which was the first long-term plan following World War II. Before the first plan, draft plans were rejected three times in the 1950s because the Ministry of Finance did not agree on them [18]. The plan aims to complete river works in 100 major rivers throughout the country within 15 years. The budget planned for 10 years accounted for 7% of NI. This amount was decided by shrinking the total estimated project costs without considering national economic growth or the share of the total government budget [33]. In 1960, the first year of the long-term plan, the budget increased by over 40%. The government created a special account to manage budgets for flood protection, which was separate from the general account (Figure 7). This special account received funds from electric companies and the local governments shared some of the costs. In addition, the general account of the national government along with the special account provided budgets for national projects and subsidies to local government.

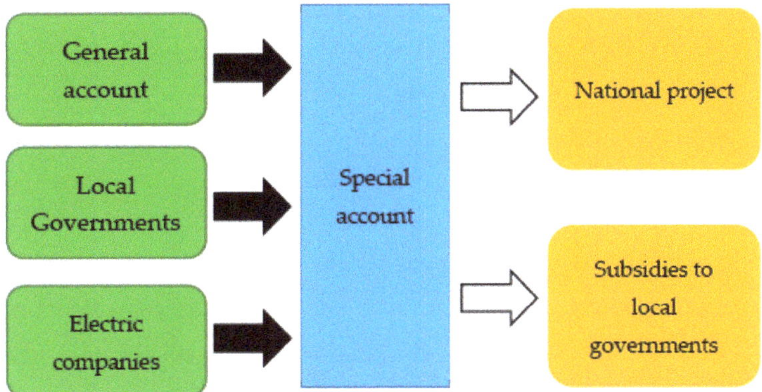

Figure 7. Mechanism of special account for flood protection. Source: Authors' elaboration.

Since 1959, the Japanese government has adopted an expansionary fiscal policy to promote economic growth and create the foundation for growth. The government rapidly expanded its fiscal scale and increased its fiscal activities, including spending on flood protection. In 1965, the government began issuing deficit covering government bonds to stimulate the economy [25].

The administration system for managing rivers was changed by revising the river law in 1964. The national government is responsible for managing major rivers from the perspective of the river basin, and can improve flood protection in a balanced manner between left and right banks, as well as upstream and downstream. Under the original law, prefectural governors were responsible for managing rivers and could implement protection work, which sometimes caused conflicts with other prefectures in the same basin [34].

Japan needed to respond to drastic socio-economic changes, such as urbanization and development activities, in the 1960s, when the average annual real growth rate of the gross domestic product was over 10%. The long-term plans for flood protection were revised before the end of the planning period to respond to financial and socio-economic changes.

Since the national budget expanded and the need for investing in urban areas increased, the government invested 18% more than the planned expenditure on flood protection for the five years between 1960 and 1965 [35]. In 1965, the Cabinet decided on a second long-term plan for five years by revising the first 10-year plan at midyear. The second plan aimed to develop flood protection facilities in a balanced manner nationwide to conserve and develop national lands in response to socio-economic development, stabilize people's lives, and strengthen the industrial foundation. The Cabinet revised the second plan and decided on a third plan, which increased the planned budget by 85% in 1968, the midyear of the second plan. The Cabinet decided on a fourth plan for five years in 1972, in which the budget was double that of the third plan [35].

The purpose of these long-term plans has evolved. Originally, their aim was to secure budgets, but this later changed to setting long-term goals for flood protection projects [36]. For example, the fourth plan in 1972 included decreasing flood damage in urban rivers and improving the river environment, and the eighth plan in 1991 included disaster management and supporting the local community's development (Figure 8).

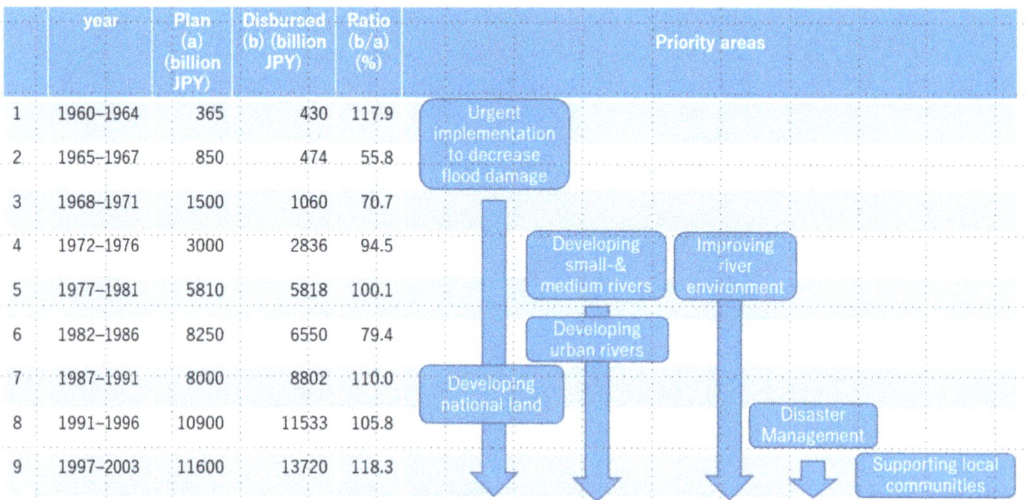

	year	Plan (a) (billion JPY)	Disbursed (b) (billion JPY)	Ratio (b/a) (%)	Priority areas
1	1960–1964	365	430	117.9	
2	1965–1967	850	474	55.8	
3	1968–1971	1500	1060	70.7	
4	1972–1976	3000	2836	94.5	
5	1977–1981	5810	5818	100.1	
6	1982–1986	8250	6550	79.4	
7	1987–1991	8000	8802	110.0	
8	1991–1996	10900	11533	105.8	
9	1997–2003	11600	13720	118.3	

Figure 8. Long-term plans. Source: Modified from Okamura (2002).

According to these long-term plans, Japan has revised its strategy for flood protection to respond to socio-economical changes. The Japanese government started new programs, increasing investment in urban areas, to mitigate the damage accelerated by urbanization in the 1970s. These urban programs cover software measures, such as evacuation and early warning, and land use regulation, in addition to conversional structural measures [37]. The government began its nature-oriented river management program in 1990 to respond to growing public environmental awareness. Further, the government expanded this program to include green infrastructure programs in response to the Great East Japan Earthquake and Tsunami in 2011 [38]. To adapt to the adverse effects of climate change, the government initiated the new strategy "River Basin Disaster Resilience and Sustainability by All" in 2021, which engages the whole society, including the central government, local government, private sector, civil societies, and local communities. This strategy comprehensively covers various measures: (a) conventional structure of flood protection; (b) exposure reduction by relocation from risk areas; and (c) software measures for warning, evacuation, response, and recovery [18,39].

The government has also started to incorporate science-based planning for flood protection projects. It adapted a probability approach to decide safety levels in the 1950s and applied the flood scale of once-in 80–100-years as the targeted safety levels for major rivers [17]. The Ministry of Construction published technical criteria for designing flood scales using probability analysis in 1958. Before World War II, the government decided on safety levels according to recorded flood volumes.

The safety levels of flood scales designed for protection have increased as the flood protection budget has expanded. The flood scale in major rivers, throughout the country, increased substantially after 1975 [16]. For example, the government is currently promoting works for securing the safety levels of major rivers in metropolitan areas at the safety levels of once in 200- or 150-year scale of floods, following several revisions of design flood volumes. The design flood volume almost doubled in the Yodogawa River, flowing through the Osaka metropolitan area [35].

The development of technology for managing dams has also enabled an increase in safety levels. The Law of Specified Multipurpose Dams was enacted in 1957 to coordinate multiple ministries and water user organizations for multipurpose dam construction [40]. The purpose of multipurpose dams is to supply urban water and protect against flooding. This law aims to promote multi-purpose dam construction by designating the construction minister to implement and manage dams and by clarifying the cost-sharing mechanisms among water users.

Decreasing Damage (1982–2000)

The flood damage decreased to less than 0.4% of the NI during this period. The protected area increased from 32% in 1980 to over 50% in 2000.

As the economy continued to stagnate in the early 1990s due to the collapse of the bubble economy, the government began to stimulate the economy through fiscal policy. In 1995, the budget for flood protection reached 0.93% of the NI, the highest on record.

Decreasing Budget (2000–Present)

The government has decreased the budget for flood protection since 2000 because of its severe financial situation. The government applied financial policies to stimulate the economy in the 1990s, but changed to tight fiscal expenditure in 2000. Since the formation of the Koizumi cabinet in 2001, the government has promoted austerity policies focusing on the reduction of public works [41]. The share of the NI in the flood protection budget halved from over 0.8% in 2000 to 0.4% in 2010. Damage was limited to less than 0.2%, except for 0.5% in 2004. In 2003, the Cabinet formulated the Priority Plan for Social Infrastructure Development by integrating nine long-term sectoral plans, including flood protection [42]. The cabinet recognized that the long-term plans for each sector lack flexibility in allocating budgets among sectors and in responding to economic and financial situations [43]. Since the new infrastructure development plans do not include targeted budgets, the government has lost sight of its long-term commitments.

3.2. Mechanisms for Securing Budgets for Flood Protection

Factors affecting flood damage and investment for flood protection can be summarized in Figure 9. Major disasters triggered the establishment of financing mechanisms for flood protection. The River Law was enacted in 1896 following major disasters in the 1880s and the 1890s. The law enabled the national government to directly implement flood protection works that local communities and the prefectural government were responsible for initially. The 1910 flood drove the formulation of the first long-term plan for flood protection that mentions multiyear budget targets. Following the Isewan typhoon disaster in 1959, the cabinet began formulating a long-term plan in 1960.

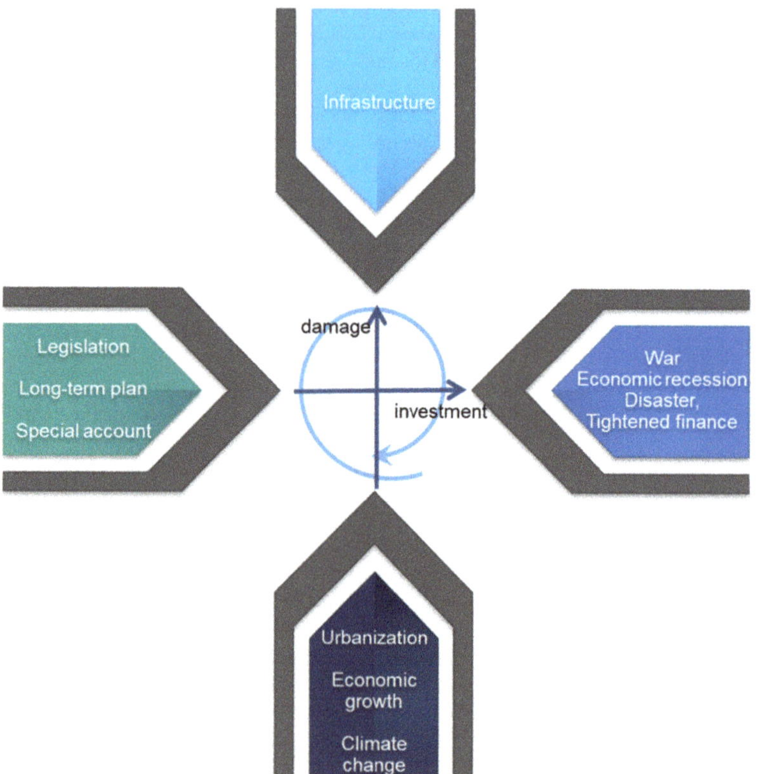

Figure 9. Factors affecting flood damage and investment for flood protection. Source: Authors' elaboration.

Legislation can provide a foundation for increasing investment in flood protection. Until the enactment of the River Law in 1896, the government could not increase budgets for flood protection, even when severe floods damaged more than 4% of the NI. After enacting this law, the government could increase budgets to develop infrastructure for flood protection following large-scale disasters that caused damage equivalent to more than 4% of the NI with the exception of the 1934 flood during the military expansion period.

The robust mechanism of financing investment enabled stable budgets from 1958 until 2000, even in times of economic recession and other shocks. This is why the last cycle shows longer periods than the other four cycles. The government has secured multi-year financing commitments by developing long-term plans and a special account for flood protection. As the cabinet, including the finance minister, decide on long-term plans involving budget amounts planned for multiple years, annual fluctuations in budgets could be avoided. Long-term plans continued to exist, being formulated nine times from 1960 to 2003 even in periods of worsening national economic and financial situations. The focus of these plans changed from the urgent need to decrease flood damage to environmental improvement and support for the local economy. Thus, Japan succeeded in increasing and maintaining its budget. The budget increased from 0.5% of NI in the late 1950s to approximately 0.9% of NI in the 1990s. Damage was limited to less than 1% of NI in 1962 and further decreased to less than 0.2% of NI in the 2000s, except for 0.5% in 2004. Flood damage decreased continuously during this period.

Securing commitment to a sector means losing flexibility in budget arrangements from the perspective of national finance. The administration led by Prime Minister Koizumi recognized that these mechanisms disturb the restructuring of the national budget and abolished the long-term plans and special accounts for specific sectors, including flood protection, in 2003 [42]. After a high-growth period, Japan has faced severe financial constraints because of economic stagnation since the 1990s.

While legislation and financial schemes can contribute to securing budgets for flood protection, major external shocks have affected the financing of investments. Before World War II, the country established a mechanism for long-term planning and special accounts. However, because of the shocks of wars, large-scale disasters, economic recession, and inflation, the government could not finance flood protection investments as planned. Also, the government is currently reducing flood protection budgets, since the Japanese economy is experiencing a long period of stagnation known as the "lost decades" following the collapse of the bubble economy in 1991 [44].

3.3. Implications of Fulfilling SFDRR Targets for Japan

While Japan achieved Target E of SFDRR, formulating DRR strategies at the national and local levels, the country has not reported its baseline and achievements on the other four targets: Target A. mortality; B. people affected; C. economic loss; and D. critical infrastructure [45]. This is because the country has not established definitions and a database. As discussed in this paper, Japan has developed a database for economic damage and mortality caused by flooding for the whole of the previous century. However, databases of damage caused by other natural hazards, such as earthquakes and volcanic eruptions, were not developed properly. Furthermore, there are no clear definitions for people affected and critical infrastructure. We have found a major research gap, which should be filled to develop the database and report its progresses for Target A–D. Furthermore, we consider that national and local government should improve their strategies as Target E of the SFDRR stipulates, by including the issue of financing investment in DRR. Currently, strategies are ignoring this important issue. As this paper discusses, securing budgets for the long-term perspective is crucial for mitigating disaster damage, leading to promotion of SDGs. It is our hope that this study can contribute to finding a clue for further improvement of Japanese DRR strategies in the future.

4. Conclusions

This study examines 150 years of flood protection and investment cycles that helped reduce damages in Japan. It examines the factors that can influence investments in flood protection and approaches that can be taken to secure such investments.

The study identified five investment cycles for flood protection and clarified the definitive triggers of major flood disasters in terms of increasing the budget for flood protection in these cycles. Japan has established mechanisms for securing budgets from a long-term perspective following major flood disasters. These mechanisms include legislation for implementing national work, long-term plans decided by the cabinet, and special accounts specifically managed for flood protection. By increasing investment, Japan could substantially reduce flood damage. However, these investments were also influenced by external shocks such as wars, economic recessions, disasters, and tightened national finance. Through these cycles, the country increased its budget to 0.9% of NI in the 1990s. It often experienced flood damage of over 1% of NI until 1961, but succeeded in decreasing damages to less than 1%, and currently this limit is less than 0.2%.

Developing countries, which are suffering from increasing flood damage because of the socioeconomic and climatic changes, apply these mechanisms to secure investment in flood protection. These mechanisms could support sustaining budgets at a certain level, even though the budgets are affected by various shocks.

The Japanese case also shows that established mechanisms lose flexibility due to the overall financial arrangement of the country. Once a national agency secures a multiyear budget for flood protection, the government cannot immediately reduce it even if the national financial situation worsens.

Future studies should examine investment cycles of DRR in other countries. Some developing countries, such as the People's Republic of China and the Philippines, are increasing their investments in DRR following recent major disasters. Comparing the factors affecting investments in these countries with those in Japan could provide useful lessons for other countries for establishment of financing mechanisms for DRR.

Author Contributions: Conceptualization, M.I. and D.S.; methodology, M.I.; validation, M.I. and D.S.; formal analysis, M.I.; investigation, M.I. and D.S.; resources, M.I.; data curation, M.I.; writing—original draft preparation, M.I.; writing—review and editing, D.S.; visualization, D.S.; supervision, M.I. All authors have read and agreed to the published version of the manuscript.

Funding: This work was supported by the collaborative research project of the International Research Institute of Disaster Science (IRIDeS), Tohoku University.

Institutional Review Board Statement: Not applicable.

Informed Consent Statement: Not applicable.

Data Availability Statement: The data that support the findings of this study are available from the corresponding author upon reasonable request.

Conflicts of Interest: The authors declare no conflict of interest. The funders had no role in the design of the study; in the collection, analyses, or interpretation of data; in the writing of the manuscript or in the decision to publish the results.

References

1. Chatterjee, R.; Shiwaku, K.; Gupta, R.D.; Nakano, G.; Shaw, R. Bangkok to Sendai and beyond: Implications for disaster risk reduction in Asia. *Int. J. Disaster Risk Sci.* **2015**, *6*, 177–188. [CrossRef]
2. Ishiwatari, M.; Surjan, A. Good enough today is not enough tomorrow: Challenges of increasing investments in disaster risk reduction and climate change adaptation. *Prog. Disaster Sci.* **2019**, *1*, 100007. [CrossRef]
3. Ishiwatari, M.; Sasaki, D. Investing in flood protection in Asia: An empirical study focusing on the relationship between investment and damage. *Prog. Disaster Sci.* **2021**, *12*, 100197. [CrossRef]
4. The Government of Japan. Voluntary National Review 2021 Report on the Implementation of 2030 Agenda: Toward Achieving the SDGs in the Post-COVID19 Era. 2021. Available online: https://sustainabledevelopment.un.org/content/documents/289572 10714_VNR_2021_Japan.pdf (accessed on 20 February 2022).
5. UNISDR. *Sendai Framework for Disaster Risk Reduction 2015–2030*; UNSIDR: Geneva, Switzerland, 2015.
6. Adeniyi, O.; Perera, S.; Collins, A. Review of finance and investment in disaster resilience in the built environment. *Int. J. Strateg. Prop. Manag.* **2016**, *20*, 224–238. [CrossRef]
7. Henstra, D.; Thistlethwaite, J. *Overcoming Barriers to Meeting the Sendai Framework for Disaster Risk Reduction*; Center for International Governance Innovation: Waterloo, ON, Canada, 2017.
8. Mizutori, M. Reflections on the Sendai Framework for disaster risk reduction: Five years since its adoption. *Int. J. Disaster Risk Sci.* **2020**, *11*, 147–151. [CrossRef]
9. Korotayev, A.V.; Tsirel, S.V. A spectral analysis of world GDP dynamics: Kondratieff waves, Kuznets swings, Juglar and Kitchin cycles in global economic development, and the 2008–2009 economic crisis. *Struct. Dyn.* **2010**, *4*, 1. [CrossRef]
10. Ishiwatari, M.; Sasaki, D. *Investments in Flood Protection: Trends in Flood Damage and Protection in Growing Asian Economies*; JICA Institute: Tokyo, Japan, 2021.
11. Ishiwatari, M. Government Roles in Community-Based Disaster Risk Reduction. In *Community-Based Disaster Risk Reduction*; Shaw, R., Ed.; Emerald Group Publishing Limited: Bingley, UK, 2012; pp. 19–33.
12. Takahasi, Y. History of water management in Japan from the end of world war II. *Int. J. Water Resour. Dev.* **2009**, *25*, 547–553. [CrossRef]
13. Takei, A. *Research on Relationship between Technology and Institution in Flood Control in Japan*; Kishimoto Syuppan: Kobe, Japan, 2017. (Original Work Published 1961); (In Japanese)
14. Takahasi, Y.; Uitto, J.I. Evolution of river management in Japan: From focus on economic benefits to a comprehensive view. *Glob. Environ. Change* **2004**, *14*, 63–70. [CrossRef]
15. Kajiwara, K. *River Administration in Modern Japan: Deployment Policy and Legislation from 1868 until 2019*; Horitsubunkasha: Kyoto, Japan, 2021. (In Japanese)
16. Nakamura, S.; Oki, T. Paradigm shifts on flood risk management in Japan: Detecting triggers of design flood revisions in the modern era. *Water Resour. Res.* **2018**, *54*, 5504–5515. [CrossRef]
17. Nakamura, S. *Floods and Probability: Modern History of Technology and Society for Designed Flood Volumes*; The University of Tokyo Publication: Tokyo, Japan, 2021.

18. Koike, T. Evolution of Japan's flood control planning and policy in response to climate change risks and social changes. *Water Policy* **2021**, *23*, 77–84. [CrossRef]
19. Ranghieri, F.; Ishiwatari, M. *Learning from Megadisasters: Lessons from the Great East Japan Earthquake*; World Bank Publication: Washington, DC, USA, 2014.
20. Matsuura, S.; Fujii, M. A study on the progress of the river administration from the consideration of Embankment Law in 1875 to the institution of River Law of 1896. *Res. Civ. Eng. Hist.* **1994**, *14*, 61–76. (In Japanese)
21. Ishiwatari, M.; Sasaki, D. *Bridging the Gaps in Infrastructure Investment for Flood Protection in Asia*; JICA Research Institute: Tokyo, Japan, 2020.
22. Shinohara, O. *What Do Three Generations of River Engineers Have Seen? Koichi Ando, Yutaka Takahashi, Takashi Okuma, and Modern River Administration for 150 Years*; Nobunkyo Production: Tokyo, Japan, 2018.
23. Institute for International Cooperation. *Disaster Management and Development: Improving Disaster Management Capacity of Society*; JICA: Tokyo, Japan, 2003. (In Japanese)
24. Yamamoto, S.; Matsuura, S. Establishing the former River Law and River Administration (2). *Suiri Kagaku* **1996**, *40*, 51–78. (In Japanese)
25. Editing Office of 100-Year History of Ministry of Finance. *100-Year History of Ministry of Finance*; Ministry of Finance: Tokyo, Japan, 1969.
26. Matsuura, S. The History of Making Long Term Flood Control Program and The Transition of its Basic Concept. *Civ. Eng. Hist. Jpn.* **1986**, *6*, 147–155. [CrossRef]
27. Matsuura, S. Severe Flood Damage in 1910 and Settle Process of the First Long-Term Flood Control Program. *J. Reg. Dev. Stud.* **2008**, *11*, 149–173.
28. Umeda, M. *The Background to Japan's Overcoming Deflation in the Early 1930s: Exchange Rate Policy, Monetary Policy, and Fiscal Policy*; Institute for Monetary and Economic Studies; Bank of Japan: Tokyo, Japan, 2005. (In Japanese)
29. Forestry Agency. *White Paper on Forest and Forestry*; Forestry Agency: Tokyo, Japan, 2014.
30. Nishikawa, T. The History of Long-term plans for flood protection 4. *Water Sci.* **1965**, *44*, 128–151. (In Japanese)
31. Nishikawa, T. The long-term plans of river administration. *Water Sci.* **1961**, *4*, 122–138. (In Japanese)
32. Ministry of Land, Infrastructure, Transport and Tourism (MLIT). *White Paper on Land, Infrastructure, Transport and Tourism*; MLIT: Tokyo, Japan, 2016.
33. Nakayasu, Y. Long-term plan for flood protection project. *River* **1962**, *197*, 2–3. (In Japanese)
34. Takemura, K. The historical process of river administration during the period of modernization in Japan: Dam construction and environmental improvement. *Nippon Suisan Gakkaishi* **2007**, *73*, 103–107. (In Japanese) [CrossRef]
35. Matsuura, S. The River Policy for the Age of Rapid Economic Growth in Japan. *Int. Reg. Study* **2010**, *13*, 57–76.
36. Okamura, J. History and importance of the river improvement and management long-term plan. *River* **2002**, *673*, 7–10. (In Japanese)
37. Ishiwatari, M. What are crucial issues in promoting an integrated approach for flood risk management in urban areas? *Jpn. Soc. Innov. J.* **2016**, *6*, 15–26. [CrossRef]
38. Nakamura, K. Nature-based solutions for river restoration in Japan. In *Financing Investment in Disaster Risk Reduction and Climate Change Adaptation: Asian Perspective*; Ishiwatari, M., Sasaki, D., Eds.; Springer: Singapore, 2022.
39. Ishiwatari, M. Disaster Risk Reduction. In *Handbook of Climate Change Mitigation and Adaptation*; Lackner, M., Sajjadi, B., Chen, W.Y., Eds.; Springer: New York, NY, USA, 2020. [CrossRef]
40. Japan Commission on Large Dams. *Dams in Japan: Past, Present and Future*; CRC Press: Leiden, The Netherlands, 2009.
41. Lincoln, E.J. Japan in 2001: A depressing year. *Asian Surv.* **2002**, *42*, 67–88. [CrossRef]
42. MLIT. *White Paper on Land, Infrastructure and Transport*; MLIT: Tokyo, Japan, 2003.
43. Prime Minister's Office. Structural Reform and the Medium-Term Economic and Fiscal Perspectives. 2002. Available online: http://www.kantei.go.jp/jp/kakugikettei/2002/0125tenbou.html (accessed on 20 February 2022).
44. Hasumi, R.; Iiboshi, H.; Nakamura, D. Trends, cycles and lost decades: Decomposition from a DSGE model with endogenous growth. *Jpn. World Econ.* **2018**, *46*, 9–28. [CrossRef]
45. UNDRR. Measuring Implementation of the Sendai Framework, Undated. Available online: https://sendaimonitor.undrr.org/ (accessed on 20 February 2022).

International Journal of
*Environmental Research
and Public Health*

MDPI

Article

Analysis in the Influencing Factors of Climate-Responsive Behaviors of Maize Growers: Evidence from China

Hongpeng Guo, Yujie Xia, Chulin Pan, Qingyong Lei and Hong Pan *

College of Biological and Agricultural Engineering, Jilin University, 5988 Renmin Street, Changchun 130022, China; ghp@jlu.edu.cn (H.G.); xiayj1997@163.com (Y.X.); pancl@jlu.edu.cn (C.P.); leiqingyong@sina.com (Q.L.)
* Correspondence: yy991205@163.com

Abstract: Due to the natural production properties, agriculture has been adversely affected by global warming. As an important link between individual household farmers and modern agriculture, it is crucial to study the influence of agricultural productive services on farmers' climate-responsive behaviors to promote sustainable development and improve agricultural production. In this paper, a questionnaire survey has been conducted among 374 maize farmers by using the combination of typical sampling and random sampling in Jilin Province of China. Moreover, the Poisson regression and the multi-variate Probit model have been used to analyze the effects of agricultural productive services on the choices of climate-responsive behaviors as well as the intensity of the behaviors. The results have shown that the switch to suitable varieties according to the frost-free period have been mostly common among maize growers in Jilin province. Agricultural productive services have a significant effect on the adoption intensity of climate- responsive behaviors, at the 1% level. Based on this conclusion, this paper proposes policy recommendations for establishing a sound agricultural social service system and strengthening the support for agricultural productive services. It has certain reference significance for avoiding climate risk and reducing agricultural pollution in regions with similar production characteristics worldwide.

Keywords: multi-variate Probit model; Poisson regression model; agricultural productive services

Citation: Guo, H.; Xia, Y.; Pan, C.; Lei, Q.; Pan, H. Analysis in the Influencing Factors of Climate-Responsive Behaviors of Maize Growers: Evidence from China. *Int. J. Environ. Res. Public Health* **2022**, *19*, 4274. https:// doi.org/10.3390/ijerph19074274

Academic Editors: Mikio Ishiwatari and Daisuke Sasaki

Received: 16 February 2022
Accepted: 29 March 2022
Published: 2 April 2022

Publisher's Note: MDPI stays neutral with regard to jurisdictional claims in published maps and institutional affiliations.

1. Introduction

Climate change has become a serious constraint on social and economic development. Many studies have zoomed into this issue in agriculture area, which is highly dependent on climate-sensitive resources [1,2]. Climate change will intensify the existing pressure on land and water resources. Due to the vulnerability of agriculture, climate change will affect poverty levels, especially those in Africa and Asia [1,3,4]. As a country with a population of 1.4 billion, food security is a crucial part of the national livelihoods of China. It is necessary to study the impacts of climate change on agricultural production and hence to come up with corresponding strategies to adapt to climate change to ensure food security [5,6]. The mode of agricultural production affects the severity of climate change in the future. Therefore, reducing greenhouse gas emissions in agriculture will mitigate the degree of climate change. The purpose of climate-smart agriculture is to solve the problem of poverty and food security under climate change. At the same time, it will mitigate the impacts of climate change on agricultural production [7,8]. The popularity of climate-smart agriculture has led to a growing number of the related literatures. Climate-smart agriculture has three major objectives: to increase agricultural yields, to enhance adaptation to climate change, and to mitigate greenhouse gas emissions [5,9–11]. Therefore, the analysis of the influencing factors of farmers' low-carbon production adaptation practice plays an important role in achieving the goal of climate-smart agriculture.

In response to climate change, farmers will adopt some adjustments in agricultural production to avoid risks and to maintain the ecological environment, including low-carbon production practice and climate adaptation practice. Emission reduction refers to a series of measures to reduce the greenhouse gas emissions in agricultural production without affecting the economic efficiency of agricultural production. Through certain technologies, the material inputs and agricultural pollution have been reduced while ensuring agricultural production so as to achieve the joint development of economic, social and ecological benefits [12–14]. At present, scholars have conducted relevant studies and they have found that gender, age, and education level are governing factors affecting the low-carbon production practice [15,16]. Among household characteristics, family income, household size, and family business scale affect farmers' low-carbon production practice greatly [16–19]. Moreover, policy and socio-cultural level also have certain impacts on farmers' low-carbon production practice and fiscal incentives affect farmers' emission reduction behaviors [20–22]. Social norms affect farmers' soil conservation behaviors while the cost of low-carbon production practice affects farmers' willingness for adaptation [16]. Farmers' concerns for the environment and their technological awareness have certain impacts on farmers' emission reduction behaviors. When considering promoting farmers' emission reduction behaviors, farmers should be allowed to fully understand the costs and the risks of changing agricultural production methods [22–26]. Technical guidance and social service will reduce farmers' resistance to low-carbon production practice, therefore promoting farmers' low-carbon production behaviors [5,22,27,28]. The promotion of low-carbon agricultural production practice can effectively reduce greenhouse gas emissions, thus realizing the goal of climate-smart agriculture [14,22].

Adaptation is a deliberate change process of farmers themselves, which can cope with various pressures and changes that affecting people's lives, to reduce the vulnerability of agriculture and improve its resilience [22,29]. Climate adaptation practice refers to the process of coping with the impacts of known or unknown climate change on agriculture by adjusting agricultural production mode [30]. Farmers are the microscopic main body of agricultural production and they are also the smallest decision-making unit for implementing climate adaptation practice. When farmers encounter or anticipate climate change during agricultural production, they will take corresponding measures to reduce agricultural production losses to maintain their normal yields. Past survey has shown that farmers will make corresponding changes in response to climate change [31]. Family characteristics also have significant impacts on the intensity of farmers' climate adaptation practice and it is reflected in the form of human capital, social capital, economic capital and natural endowments. There's a positive correlation shown between climate adaptation practice and family size, household income and relationship with neighbors [32]. There is also a positive correlation between climate adaptation practice and ecological environment [32]. The readiness of technology and resources and the availability of information affect farmers' climate adaptation practice deeply [29,32]. Therefore, the level of technical service is a key factor affecting farmers' climate adaptation practice [33]. The climate-adaptive behavior of farmers is also affected by pressure and institutions [29,34,35].

According to previous studies, climate-responsive behaviors can reduce agricultural production losses, maintain food security. Meanwhile, it can increase farmers' income and promote stable global food production. It helps realize the goal of climate-smart agriculture. These behaviors are a form of "sustainable intensification", which is an approach of agriculture that increases production without increasing adverse effects on the environment [36]. In this paper, climate response behaviors are divided into two categories: one is the low-carbon practice actively implemented by farmers, which is generally due to farmers' awareness of the deterioration of the natural environment, and the purpose of which is to reduce agricultural pollution and agricultural emissions. The second is the climate adaptation practice that is passively implemented by farmers. These behaviors are the changes in production behaviors that farmers have to make due to climate change and climate disasters. Studying and evaluating the influencing factors of climate-responsive behaviors will help achieve the goals of climate-smart agriculture. Farmers are the main research object in this paper. Exploring the degrees of influence of individual characteristics, family characteristics, social characteristics, and access to productive agricultural services on farmers' behaviors will help in the establishment of related policies around the world. They can take locally targeted measures to protect agro-ecosystems, mitigate climate change and increase yields. Moreover, it will drive the development of climate-smart agriculture, contributing to the realization of modern agricultural goals and agricultural sustainability [37].

1.1. Theoretical Analysis and Research Hypothesis

1.1.1. Theoretical Analysis

The continuous global climate change will intensify the risks related to farmers' agricultural production. As a rational economic entity, farmers will adopt various measures to ensure the yields and the product quality to maximize their benefits. At present, there are few studies on the impacts of agricultural productive services on farmers' climate-responsive behaviors. However, there is a lack of systematic theoretical support for it. Therefore, we can treat the climate-responsive behaviors due to climate change as the farmers' technology adoption behaviors. Hence the theory of farmers' technology adoption behaviors can be used to analyze the adoption of farmers' climate-responsive behaviors.

Drawing on the research ideas of A. Saha and other scholars, this paper constructs the decision model of farmers' climate-responsive behaviors below [38].

$$maxH = E_{i*}\left[U(\widetilde{W})\right] = E_{i*}\{U[p(f(m) + g(z)\widetilde{e}) - w(m + z) - rz]\}$$
$$m + z = x, \widetilde{Q} = f(m) + g(z)\widetilde{e} \tag{1}$$

In this formula: H represents the net income of farmers, E_{i*} represents the income expectation of farmers when the amount of information is i^*, U represents the income function, $\widetilde{W}$ represents a series of factors that affect the net income of farmers, $\widetilde{Q}$ means that when there are m units of arable land area without climate-responsive behaviors and z units of arable land area adopting climate-responsive behaviors, where the total output of crops is represented by $m + z = x$. x is the total arable land area. Generally, the risks of farmers adopt climate-responsive behaviors is greater than the risks when they do not adopt climate-responsive behaviors. So, when climate-responsive behaviors are not adopted, the production function is non-random $f(^{\circ})$. When the climate-responsive behaviors are adopted, the production function is $g(z)\widetilde{e}$ and $\widetilde{e}$ is a random variable. w represents the costs incurred by farmers when climate-responsive behaviors are not adopted, r represents the additional costs incurred by farmers when adopting climate-responsive behaviors, and p represents the product price.

Since the main objective is to maximize benefits, whether to adopt climate-responsive behaviors depends on if there's any changes to their expected benefits. With other conditions and factors remain unchanged, if the expected benefits of the farmers when adopting climate-responsive behaviors are greater than that of not adopting climate-responsive behaviors, they will make the adoption for their own good. The conditions for the farmers to adopt the climate-responsive behaviors are below:

$$p_1 g(m)\widetilde{e}(Z) - (w + r)m \geq p_0 f(m) - wm \tag{2}$$

$g(^{\circ})$ is the production function after the climate-responsive behaviors are adopted. p_1 represents the price of crops after the climate-responsive behaviors are adopted. p_0 represents the crop price when the climate-responsive behaviors are not adopted. m is the decision scale. $f(^{\circ})$ is the production function without the climate-responsive behaviors. $\widetilde{e}(Z)$ represents the subjective risk function which is determined by Z. Z is a factor that affects decision-making and $\widetilde{e}(Z) \in [0,1]$. The price of crops changes little before and after the farmers adopt the climate-responsive behaviors, so it can be assumed that $p_0 = p_1$, then,

$$\widetilde{e}(Z) \geq \frac{p_0 f(m) + rm}{p_0 g(m)} \tag{3}$$

Price, production function and costs are relatively easy to determine in the above formula, while the subjective risk function $\tilde{e}(Z)$ is difficult to be determined. It depends on the adoption of agricultural productive services and the ambient environment in which they are located at. Hence, we can change the mathematical expression to:

$$\tilde{e}(Z) = F\{G(I), H(O)\} \tag{4}$$

In (4), $G(I)$ indicates the resource endowment that affect farmers' adoption of climate-responsive behaviors, including but not limited to personal characteristics, family characteristics, and social capital. $H(O)$ indicates other external factors that affect farmers' adoption of climate-responsive behaviors, such as the adoption of agricultural productive services. The climate-responsive behaviors of farmers are affected by both.

1.1.2. Research Hypothesis
Individual Characteristics of Farmers

Studies have shown significant differences between male and female farmers in terms of technology adoption in agricultural production [39]. At present, most agricultural production have been carried out by women. Compared with male farmers, females know better about how to take corresponding actions or measures to make adjustments for their agricultural production [33]. These measures help reduce the economic losses caused by climate change. However, other studies have also shown that, due to the influence of traditional concepts, women receive fewer resources compared with men, leading to the fact that women have been less empowered to make adjustment in their farming practice [15]. In Northeast China, men are still in the dominating position in their households. They make more decisions in practice, and they have more resource advantages. So, the male farmers in Northeast China tend to adopt climate-responsive behaviors in agricultural production. Some studies have shown that older people have more experience in terms of adjustments of agricultural production. As a result, these adjustment are more flexible, and their climate-responsive behaviors are more practical [33,40]. However, other studies have also shown that the age factor has a negative correlation with the adoption of new technology [40]. Education level can improve farmers' ability to deal with climate change and adopt new technology used in future agricultural production [1,16,29].

Hypothesis 1. *In comparison with females, male farmers are more likely to adopt climate-responsive behaviors. Age, experience, and education level have significant positive impacts on climate-responsive behaviors.*

Characteristics of Farmer Households' Management

Farmer households with more family members can provide more labor resources for agricultural production. At the same time, their family members can do non-agricultural works, which will increase their overall family income and help relieve monetary pressure. This ensures better planning and conduction of agricultural production [17,38,41]. Moreover, households with more family members use more conservation tillage methods, and they tend to change more often on fertilizer usage [32,42]. Farmers with larger planting areas are generally more professional; hence they are more likely to adopt new technology [38]. Social capital can accelerate the delivery of information, which helps promote of agricultural technology adoption. In the meantime, it can also increase opportunities for collaboration and the successful application of loans. As a result, social capital can improve farmers' awareness of climate, thereby affecting their climate-responsive behaviors [32].

Hypothesis 2. *Household size, planting areas, and social capital have a significant positive impact on the adoption of climate-responsive behaviors.*

Government Subsidies and Loans

Different policy backgrounds affect farmers' perception of climate change differently, and available subsidies from the government can reduce farmers' monetary pressure, increase their production motivation and enhance farmers' cooperation. This can help reduce the operating costs incurred, and farmers will be more likely to change their corresponding planting behaviors under climate change [20,22,23,25,35]. Loans can also reduce the monetary pressure farmers face in the agricultural production process, helping them to adopt new technology faster. Moreover, it encourages farmers to easily purchase fertilizers, pesticides, and new planting varieties. At the same time, it increases the possibility of improving the basic infrastructures of agricultural production. Generally, the availability of agricultural loans has a positive correlation with climate-responsive behaviors [32].

Hypothesis 3. *Government subsidies and loans have a significant positive impact on the adoption of climate-responsive behaviors.*

Agricultural Productive Services

Agricultural productive services can assist agricultural production at all stages, from pre-production, production to post-production. It makes agricultural production activities more professional and well structured. At the same time, it also improves agricultural production efficiency greatly. Moreover, agricultural productive services improve labor-shortage problems, introduce advanced agricultural production technology to farmers [43,44]. It can improve farmers' understanding of climate change and increase the intensity of farmers' climate-responsive behaviors [32]. Furthermore, agricultural productive services can increase farmers' awareness of advanced technology, helping them adopt low-carbon production practices easily. As a result, the promotion of related production behaviors can be improved [5,15,22,26–28].

Hypothesis 4. *Agricultural productive services have a positive impact on the adoption of climate-responsive behaviors.*

2. Materials and Methods

2.1. Model Setting

2.1.1. Poisson Regression

When farmers perform climate-responsive behaviors, there will be differences in their behaviors' intensity. The number of behaviors implemented is not a binary variable, whereas it is a limited dependent variable that takes only positive integer values. Moreover, the dependent variable is the sum of the types of farmers' climate-responsive behaviors, which is a counting variable. Linear models are not suitable. For such dependent variables, the Poisson distribution has been chosen [45,46].

$$P(Y_i = y_i | x_i) = \frac{e^{-\lambda_i} \lambda_i^{y_i}}{y_i!} (y_i = 0, 1, 2, \cdots)$$

Since the number of behaviors is a counting variable, such discrete variables obey the Poisson distribution. The model can be set as below:

For individual i, this article assumes that the explained variable is Y_i. $\lambda_i > 0$, which is the "Poisson arrival rate", representing the average number of events. It is determined by x_i. The expectation and variance of the variable distributions are the same in the Possion regression and they are equal to the Poisson arrival rate. In order to ensure that the Poisson arrival rate is negative, this paper makes the assumptions:

$$E(Y_i | x_i) = \lambda_i = \exp(x_i'\beta)$$

In this way, the intensity of the farmers' climate-responsive behaviors can be linked to the explanatory variables. In this paper, the Poisson regression has been used to calculate the impacts of farmers' use of productive services on the intensity of the implementation of climate-responsive behaviors.

2.1.2. Multi-Variate Probit Model

The Probit model is a discrete model and generally it's used to fit a $0 - 1$ dependent variable regression. ε is an error term that obeys the standard normal distribution [47].

Ordinary multiple linear regression equation:

$$Y = \beta_0 + \sum_{i=1}^{n} \beta_i x_i + \varepsilon$$
$$E(y|x) = 1 \cdot P(y = 1|x) + 0 \cdot P(y = 0|x) = P(y = 1|x)$$

If $F(x, \beta)$ is the standard normal cumulative distribution function:

$$P(y = 1|x) = F(x, \beta) = \Phi(x', \beta) = \int_{-\infty}^{x'\beta} \varphi(t)dt$$

When studying the choice of n types of production behaviors of farmers, it is necessary to estimate the n Probit models. The implicit assumption is that the error term ε between the n Probit models are not related to each other. However, in actual agricultural production, farmers may choose a variety of climate-responsive behaviors and these behaviors are not mutually exclusive [48]. Therefore, some variables that cannot be observed may affect farmers to adopt different climate-responsive behaviors at the same time. The above-mentioned error term may be correlated. Therefore, this article uses a multi-variate Probit model for the estimation. It contains multiple binary dependent variables:

$$y_{im}^* = \alpha_i S + \beta_i Control + \varepsilon_{im}$$
$$y_{im} = \begin{cases} 1, & if\ y_{im}^* > 0 \\ 0, & x \end{cases}$$

Among them, y_{im}^* is a potential variable for the individual i to implement m kinds of climate-responsive behaviors. $m = 1, 2, 3, 4, 5, 6, 7$, indicate the adoption of planting-breeding combination methods, the conservation tillage methods, the switch-over to organic fertilizers, the change of suitable varieties according to the frost-free periods, the adjustment of pesticides and fertilizer usage, the adjustment of the time of sowing and harvesting, and the replenishing seedlings, respectively. y_{im} is the final result variable and ε_{im} is a random disturbance term, which obeys a multi-variate normal distribution with a mean value of 0 and a covariance of K, $\varepsilon_{im} \sim (0, k)$.

2.2. Data Sources

Data used in this article comes from questionnaire surveys conducted on maize farmers by the research team from June to September in 2021 in Jilin province in Northeast China. Based on existing documents and interviews with farmers, combined with the characteristics of agricultural production in Jilin province. The area has an average annual rainfall of 400–800 mm, 3000 h of sunshine and high-quality soil. The farmers come from Changchun, Songyuan, Jilin, Siping and Liaoyuan cities in Jilin Province. This area is characterized by rain hot during the same period. It is flat, fertile and has four distinct seasons, making it a good place to grow corn in China. Questionnaires on agricultural productive services and climate-responsive behaviors have been made. It includes the basic information of farmer households, the adoption of agricultural productive services, and their awareness of climate-responsive behaviors. A combination of typical sampling and random sampling has been used for sample selection. A total of 389 questionnaires have been conducted with, 374 of them were being valid. The overall effective rate was 96%.

2.3. Variable Setting and Descriptive Statistics

To determine the impacts of agricultural productive services on climate-responsive behaviors, it is necessary to determine what climate-responsive behaviors are. There are various types of low-carbon practice and climate adaptation practice. However, not all of these methods are suitable for the farmers in Jilin province of China. After literature review and in-person interviews, related information has been obtained. At present, the main low-carbon production practice aiming to reduce agricultural pollution used by farmers in Jilin province includes planting-breeding combination methods, conservation tillage methods, and organic fertilizers. Climate adaptation practice aiming at reducing yield losses and increasing profits includes the change of suitable varieties according to the frost-free periods, the adjustment of pesticides and fertilizer usage, the adjustment of the time of sowing and harvesting, and the filling of gaps with seedlings.

Figure 1 shows the adoption of farmers' climate-responsive behaviors. From the left to the right: planting-breeding combination methods (b1), conservation tillage methods (b2), use of organic fertilizers (b3), the change of suitable varieties according to the frost-free periods (b4), adjustment of pesticides and fertilizer usage (b5), adjustment of the time of sowing and harvesting (b6), filling the gaps with seedlings (b7), respectively.behaviors. Among the farmers being surveyed, changing suitable planting varieties according to the frost-free periods is the most used method, followed by adjusting pesticides and fertilizers usage. Conservation tillage is commonly used in low-carbon production practice, which also shows that the promotion of conservation tillage is effective.

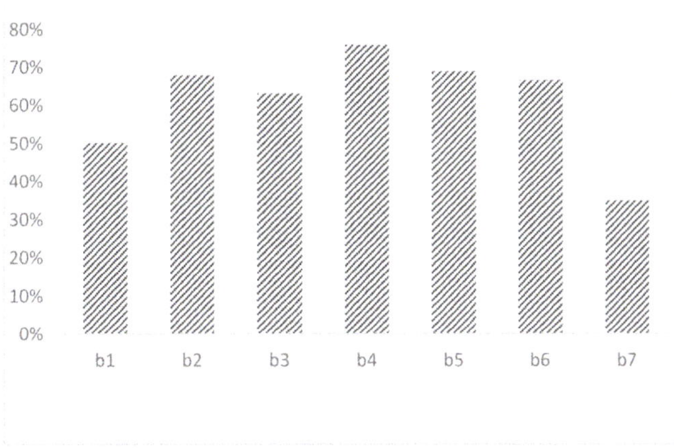

Figure 1. Farmers' behaviors.

The climate-responsive methods adopted by the farmers will be assigned with a value of 1, otherwise it is set to be 0. When performing multi-variate Probit test, regression is performed on each behavior separately. When performing the Poisson regression, the dummy variables of corresponding variables are summed. The more types of method being used, the higher the intensity of climate-adaptive behaviors would be.

Based on the hypothesizes made in Section 2.2, a total of 9 variables in 4 aspects namely farmers' individual characteristics, farmer households' characteristics, government subsidies and loans, and agricultural productive services are selected for analysis. Detailed definitions and explanations of the variables can be seen in Table 1.

Table 1. Definition and Descriptive Statistics of Variables.

Variables	Value	Mean Value	Standard Deviation	Min	Max
Gender	1 = Male, 2 = Female	1.353	0.479	1	2
Age	Age of farmers	46.385	7.5	28	68
Education level	1 = Primary school education and below, 2 = Junior high school education, 3 = High school education, 4 = University degree	2.414	0.766	1	4
Household size	Population of peasant households (people)	3.882	1.230	1	10
Planting area	Operating land area of peasant household(ha)	16.624	35.398	0.053	500
Social capital	Neighbors help each other during the busy period, 1 = Yes, 0 = No	0.882	0.323	0	1
Government subsidies	Receive agricultural subsidies, 1 = Yes, 0 = No	0.802	0.399	0	1
Loans	Take loans, 1 = Yes, 0 = No	0.537	0.499	0	1
Agricultural productive services	Number of agricultural productive services received, 1–5 (Agricultural resources service, 1 = Yes, 0 = No. Rural insurance service, 1 = Yes, 0 = No. Service of agricultural machinery, 1 = Yes, 0 = No. Hires labor, 1 = Yes, 0 = No. Technical training, 1 = Yes, 0 = No)	3.856	1.149	0	5

3. Results and Analysis

With the survey data obtained, the nine independent variables are tested for multicollinearity before regression. It is showed in Table 2 that the variance inflation factor (VIF) value of all variables is less than 10 and the average value is only 1.13. It means that the selected explanatory variables all meet the principle of independence, so there is no multicollinearity problem anticipated and regression analysis can be performed.

Table 2. Multicollinearity.

Variables	VIF
Gender	1.04
Age	1.18
Education level	1.21
Household size	1.08
Planting area	1.17
Social capital	1.07
Government subsidies	1.08
Loans	1.14
Agricultural productive services	1.19
Mean VIF	1.13

3.1. Analysis of the Factors Affecting the Intensity of Low-Carbon Production Practice and Climate Adaptation Practice

Climate response behavior is divided into low-carbon production practice aiming at protecting the environment and mitigating climate change and climate adaptation practice aiming at reducing climate-induced agricultural production loss. The influencing factors of low-carbon production behavior intensity and climate adaptation behavior intensity were analyzed, respectively. Stata15.1 software (Statacorp LLC, College Station, TX, USA) has been used to perform the Poisson regression on the survey data collected, and Table 3 shows that the overall fitting effect is good. Table 3 also shows the estimation results of the general least squares (OLS). No matter which model is used, the regression is significant overall, and the significance of the estimated coefficients is the same.

Gender is significant at the 1% level in the adoption of low-carbon production practice and climate-responsive practice, and it has a remarkable negative impact on them. The probability of female farmers adopting low-carbon production behaviors is 25.4% lower than that of male farmers. The probability of female farmers adopting climate adaptation practice is 16.0% lower than that of men. With other conditions unchanged, men tend to adopt climate-responsive behaviors compared with women farmers. This shows that women are generally not willing to change their agricultural production behaviors, hence changes in production behaviors caused by climate are not as good as men. Influenced by traditional concepts, men, as the dominant force in family production, receive better resources than women and men are more inclined to switch to new production behaviors. In this part, hypothesis 1 was verified. Age has an insignificant influence on the intensity of climate-responsive behaviors, so the hypothesis is not valid. Education level has a significant positive impact on farmers' low-carbon production practice. With each education level up, the possibility of adopting low-carbon production behaviors increases by 12.2%. This shows that those who have higher education level has a stronger awareness of global warming. At the same time, they are more inclined to emission reduction and they tend to adopt low-carbon production behaviors more compared with the rest. It's also worth to mention that education level has an impact on climate adaptation practice although it's not significant. With other conditions remain unchanged, farmers with higher education levels are more cognizant and they are more inclined to adopt climate-responsive behaviors. People with higher education levels are more likely to learn new production techniques and they often understand the climate better.

Household size and planting area do not significantly affect the intensity of climate-responsive behaviors. In Northeast China, families with more family members are generally old-fashioned with average or below-average household income. Hence, these farmers are not sensitive to climate-responsive agricultural production behaviors.

The impact of planting area on climate-responsive behaviors is not significant, and the hypothesis cannot be verified.

Government subsidies have positive impacts on climate adaptation practice and it is significant at the 5% level. At the same time, the impact of government subsidies and loans are not significant on low-carbon production practice. Government subsidies can reduce catastrophic losses caused by climate change and reduce the costs incurred during agricultural production. So, farmers are more inclined to carry out technological transformation. Loans have an insignificant negative correlation with the adoption of climate-responsive behaviors.

Agricultural productive services have a clear positive impact on climate-responsive behaviors. The impact of agricultural productive services on the intensity of climate adaptation practice and climate adaptation practice are both significant at the 1% level in both OLS and Poisson. The possibility of adopting low-carbon production practice increases by 7.8%, and that of adopting climate-adaptive services increases by 9.4% for each additional productive service used by farmers. The adoption of agricultural productive services helps to achieve the emission reduction goal advocated by climate-smart agriculture. It improves the level of farmers' adaptive behaviors at the same time. Agricultural productive services enable farmers to obtain climate change information promptly. The agricultural production technology level of farmers continues to improve, and the relevant agricultural materials can be obtained by farmer households faster. The entire agricultural activities are more specialized and properly divided, and the efficiency of agricultural production is improved. Farmers tend to adopt flexible production methods to promote the occurrence of climate response behaviors.

Table 3. Estimation of the intensity of low-carbon production practice and climate adaptation practice of farmers.

	Low-Carbon Production Practice			Climate Adaptation Practice		
	Poisson Regression	OLS	Incidence-Rate Ratios	Poisson Regression	OLS	Incidence-Rate Ratios
Gender	−0.293 ***	−0.499 ***	0.746	−0.175 ***	−0.417 ***	0.840
	(−4.51)	(−4.75)		(−3.21)	(−3.18)	
Age	0.046	0.093	1.047	0.183	0.483	1.201
	(−0.29)	(−0.33)		(−1.13)	(−1.36)	
Education level	0.115 ***	0.216 ***	1.122	0.043	0.108 *	1.044
	(−3.19)	(−3.05)		(−1.29)	(−1.23)	
Household size	0.003	0.008	1.003	−0.023	−0.046	0.977
	(−0.54)	(−0.39)		(−1.3)	(−1.70)	
Planting area	0.079	0.123	1.082	0.143	0.309	1.154
	(−0.90)	(−0.78)		(−1.46)	(−1.57)	
Social capital	0.000	0.000	1.000	−0.001	−0.002	0.999
	(−0.24)	(−0.15)		(−0.92)	(−0.87)	
Government subsidies	0.061	0.116	1.063	0.174 **	0.403 **	1.190
	(−0.80)	(−0.90)		(−2.43)	(−2.52)	
Loans	−0.049	−0.096	0.953	−0.022	−0.061	0.978
	(−0.84)	(−0.91)		(−0.42)	(−0.46)	
Agricultural productive services	0.0748 ***	0.129 ***	1.078	0.0896 ***	0.210 ***	1.094
	(−2.66)	(−2.75)		(−3.32)	(−3.60)	
_cons	0.280	1.247 ***	1.323	0.440 **	1.382 ***	1.552
	(−1.38)	(−3.27)		(−2.07)	(−2.91)	
N	374	374		374	374	
P	0.0000	0.0000		0.0000	0.0000	

OLS: Ordinary Least Square; Cons: constant terms; z statistics in parentheses; N sample size; * $p < 0.1$, ** $p < 0.05$, *** $p < 0.01$; Indicate that the coefficients of the explanatory variables are significant at the 10%, 5%, and 1% levels, respectively.

3.2. Heterogeneity Analysis

The result in Table 3 describes how various variables influence the climate-responsive behaviors. Farmers' adoption of climate-responsive behaviors is significantly affected by productive services. To exploit the difference between the influence brought by productive services, this part does analysis from three aspects: age, gender, and planting scale.

3.2.1. Age

With the continuously growing social development in recent years, generation gap has become a more critical influencing factor in the ways of thinking of individuals. Similarly, this effect can be observed when people accept fresh viewpoints, adopt agricultural productive services and put theoretical knowledge into practices. After conducting research in Chinese rural regions, the farmer in this paper can be categorized into three types: the young generation (born after 1980), the middle-age generation (born between 1965 and 1979), and the old generation (born before 1965) [49].

According to Table 4, after controlling the variable, the young and the middle-age are more tendentious to adopt low-carbon production practice and climate adaptation practice under the influence of agricultural productive services. Oppositely, the minor response in the old indicates that the willingness of elder farmer on learning and adopting advanced technologies, and they have less tendency on changing the conventional methods they used.

Table 4. Regression results of different generations of farmers.

Low-Carbon Production Practice	The Young Generation	The Middle-Age Generation	The Old Generation
Agricultural productive services	1.1258 ***	1.0713 ***	0.9027
	(2.04)	(2.09)	(−0.76)
Control variable	Controlled	Controlled	Controlled
Log likelihood	−128.90164	−371.99202	−41.994357
N	93	251	30
P	0.0000	0.0000	0.0230
Pseudo R^2	0.0645	0.0223	0.0245
Climate Adaptation Practice	**The Young Generation**	**The Middle-Age Generation**	**The Old Generation**
Agricultural productive services	1.1060 *** 2.93	1.0933 *** 2.81	0.0605 0.1370
Control variable	Controlled	Controlled	Controlled
Log likelihood	−156.0848	−410.3360	−48.7649
N	93	251	30
P	0.0000	0.0032	0.0607
Pseudo R^2	0.0545	0.0223	0.0729

z statistics in parentheses. N sample size. *** $p < 0.01$. Indicate that the coefficients of the explanatory variables are significant at the 10%, 5%, and 1% levels, respectively.

3.2.2. Gender

The analysis conducted in Section 2.2 is based on two groups, female and male. According to Table 5, the level of female farmers who adopt climate-responsive behaviors is not remarkably influenced by the extent of they accept productive services. However, a significant feedback can be found in the analysis of another group. Therefore, it seems that agricultural productive services have brought great impact to male farmers' adoption of climate-responsive behaviors. At the same time, the analysis tells that the attitude of the females are more conventional when embracing advanced technologies. The males, on the contrary, are more likely to improve the production efficiency by changing technical practices.

Table 5. Regression results of different gender.

Low-Carbon Production Practice	Female	Male
Agricultural productive services	1.0570	1.0828 ***
	(0.97)	(2.64)
Control variable	Controlled	Controlled
Log likelihood	−181.3164	−362.0998
N	132	242
P	0.0053	0.0095
Pseudo R^2	0.0341	0.0129
Climate Adaptation Practice	**Female**	**Male**
Agricultural productive services	1.0829	1.0826 ***
	(1.55)	(2.62)
Control variable	Controlled	Controlled
Log likelihood	−208.8982	−409.3071
N	132	242
P	0.0000	0.0000
Pseudo R^2	0.0473	0.0223

z statistics in parentheses. N sample size. *** $p < 0.01$. Indicate that the coefficients of the explanatory variables are significant at the 10%, 5%, and 1% levels, respectively.

3.2.3. Planting Scale

In the third national agricultural census, the planting area in the region where is one crop per annum is categorized into two types: small scale (6.7 hectares) and large scale (greater than 6.7 hectares) [50]. The data, according to this regulation, is separated into the small-scale farmer and the large-scale farmer group. Analyses are made independently in two groups to see the impact of agricultural productive services on farmers' adoption of climate-responsive behaviors.

The result in Table 6 describes that the productive services increase the acceptance of climate-responsive behaviors in both groups. However, the likelihood of those farmers who have a smaller scale planting area is relatively low. Individuals in the large-scale farmer group trust more in local technicians and agricultural extension personnel. At the same time, they earn more attention from the local government. Hence, they have more opportunities to attend training activities in agricultural technologies, and they are more likely to accept advanced knowledge, which develops the level of climate-responsive behaviors.

Table 6. Regression results of different scale.

Low-Carbon Production Practice	Small Scale	Large Scale
Agricultural productive services	1.0368 0.95	1.1019 ** 2.41
Control variable	Controlled	Controlled
Log likelihood	−293.7519	−252.1136
N	199	242
P	0.0002	0.0004
Pseudo R^2	0.0226	0.0314
Climate Adaptation Practice	**Small Scale**	**Large Scale**
Agricultural productive services	1.0658 * (1.70)	1.0861 ** (2.17)
Control variable	Controlled	Controlled
Log likelihood	−326.3521	−291.1609
N	132	242
P	0.0066	0.0003
Pseudo R^2	0.0271	0.0403

z statistics in parentheses. N sample size. * $p < 0.1$, ** $p < 0.05$. Indicate that the coefficients of the explanatory variables are significant at the 10%, 5%, and 1% levels, respectively.

3.3. Analysis of Factors Affecting Farmers' Choices for Climate-Responsive Behaviors

Reference to Stata 15.1, a multi-variate Probit model has been used to estimate the choices of climate adaptation practice of the maize farmers in Jilin province. In the correlation matrix, 19 correlation coefficients ρ are significant at the 1% level and all correlation coefficients are significant at the 5% level. This result shows that, farmers' choices of climate-responsive behaviors are influenced by other climate-responsive behaviors. The planting-breeding combination methods, the conservation tillage methods, the use of organic fertilizers, the changes of planting varieties according to the frost-free periods, the adjustment of using pesticides and fertilizers, the adjustments of sowing and harvesting time and the filling of the gaps with seedlings are complementary to one another. The probability of using other methods will grammatically increase after adopting a single climate-responsive method. Table 7 shows the result of parametric regression and the model is significant at all levels. Detailed explanations of the result according to different variables are discussed as follows:

Gender is significant at the 1% level in the adoption of the planting-breeding combination methods, the conservation tillage methods and the changing of planting varieties according to the frost-free periods' methods. It is significant at the 5% level in the adoption of using organic fertilizers and the adjustment of fertilizers and pesticides usage. It is significant at the 10% level in the adoption of filling the gaps with seedlings.

Table 7. Estimation of farmers' choices of climate-responsive behaviors.

	Planting-Breeding Combination Methods	Conservation Tillage Methods	Use of Organic Fertilizers	The Change of Suitable Varieties according to the Frost-Free Periods	Adjustment of Pesticides and Fertilizers Usage	Adjustment of the Time of Sowing and Harvesting	Filling the Gaps with Seedlings
Gender	−0.540 ***	−0.560 ***	−0.294 **	−0.480 ***	−0.360 **	−0.204	−0.257 *
	(−3.81)	(−3.87)	(−2.08)	(−3.24)	(−2.5)	(−1.44)	(−1.8)
Age	0.000	−0.003	0.517	0.884 **	0.737 *	0.737 *	−0.233
	(0.00)	(−0.01)	(1.29)	(2.08)	(1.83)	(1.86)	(−0.58)
Education level	0.178 *	0.159	0.312 ***	0.048	0.095	0.412 ***	−0.099
	(1.81)	(1.52)	(3.05)	(0.45)	(0.92)	(3.84)	(−1.02)
Household size	0.063	−0.021	0.086	0.079	0.016	−0.018	−0.011
	(1.07)	(−0.35)	(1.44)	(1.21)	(0.27)	(−0.3)	(−0.19)
Planting area	−0.165	0.172	0.345 *	0.240	0.311	0.346 *	−0.024
	(−0.77)	(0.81)	(1.68)	(1.07)	(1.48)	(1.65)	(−0.11)
Social capital	0.003	−0.001	−0.002	−0.001	−0.001	0.001	−0.004
	(1.06)	(−0.32)	(−0.78)	(−0.39)	(−0.67	(0.37	(−1.57
Government subsidies	0.275	0.003	0.022	0.228	0.332*	0.162	0.406 **
	(1.56)	(0.02)	(0.13)	(1.26)	(1.9)	(0.92)	(2.23)
Loans	−0.046	−0.279 *	0.008	0.037	−0.192	−0.062	−0.024
	(−0.32)	(−1.85)	(0.06)	(0.24)	(−1.31)	(−0.43)	(−0.17)
Agricultural productive services	0.090	0.210 ***	0.082	−0.027	0.237 ***	0.207 ***	0.186 ***
	(1.42)	(3.21)	(1.3)	(−0.38)	(3.65)	(3.23)	(2.75)
_cons	−0.362	0.162	−1.184 **	0.257	−0.924	−1.702 ***	−0.605
	(−0.66)	(0.28)	(−2.08)	(0.43)	(−1.61)	(−2.97)	(−1.05)
N				374			
P				0.0000			

pCorrelation coefficient matrix: ρ21 = 0.2786 *** ρ31 = 0.2479 *** ρ41 = 0.2907 *** ρ51 = 0.2926 *** ρ61 = 0.3155 *** ρ71 = 0.3043 *** ρ32 = 0.1273 ** ρ42 = 0.1758 *** ρ52 = 0.3688 *** ρ62 = 0.3387 *** ρ72 = 0.1325 ** ρ43 = 0.1788 *** ρ53 = 0.2539 *** ρ63 = 0.2100 *** ρ73 = 0.3059 *** ρ54 = 0.2581 *** ρ64 = 0.2774 *** ρ74 = 0.2035 *** ρ65 = 0.4562 *** ρ75 = 0.1773 *** ρ76 = 0.2826 ***

Wald χ²				140.13			
Log likelihood				−1411.93			

OLS: Ordinary Least Square. Cons: constant terms. z statistics in parentheses. N sample size. * $p < 0.1$, ** $p < 0.05$, *** $p < 0.01$ indicate that the coefficients of the explanatory variables are significant at the 10%, 5%, and 1% levels, respectively.

Age has a remarkable positive impact on the changing of planting varieties according to the frost-free periods' methods, the adjustment of pesticides and fertilizers usage and the adjustment of sowing and breeding time. These three climate-responsive behaviors are the most empirical ones among the seven behaviors discussed in this article. Farmers who have been working in the agricultural production industry for a long time can adjust on the basis of their previous experiences. Education level is significant at the 1%, 5%, 10% level in the adoption of the planting-breeding combination methods, the usage of organic fertilizers and the adjustment of the time of sowing and breeding, respectively. At the same time, education level has a significant positive impact on these three climate-responsive behaviors.

Social capital has a 10% significance in the method of switching over to organic fertilizers and the adjustment of sowing and harvesting time.

Government subsidies are significant at the 5% and 10% level in the adjustment of the pesticides and fertilizers usage and the filling of the gaps with seedlings, respectively. This shows that, government subsidies help reduce farmers' monetary pressure and the related risky behaviors in great deal. This is especially obvious in the filling of the gaps with seedlings behaviors.

Agricultural productive services have positive impacts on climate-responsive behaviors. It is significant at the 1% level in the adoption of conservation tillage methods, the adjustment of sowing and harvesting time and replanting methods. Agricultural productive services also have positive impacts on the planting-breeding combination methods and the usage of organic fertilizers although it's not significantly.

Agricultural productive services have positive impacts on climate response behaviors. Among them, the conservation tillage methods, the adjustment of pesticides and fertilizers usage, the adjustment of sowing and harvesting time and the behaviors of replenishing seedlings have significance at 1%.

4. Discussion

Based on the survey conducted on 374 maize farmers in Jilin province of China, an empirical analysis of the low-carbon production practice and climate adaptation practice due to climate change have been carried out, with findings.

Firstly, most maize farmers adjust their production methods promptly according to the changes of climate. Conservation tillage methods have been used the most by farmers among all the low-carbon production practice. Among all of the adaptation methods practiced, 75.9% of the farmers will choose suitable varieties according to the frost-free periods and 69.0% of the farmers will adjust their usage of pesticides and fertilizers according to different climate conditions. Secondly, gender and education level have a remarkable impact on the intensity of farmers' climate-responsive behaviors. Thirdly, agricultural productive services have crucial impacts on farmers' climate-responsive behaviors. This part validates the hypothesis. Lastly, in actual operation, farmers have used methods according to local needs.

5. Suggestion

Based on the data and the analysis conducted, policy recommendations related to the existing problems are proposed as below.

Firstly, agricultural socialization service system needs to be improved, different service suppliers need to be developed, agricultural market needs to be standardized, and the construction of market needs to be strengthened.

Secondly, the government should continue to support on the provision of information and technology related to agricultural production. The government can set up propaganda boards in the village, distribute leaflets, and provide guidance for technical personnel to enter the households, so that the villagers know its benefits and effects and increase the acceptance of agricultural productive services. Agricultural productive services need to be diversified. It will not only help with the realization of climate-smart agriculture and the adaptation of farmers to climate change, but will also help to reduce external risks during production processes.

Thirdly, as an essential and highly specialized portion, emission reductions can be outsourced to external business entities. It will also help to realize the goal of low-carbon production as those related technology does not need to be promoted to farmers.

On the fourth, farmers' education level needs to be improved and the financial supports on farmers' education need to be strengthened. These measures will help to improve the farmers' understanding of climate change and their adaptive methods.

On the fifth, the availability and readiness of loans and subsidies for farmers should be improved. Risks coming from climate change can be avoided with the guidance and control of agricultural subsidies.

In general, agricultural productive services have a great impetus to farmers' climate-responsive behaviors. Globally, agricultural production characteristics with similar climatic characteristics may have similar effects.

6. Conclusions

In contrast, the males are willing to try and use new ideas and technologies, while females tend to be conservative. This part validates the hypothesis. In addition, farmers with higher education level have stronger understanding and ability to accept new things. They have more positive attitudes towards environmental protection, and they are more likely to adopt climate-responsive behaviors. In previous literatures, some scholars have also verified the influencing factors of these factors. Their proof is similar to the conclusions of this study [15,16,48]. Agricultural productive services help households to participate in the division and proper allocation of labor. It helps farmers to obtain information related to climate change, government support on technology and resources promptly. Therefore, agricultural productive services have positive impacts on climate-responsive behaviors. This conclusion has also been confirmed by scholars in previous studies. They believe that training services and information disclosure can help rainfed farmers develop reasonable adaptation measures to reduce climate risks to agriculture [48,51]. Different farmer groups have different performances in agricultural productive services promoting climate-responsive behaviors. It has a more significant effect on farmers with male, large-scale planting and younger characteristics. Males are more receptive to new technologies than women. Younger farmers have stronger learning ability and can translate new ideas into practice well. Large-scale planting farmers are valued by local agricultural associations. They received more information and technical training, resulting in higher adoption of climate-responsive behaviors by farmers.

The method used in this paper fails to study the internal mechanism of agricultural productive services on farmers' behaviors. In addition, the consistency of farmers' willingness and behavior needs to be further explored. According to the effects of different types of agricultural productive services, a comparative analysis can be carried out to find the most effective method in the region. In the future, we can also discuss the specific relationship between gender and farmers' behavior so as to contribute to promoting equality between males and females and solving social conflicts.

Author Contributions: Conceptualization, H.G. and H.P.; methodology, H.G. and Y.X.; software, Y.X.; validation, H.G. and Y.X.; formal analysis, Y.X.; resources, H.G.; data curation, Y.X.; writing—original draft preparation, H.G.; writing—review and editing, Q.L. and H.P.; visualization, Q.L.; supervision, C.P.; project administration, H.G.; funding acquisition, H.G. All authors have read and agreed to the published version of the manuscript.

Funding: This work was supported by the Jilin Province Science and Technology Development Plan Project under Grant [20190301080NY].

Institutional Review Board Statement: Not applicable.

Informed Consent Statement: Informed consent was obtained from all subjects involved in the study.

Data Availability Statement: Not applicable.

Conflicts of Interest: The authors declare no conflict of interest.

References

1. Le Dang, H.; Li, E.; Nuberg, I.; Bruwer, J. Factors influencing the adaptation of farmers in response to climate change: A review. *Clim. Dev.* **2019**, *11*, 765–774. [CrossRef]
2. Karki, S.; Burton, P.; Mackey, B. The experiences and perceptions of farmers about the impacts of climate change and variability on crop production: A review. *Clim. Dev.* **2019**, *12*, 80–95. [CrossRef]
3. Shirsath, P.B.; Aggarwal, P.; Thornton, P.; Dunnett, A. Prioritizing climate-smart agricultural land use options at a regional scale. *Agric. Syst.* **2017**, *151*, 174–183. [CrossRef]
4. Mahmood, N.; Arshad, M.; Kächele, H.; Ma, H.; Ullah, A.; Müller, K. Wheat yield response to input and socioeconomic factors under changing climate: Evidence from rainfed environments of Pakistan. *Sci. Total Environ.* **2019**, *688*, 1275–1285. [CrossRef]
5. Long, T.B.; Blok, V.; Coninx, I. Barriers to the adoption and diffusion of technological innovations for climate-smart agriculture in Europe: Evidence from the Netherlands, France, Switzerland and Italy. *J. Clean. Prod.* **2016**, *112*, 9–21. [CrossRef]
6. Regan, P.M.; Kim, H.; Maiden, E. Climate change, adaptation, and agricultural output. *Reg. Environ. Chang.* **2019**, *19*, 113–123. [CrossRef]
7. Kpadonou, R.A.B.; Owiyo, T.; Barbier, B.; Denton, F.; Rutabingwa, F.; Kiema, A. Advancing climate-smart-agriculture in developing drylands: Joint analysis of the adoption of multiple on-farm soil and water conservation technologies in West African Sahel. *Land Use Policy* **2017**, *61*, 196–207. [CrossRef]
8. Ngoma, H.; Lupiya, P.; Kabisa, M.; Hartley, F. Impacts of climate change on agriculture and household welfare in Zambia: An economy-wide analysis. *Clim. Chang.* **2021**, *167*, 55. [CrossRef]
9. Andrieu, N.; Sogoba, B.; Zougmore, R.; Howland, F.; Samake, O.; Bonilla-Findji, O.; Lizarazo, M.; Nowak, A.; Dembele, C.; Corner-Dolloff, C. Prioritizing investments for climate-smart agriculture: Lessons learned from Mali. *Agric. Syst.* **2017**, *154*, 13–24. [CrossRef]
10. Arslan, A.; McCarthy, N.; Lipper, L.; Asfaw, S.; Cattaneo, A.; Kokwe, M. Climate Smart Agriculture? Assessing the Adaptation Implications in Zambia. *J. Agric. Econ.* **2015**, *66*, 753–780. [CrossRef]
11. Tong, Q.; Swallow, B.; Zhang, L.; Zhang, J. The roles of risk aversion and climate-smart agriculture in climate risk management: Evidence from rice production in the Jianghan Plain, China. *Clim. Risk Manag.* **2019**, *26*, 100199. [CrossRef]
12. Grottera, C.; La Rovere, E.L.; Wills, W.; Pereira, A.O., Jr. The role of lifestyle changes in low-emissions development strategies: An economy-wide assessment for Brazil. *Clim. Policy* **2020**, *20*, 217–233. [CrossRef]
13. Hamdi-Cherif, M.; Li, J.; Broin, E. Leveraging the transport sector to mitigate long-term climate policy costs in China: A behavioural perspective. *Clim. Policy* **2021**, *21*, 475–491. [CrossRef]
14. Harada, H.; Kobayashi, H.; Shindo, H. Reduction in greenhouse gas emissions by no-tilling rice cultivation in Hachirogata polder, northern Japan: Life-cycle inventory analysis. *Soil Sci. Plant Nutr.* **2007**, *53*, 668–677. [CrossRef]
15. Mugwe, J.; Mugendi, D.; Mucheru-Muna, M.; Merckx, R.; Chianu, J.; Vanlauwe, B. Determinants of the decision to adopt integrated soil fertility management practices by smallholder farmers in the central highlands of Kenya. *Exp. Agric.* **2009**, *45*, 61–75. [CrossRef]
16. Willy, D.K.; Holm-Müller, K. Social influence and collective action effects on farm level soil conservation effort in rural Kenya. *Ecol. Econ.* **2013**, *90*, 94–103. [CrossRef]
17. De Souza Filho, H.M.; Young, T.; Burton, M.P. Factors Influencing the Adoption of Sustainable Agricultural Technologies: Evidence from the State of Espírito Santo, Brazil. *Technol. Forecast. Soc. Chang.* **1999**, *60*, 97–112. [CrossRef]
18. Mazvimavi, K.; Twomlow, S. Socioeconomic and institutional factors influencing adoption of conservation farming by vulnerable households in Zimbabwe. *Agric. Syst.* **2009**, *101*, 20–29. [CrossRef]
19. Somda, J.; Nianogo, A.; Nassa, S.; Sanou, S. Soil fertility management and socio-economic factors in crop-livestock systems in Burkina Faso: A case study of composting technology. *Ecol. Econ.* **2002**, *43*, 175–183. [CrossRef]
20. Costa, N.B.d., Jr.; Baldissera, T.C.; Pinto, C.E.; Garagorry, F.C.; de Moraes, A.; Carvalho, P.C.d.F. Public policies for low carbon emission agriculture foster beef cattle production in southern Brazil. *Land Use Policy* **2019**, *80*, 269–273. [CrossRef]
21. Paustian, K. Agriculture, farmers and GHG mitigation: A new social network? *Carbon Manag.* **2012**, *3*, 253–257. [CrossRef]
22. Sánchez, B.; Álvaro-Fuentes, J.; Cunningham, R.; Iglesias, A. Towards mitigation of greenhouse gases by small changes in farming practices: Understanding local barriers in Spain. *Mitig. Adapt. Strat. Glob. Chang.* **2016**, 995–1028. [CrossRef]
23. D'Emden, F.H.; Llewellyn, R.S.; Burton, M.P. Adoption of conservation tillage in Australian cropping regions: An application of duration analysis. *Technol. Forecast. Soc. Chang.* **2006**, *73*, 630–647. [CrossRef]
24. Knowler, D.; Bradshaw, B. Farmers' adoption of conservation agriculture: A review and synthesis of recent research. *Food Policy* **2007**, *32*, 25–48. [CrossRef]
25. Kragt, M.; Dumbrell, N.; Blackmore, L. Motivations and barriers for Western Australian broad-acre farmers to adopt carbon farming. *Environ. Sci. Policy* **2017**, *73*, 115–123. [CrossRef]
26. Wandel, J.; Smithers, J. Factors affecting the adoption of conservation tillage on clay soils in southwestern Ontario, Canada. *Am. J. Altern. Agric.* **2009**, *15*, 181–188. [CrossRef]

27. Liu, Y.; Ruiz-Menjivar, J.; Zhang, L.; Zhang, J.; Swisher, M.E. Technical training and rice farmers' adoption of low-carbon management practices: The case of soil testing and formulated fertilization technologies in Hubei, China. *J. Clean. Prod.* **2019**, *226*, 454–462. [CrossRef]
28. Moges, D.M.; Taye, A.A. Determinants of farmers' perception to invest in soil and water conservation technologies in the North-Western Highlands of Ethiopia. *Int. Soil Water Conserv. Res.* **2017**, *5*, 56–61. [CrossRef]
29. Li, C.; Ting, Z.; Rasaily, R.G. Farmer's Adaptation to Climate Risk in the Context of China. *Agric. Agric. Sci. Procedia* **2010**, *1*, 116–125. [CrossRef]
30. Field, C.; Barros, V.; Change, I. *Climate Change 2014: Impacts, Adaptation, and Vulnerability*; Working Group II Contribution to the Fifth Assessment Report of the Intergovernmental Panel on Climate Change; Cambridge University Press: Cambridge, UK, 2014.
31. Woods, B.A.; Nielsen, H.Ø.; Pedersen, A.B.; Kristofersson, D. Farmers' perceptions of climate change and their likely responses in Danish agriculture. *Land Use Policy* **2017**, *65*, 109–120. [CrossRef]
32. Deressa, T.T.; Hassan, R.M.; Ringler, C.; Alemu, T.; Yesuf, M. Determinants of farmers' choice of adaptation methods to climate change in the Nile Basin of Ethiopia. *Glob. Environ. Chang.* **2009**, *19*, 248–255. [CrossRef]
33. Nhemachena, C.; Hassan, R.M. Micro-level analysis of farmers' adaptation to climate change in Southern Africa. *IFPRI Discuss. Pap.* **2007**, *7778*, 5–15.
34. Bechtoldt, M.N.; Götmann, A.; Moslener, U.; Pauw, W.P. Addressing the climate change adaptation puzzle: A psychological science perspective. *Clim. Policy* **2021**, *21*, 186–202. [CrossRef]
35. Prokopy, L.S.; Arbuckle, J.G.; Barnes, A.P.; Haden, V.R.; Hogan, A.; Niles, M.T.; Tyndall, J. Farmers and Climate Change: A Cross-National Comparison of Beliefs and Risk Perceptions in High-Income Countries. *Environ. Manag.* **2015**, *56*, 492–504. [CrossRef]
36. Simon winter Sustainable Intensification. Available online: https://www.syngentafoundation.org/sustainable-intensification (accessed on 15 February 2022).
37. Azadi, H.; Moghaddam, S.M.; Burkart, S.; Mahmoudi, H.; Van Passel, S.; Kurban, A.; Lopez-Carr, D. Rethinking resilient agriculture: From Climate-Smart Agriculture to Vulnerable-Smart Agriculture. *J. Clean. Prod.* **2021**, *319*, 128602. [CrossRef]
38. Atanu, S.; Love, H.A.; Schwart, R. Adoption of Emerging Technologies under Output Uncertainty. *Am. J. Agric. Econ.* **1994**, *76*, 836–846. [CrossRef]
39. Ngigi, M.W.; Mueller, U.; Birner, R. Gender Differences in Climate Change Adaptation Strategies and Participation in Group-based Approaches: An Intra-household Analysis from Rural Kenya. *Ecol. Econ.* **2017**, *138*, 99–108. [CrossRef]
40. Tang, J.; Folmer, H.; Xue, J. Adoption of farm-based irrigation water-saving techniques in the Guanzhong Plain, China. *Agric. Econ.* **2016**, *47*, 445–455. [CrossRef]
41. Tizale, C.Y. The Dynamics of Soil Degradation and Incentives for Optimal Management in the Central Highlands of Ethiopia. Ph.D. Thesis, Department of Agricultural Economics, Extension and Rural Development, University of Pretoria, Pretoria, South Africa, 2007.
42. Gbetibouo, G.A.; Hassan, R.M.; Ringler, C. Modelling farmers' adaptation strategies for climate change and variability: The case of the Limpopo Basin, South Africa. *Agrekon* **2010**, *49*, 217–234. [CrossRef]
43. Lewis, B.D.; Pattinasarany, D. Determining Citizen Satisfaction with Local Public Education in Indonesia: The Significance of Actual Service Quality and Governance Conditions. *Growth Chang.* **2009**, *40*, 85–115. [CrossRef]
44. Vecchio, Y.; Agnusdei, G.P.; Miglietta, P.P.; Capitanio, F. Adoption of Precision Farming Tools: The Case of Italian Farmers. *Int. J. Environ. Res. Public Health* **2020**, *17*, 869. [CrossRef] [PubMed]
45. Mussida, C.; Zanin, L. Determinants of the Choice of Job Search Channels by the Unemployed Using a Multivariate Probit Model. *Soc. Indic. Res.* **2020**, *152*, 369–420. [CrossRef] [PubMed]
46. Wooldridge, J.M. *Introductory Econometrics: A Modern Approach*, 2nd ed.; Thomson Learning: Boston, MA, USA, 2003.
47. Siddhartha, C.; Edward, G. Analysis of multivariate probit models. *Biometrika* **1998**, *85*, 347–361.
48. Mahmood, N.; Arshad, M.; Mehmood, Y.; Faisal Shahzad, M.; Kächele, H. Farmers' perceptions and role of institutional arrangements in climate change adaptation: Insights from rainfed Pakistan. *Clim. Risk Manag.* **2021**, *32*, 100288. [CrossRef]
49. Duan, C.; Ma, X. Analysis on the intergenerational difference of Migrant workers in China. *Labour Econ. Rev.* **2011**, *4*, 34–53.
50. National Bureau of Statistics of China. *Plan for the third National Agricultural Census*; China Statistical Publishing House: Beijing, China, 2016.
51. Mahmood, N.; Arshad, M.; Kaechele, H.; Shahzad, M.; Ullah, A.; Mueller, K. Fatalism, Climate Resiliency Training and Farmers' Adaptation Responses: Implications for Sustainable Rainfed-Wheat Production in Pakistan. *Sustainability* **2020**, *12*, 1650. [CrossRef]

International Journal of
*Environmental Research
and Public Health*

Article

Analyzing the Disaster Preparedness Capability of Local Government Using AHP: Zhengzhou 7.20 Rainstorm Disaster

Linpei Zhai and Jae Eun Lee *

Department of Public Administration & National Crisisonomy Institute, Chungbuk National University, Cheongju 28644, Republic of Korea
* Correspondence: jeunlee@chungbuk.ac.kr; Tel.: +82-43-261-2197

Abstract: This study aimed to identify factors influencing disaster preparedness capability, measure and compare the relative importance of evaluation indicators of preparedness capability in a rainstorm disaster, and analyze the impact of these factors on disaster preparedness so as to improve disaster preparedness capability. The evaluation model was proposed by constructing the target level (the first level) as an indicator system; this was divided into four indicators (the second level): planning, organization, equipment, and education and exercise, and 14 tertiary evaluation indicators (the third level). The validity of the evaluation index system was demonstrated, and the weight of each level was calculated using the Analytic Hierarchical Process and expert survey methods, taking the example of the Zhengzhou "7.20" rainstorm to conduct an empirical analysis of the proposed model. The weak points of disaster preparedness capability were identified. The empirical analysis revealed that organization scored the highest, followed by planning, equipment, and education and exercise, indicating the lack of disaster management equipment and resources, disaster management training, and exercise and public emergency safety education. These results will help in future decision-making, as they provide a clear understanding of what needs to be done to improve disaster preparedness capability.

Keywords: disaster preparedness capability; heavy rainstorm; local government; AHP; evaluation index system

Citation: Zhai, L.; Lee, J.E. Analyzing the Disaster Preparedness Capability of Local Government Using AHP: Zhengzhou 7.20 Rainstorm Disaster. *Int. J. Environ. Res. Public Health* **2023**, 20, 952. https://doi.org/10.3390/ijerph20020952

Academic Editors: Mikio Ishiwatari and Daisuke Sasaki

Received: 22 November 2022
Revised: 30 December 2022
Accepted: 2 January 2023
Published: 4 January 2023

1. Introduction

Nowadays, various types of emergencies, such as natural disasters, public health crises, and social safety disasters occur frequently, and the importance of emergency response and disaster management has increased significantly in countries around the world. A key task of emergency management is ensuring that the public is adequately prepared for an impending disaster to minimize the loss of life and property [1]. In developing countries, environmental degradation, rapid urbanization, disaster scale, population density, preparedness, and mitigation measures are the main factors affecting disaster-related damage and mortality [2]. China's new urbanization process has continued to accelerate, with cities becoming larger and attracting an increasing number of people in recent years. With the progress of urbanization, Chinese cities are constantly exposed to a variety of unexpected disasters, including geological disasters (e.g., the Wenchuan earthquake in 2008), meteorological disasters (e.g., Super Typhoon Moranti in 2016, extensive haze and heavy rainfall-related flooding in many places in recent years), fire disasters, traffic disasters, accidents (collapse of self-built houses in Changsha in 2022), and infectious diseases (e.g., SARS from winter 2002 to spring 2003 and COVID-19 in 2020). Many cities suffered catastrophic consequences, such as human casualties, property damage, urban function failure, and social order imbalance as a result of such disasters.

Between 17 and 23 July 2021, China's Henan Province was struck by an extraordinarily heavy rainstorm that caused severe flooding. The event was named the "7·20"

Zhengzhou rainstorm. The "Investigation Report of the '7·20' Extraordinary Rainstorm Disaster in Zhengzhou, Henan Province" labelled the "7.20" rainstorm as a natural disaster triggering severe floods in cities and rivers and causing multiple other disasters, such as landslides, building collapses, and subway accidents, resulting in major casualties and property damage and changing the lives of millions of people [3]. According to verified sources, 14,786,000 people were affected, with direct economic losses of 120.6 billion RMB as of September 30; in Henan Province, 398 people died or were reported missing due to the disaster. The "7.20" Zhengzhou rainstorm disaster was labelled an overall "natural disaster", and, specifically, a "man-made disaster". Zhengzhou's municipal government and relevant districts, counties, departments, and units had poor understanding, risk awareness, and preparedness, with weak preventive measures in place to tackle such a mega-disaster. Moreover, there was dereliction of duty and malfeasance in the emergency response. Local governments play an important role in disaster management because they know their communities and citizens well. When a disaster occurs, local governments should be the first on the scene but, unfortunately, they remain one of the least studied institutions in disaster management literature [4]. Cities' ability to cope with unexpected disasters need to be improved, and disaster prevention and mitigation strategies pose an urgent problem facing governments at all levels and sectors of society.

Disaster preparedness has been recognized as a critical element in reducing the impact of disasters worldwide [5]. It has been studied from different perspectives and contexts related to individuals [6,7], households [8,9], and communities [10,11]. For example, by integrating community/individual behavior for disaster preparedness [12], certain populations, households, and individuals were found to have different preparedness needs and vulnerabilities [13].

While the literature on disaster risk, disaster resilience, disaster policy and management are relatively mature, public managers are unaware of how to design effective preparedness programs [14]. Few studies explain disaster management capability or disaster governance capability as a key aspect of central and local governments' disaster management; moreover, studies examining the role of local governments, especially in terms of disaster preparedness, are scarce. It is noteworthy that the primary responsibility for preparedness planning and response, in most cases, lies with the municipality or city [15]. There are two important areas that have not been fully explored with respect to the role of local governments in disaster and emergency management. First, although current research has focused on local governments in developed countries, research on local governments in developing countries is far from adequate. Second, in recent years, many local government agencies seem to be overwhelmed, rushed, and facing difficulties in responding to disasters and reducing related losses, especially in developing countries. The preparedness of local governments to manage disasters at each stage (before, during, and after) has not yet been tested.

The purpose of this study is to identify the factors that influence disaster preparedness capability, measure and compare the relative importance of evaluation indicators of preparedness capability in a rainstorm disaster, and analyze the impact of these factors on disaster preparedness to improve disaster preparedness capability. We also analyze relevant literature, refer to previous research results, combine the expert survey and the Analytic Hierarchical Process (AHP) methods, refine and construct a comprehensive disaster preparedness assessment system, develop a disaster preparedness assessment capability model, assign weights to indicators, and finally take the example of the "7·20" rainstorm in Zhengzhou city to conduct an empirical analysis of the proposed model. Through this comprehensive assessment, we aim to determine the weak points of disaster preparedness capability.

2. Theoretical Background

2.1. Disaster Preparedness

Social scientists, emergency managers, and public policymakers generally study and guide the process of disaster occurrence around four phases: mitigation, preparedness, response, and recovery [16]. This prevention phase of disaster management can be defined as all activities that can be implemented by the population, government, and relief organizations before a disaster occurs, with the aim of reducing its potentially devastating effects [17]. Preparedness is not only a state of readiness, but also a theme throughout most aspects of emergency management. It should be a dynamic and continuous management process, directly affecting the performance of emergency response capabilities, thus determining the development and evolution of the situation [18]. Preparedness comprises measures that enable different units of analysis—individuals, households, organizations, communities, and societies—to respond effectively and recover more quickly when disasters strike [16]. It is the ability of people to (a) anticipate what they have to deal with (dangerous consequences); (b) respond to, adapt to, and recover from disaster-related consequences, especially in areas that are likely to experience repeated disaster events; and (c) learn from these experiences [19] (p. 46).

Natural disaster preparedness is generally considered the preferred mechanism to encourage proactive activities (behavioral, cultural, structural, or institutional) to mitigate the disastrous potential of these events [20]. Preparedness has dual objectives: to reduce vulnerability to a potential threat [21–23] and to increase the resilience of the public exposed to a threat [24–26]. Activities that are commonly associated with disaster preparedness include developing planning processes to ensure readiness, formulating disaster plans [27], stockpiling resources necessary for an effective response [28], and developing skills [8] and competencies to ensure effective performance of disaster-related tasks [16].

Many previous studies have revealed that preparedness factors contribute to differences in disaster preparedness levels [27,29,30], such as personal, family, and social factors, and selective measures of preparedness. Residents' personal disaster preparedness refers to the actions taken in response to disasters and loss reduction [31]; it is also known as risk perception [32,33]. Physiological activities, such as attitudes and beliefs thus alter people's hazard avoidance behavior [34], as do the previous experience and knowledge of hazards [35,36], disaster preparedness knowledge [37], and access to information sources [38]. Social factors include social networking [8] and trust in the government [30,38]. These studies considered a range of dimensions and different measures of preparedness.

In summary, disaster preparedness is a series of activities implemented to mitigate possible damage and reduce the adverse effects of a disaster. It is not only a part of the crisis and emergency management activities according to the time division, but also a fundamental action throughout the crisis and emergency management process that is performed before, during, and after the disaster.

2.2. Components of Disaster Preparedness Capability

In the case of disasters, it is critical to identify the changing needs of the disaster response environment and to foster the management capability needed to respond to disasters. Capability operational transformation is a critical success factor for disaster management [39]. Cigler [40] defines capability as the financial, technical, policy-related, institutional, leadership, and human resource capabilities that local government agencies must have in order to operate in all phases of daily emergency and disaster situations. The capability required for disaster management is related to the delegation of authority, communication, decision-making, and inter-agency coordination [41]. Common mistakes that local governments make in preventing disasters are often related to rigid institutional beliefs, ignoring external complaints, difficulties dealing with multiple sources of information, and a tendency to minimize danger [42]. Therefore, disaster preparedness often requires coordination between individuals, governments, agencies, and organizations to improve training and exercise plans, enhance and introduce technological innovations, and ensure that individuals, social organizations, and businesses in various fields support this ability.

This paper is based on the preparedness capability elements classified by the Federal Emergency Management Agency (FEMA) National Preparedness Directorate, and the components of disaster preparedness capability are shown in Table 1. It shows the planning process that begins with planning for the various hazards that exist, and then builds and improves readiness systematically. This cycle recognizes the importance of the four main components of any preparation: planning, organization, equipment, and education and exercise. This cycle represents not only readiness at all levels of government jurisdictions, but also readiness actions taken by individuals, businesses, NGOs, and other entities [18,43].

Table 1. Components of disaster preparedness capability.

Planning	Disaster management planning at the government level is a necessary and complex process. Governments must know what needs to be done, how they will do it, what equipment will be used, and how they can get other agencies or people to help them. In the event of a disaster, each level of government is required to perform a range of tasks and functions before, during, and after the event. The most comprehensive approach to disaster management planning is the development of a national Emergency Operations Plan (EOP that includes disaster risk assessment (B1), disaster response plan (B2), nd plan preparation and approval (B3). Planning may also include: demand analysis, hazards risk analysis, plan evaluation, revision and improvement, the disaster planning system, hazard identification, and comprehensive evaluation of disaster risks.
Organization	When the government responds to a disaster, it is critical to ensure that all individuals and agencies involved in the emergency management system can perform their responsibilities and have appropriate statutory disaster management organization: disaster management leading agency (B4) and disaster management grassroots working organization (B5) under laws and regulations(B6). EOPs define the actions of specific authorities, and statutory authorities give them the authority to take those actions. Agreements between neighboring communities and even countries, as well as between jurisdictions, contribute to the need for a legal and disaster response system (B7) framework in the same country before a disaster occurs. Examples of organization include: policy guidance, disaster management system, disaster management leading agency, disaster management organization, and expert groups.
Equipment	Developing tools, technologies, and other equipment to assist in disaster response and recovery helps the responding agencies significantly reduce the number of casualties and properties damaged and destroyed as a result of disasters. Disaster rescue equipment also adds to the effectiveness of responding agencies by protecting the lives of responders. This equipment is primarily driven by available disaster management resources (B8), disaster management funding (B9), disaster medical rescue supplies (B10), and disaster communication and transportation guarantee (B11).
Education and exercise	Disaster management training (B12), disaster management exercise (B13), and public emergency safety education(B14) constitute the fourth component of government disaster preparedness capability. Considering disaster management, response officials who are not adequately trained in the details of specialized responses are at serious risk. Untrained or inadequately trained responders increase the likelihood of secondary emergencies or disasters, further contributing to the shortage of response resources. Examples of education and exercise include: training of general personnel, training of disaster response team, qualification certification, public emergency safety education, evaluation of educational activities, disaster exercise, exercise planning, and exercise evaluation.

Source: Adapted from Coppola [43] (pp. 276–296).

2.3. Comprehensive AHP Evaluation Model

Based on an extensive data research and literature review, we developed a three-level AHP evaluation model with disaster preparedness capability as the target with reference to the US FEMA, as shown in Figure 1. In the AHP model, the target level is the disaster preparedness capability. The evaluation indices of disaster preparedness are divided into four second-level indicators: planning (A1), organization (A2), equipment (A3), and education and exercise (A4), and 14 tertiary evaluation indicators: disaster risk assessment (B1), disaster response plan (B2), plan preparation and approval (B3), disaster management leading agency (B4), disaster management grassroots working organization (B5), laws and regulations (B6), disaster management system (B7), disaster management resources (B8), disaster management funding (B9), disaster medical rescue supplies (B10), disaster communication and transportation guarantee (B11), disaster management training (B12), disaster management exercise (B13), and public emergency safety education (B14).

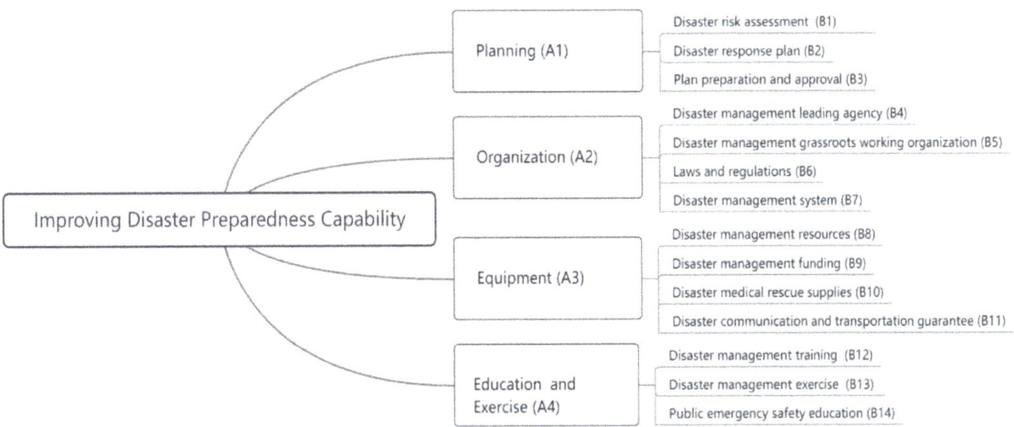

Figure 1. Comprehensive evaluation model of the disaster preparedness capability index system.

3. Materials and Methods

The Analytic Hierarchical Process, proposed by American operations researcher Saaty in the 1970s, is a comprehensive weighted decision-making method that uses mathematics and psychology to organize and analyze complex decisions, assigning weights in the process of comparing the relative importance of indicators to ensure that a logically consistent solution is reached. It is applicable to decision-making problems involving complex hierarchies and multiple indicators [44]. As a decision system, AHP is valuable for using human cognition to determine the relative importance between a set of alternatives through pairwise comparisons [45]. This approach has been used in various studies aimed at promoting development in different sectors, such as environment and natural resources [46]; disaster management, disaster resilience, and vulnerability indices [47–50]. Therefore, in this study, AHP was used to not only identify the criteria and influencing factors that best describe disaster preparedness, but also evaluate the importance and priority among indicators, and develop a tool to quantify disaster preparedness capability.

The use of an AHP analysis to determine the evaluation index system and the weights can be divided into the following steps: establishing the hierarchical structure according to the hierarchical relationship, constructing a judgment matrix, calculating the judgment matrix to obtain the relative weights of the evaluation indices, and testing the consistency of judgment to obtain the final weights of indices at each level. On this basis, we take the Zhengzhou "7·20" rainstorm as an example to conduct empirical analysis; carry out qualitative and quantitative empirical analysis of emergency preparedness ability; obtain the scores of various indicators; use EXCEL and SPSS software 26 to input and process score data, respectively; and obtain the final comprehensive evaluation results. The specific steps and process are shown in Figure 2.

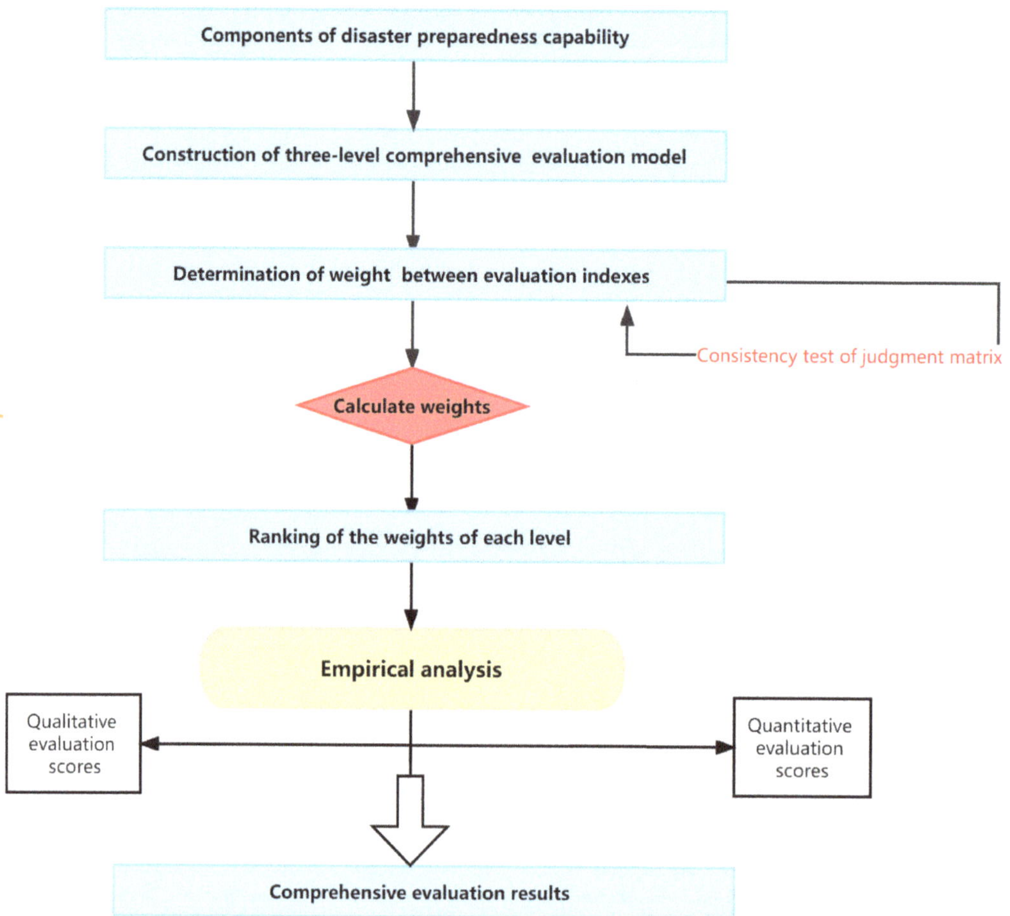

Figure 2. Calculation process chart.

To ensure the objectivity of the relevant data obtained and to scientifically determine and rank the importance of the weights of the indicators to ensure the validity of the indicator system, this paper solicited and obtained data of the weights of each indicator by issuing questionnaires to 14 experts and scholars in the field of government disaster management who were recommended by professors and contacted directly. They were viewed as decision-makers in the prioritization process, making evaluations and choices based on their experience, skill, knowledge, and practice [49]. The questionnaire data were collected from 17 to 20 September 2022. All 14 completed questionnaires were collected. The basic information of the survey respondents is shown in Table 2. The questionnaire used the scale method of 1~9 and their reciprocals. The complex problem was broken down, level by level, and the indicators in the hierarchy were compared in terms of their relative importance in determining their overall order of importance.

Table 2. Descriptive statistics of survey participants.

Characteristics		Frequency	Characteristics		Frequency
Gender	Male	8	Education	Master's degree	4
	Female	6		Doctoral degree	10
Age	30–39	6	Number of years of research or work	Less than 5 years	2
	40–49	3		5–10 years	4
	>50	5		More than 10 years	8

3.1. Determination of Weight Value between Evaluation Indices

In this paper, we used YAAHP software 12.8 to calculate the weights for each level of indicator (i.e., the degree of importance, according to the abovementioned steps). The index weights were calculated according to AHP and the weights of each hierarchical evaluation index system were also calculated; the results are shown in Table 3. Four indicators were evaluated at the second level: (A1), (A2), (A3) and (A4) with weights of 0.294, 0.220, 0.257, and 0.228, respectively.

Table 3. Weight of each index of the disaster preparedness capability evaluation index system.

Second-Level(A)	Weight	Priority	Third-Level(B)	Relative Importance	Priority	Composite Weight	Priority
Planning (A1)	0.294	1	Disaster risk assessment (B1)	0.233	3	0.068	5
			Disaster response plan (B2)	0.457	1	0.135	1
			Plan preparation and approval (B3)	0.310	2	0.091	4
Organization (A2)	0.220	4	Disaster management leading agency (B4)	0.222	4	0.049	13
			Disaster management grassroots working organization (B5)	0.253	3	0.056	10
			Laws and regulations (B6)	0.271	1	0.060	8
			Disaster management system (B7)	0.255	2	0.056	9
Equipment (A3)	0.257	2	Disaster management resources (B8)	0.393	1	0.101	3
			Disaster management funding (B9)	0.202	3	0.052	11
			Disaster medical rescue supplies (B10)	0.152	4	0.039	14
			Disaster communication and transportation guarantee (B11)	0.253	2	0.065	7
Education and exercise (A4)	0.228	3	Disaster management training (B12)	0.490	1	0.112	2
			Disaster management exercise (B13)	0.285	2	0.065	6
			Public emergency safety education (B14)	0.225	3	0.052	12

In the consistency test, the consistency ratio (CR) is generally within 0.1, suggesting that the calculation results are consistent and that the consistency of the judgment matrix is acceptable [51] (p. 287). However, in some cases, 0.2, but never more, is tolerable [52] (p. 34). According to the software's calculation results, the consistency index (CI) and the average random consistency index (RI) are derived, and the consistency ratio (CR) is finally calculated as follows: CR = CI/RI. The analysis results of the AHP model in the disaster preparedness index system revealed that the CR of disaster preparedness = 0.077 < 0.1, which meets the consistency requirement. Meanwhile, the CRs of planning (A1), organization (A2), equipment (A3), and education and exercise (A4) were 0.055, 0.079, 0.038, and 0.076, respectively, suggesting that the constructed judgment matrix had a high degree of consistency.

3.2. Composite Weight Ranking of Judgment Matrix

Figure 3a shows the weights of the different indicators at each level, including comparisons between levels. Combined with the hierarchical relationships of the indicators in the constructed model, different weight distributions of the different indicators affecting the disaster preparedness capability can be seen. The weight ranking of the criterion layer (second-level) and scheme layer (third-level) to the target layer is shown in Figure 3b,c, respectively. In Figure 3b, Planning (A1) had the largest weight in the overall disaster preparedness capability, followed by equipment (A3), education and exercise (A4), and finally, organization (A2), which accounted for the smallest weight, while training and education and exercise had the least influence on the overall disaster preparedness capability. In Figure 3c, compared with other indicators, the disaster response plan (B2) was the most important for assessing overall disaster preparedness capability, indicating that the prevention work plan before an incident was critical to overall disaster preparedness. When hazardous accidents and disasters occur, we must target our disaster preparedness efforts toward improving emergency rescue and disposal capabilities and minimize losses. Additionally, disaster management training (B12), disaster management resources (B8), and plan preparation and approval (B3) had a greater impact on disaster preparedness. The three with the least impact were public emergency safety education (B14), disaster management leading agency (B4), and disaster medical rescue supplies (B10).

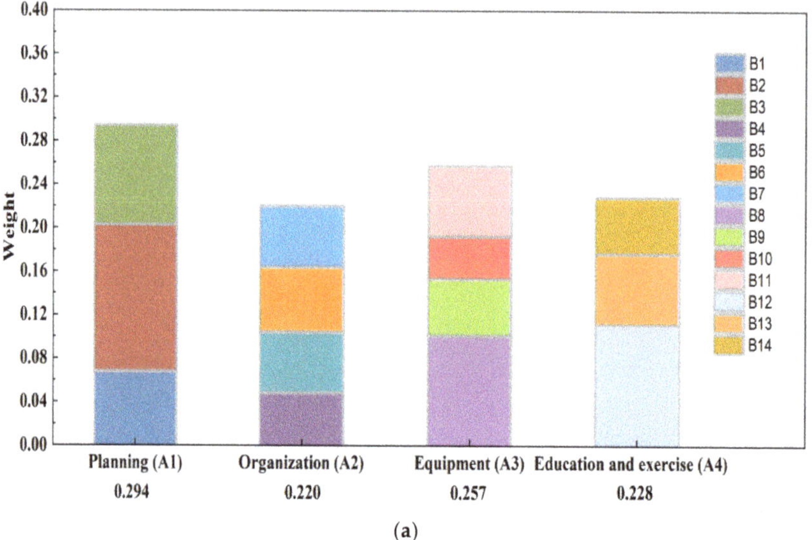

(**a**)

Figure 3. *Cont.*

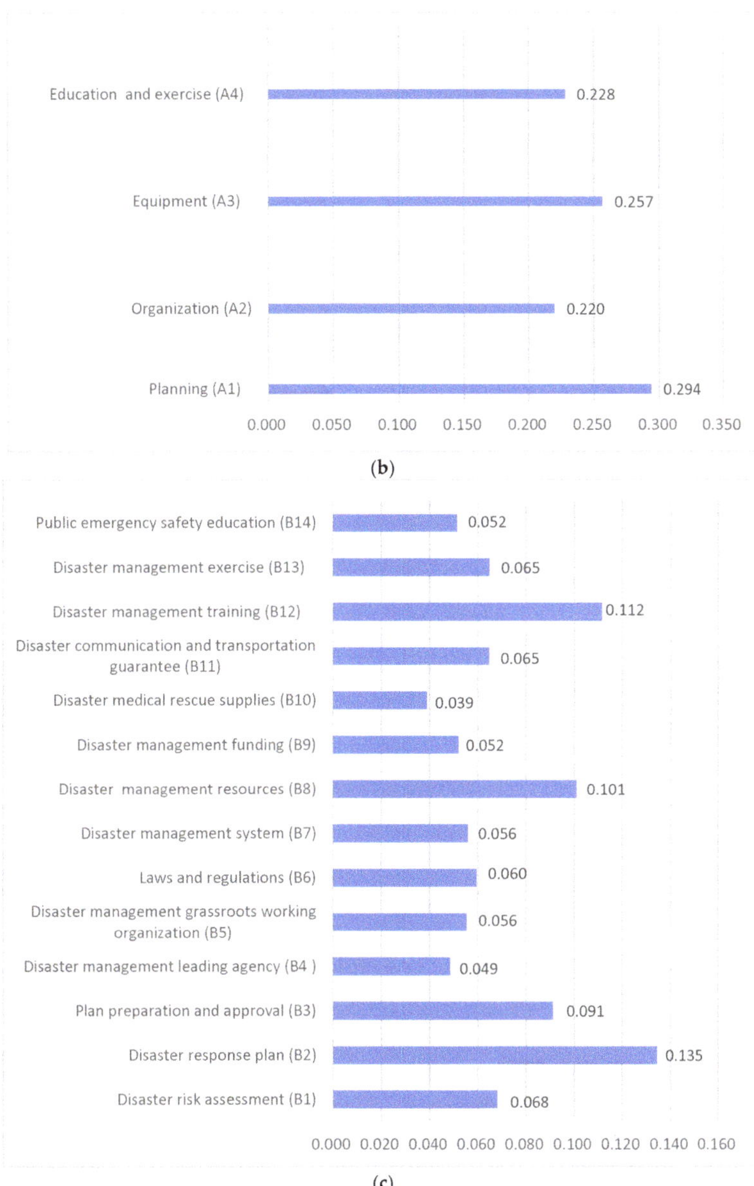

Figure 3. (**a**) Comparisons between different indicators at each level. (**b**) The weight ranking of the second level (A1~A4) to the target level. (**c**) The weight ranking of the third level (B1~B14) to the target level.

3.3. Relative Weight Ranking of Judgment Matrix

According to the judgment matrix, the relative weight ranking of the third level to the corresponding second level can be obtained separately (Figure 4). For planning (A1), disaster response plan (B2) had the largest weight (45.7%), followed by plan preparation and approval (B3) (31%), and disaster risk assessment (B1) (23.3%). Thus, the disaster response plan greatly impacted planning (see Figure 4a). For organization (A2), laws and regulations (weighted at 27.1%) and disaster management systems (weighted at 28.4%) were more important. Additionally, the proportions of disaster management grassroots working organizations and disaster management leading agencies to the organization were 25.3% and 22.2%, respectively (Figure 4b). For equipment (A3), the weight ratio of disaster resources was as high as 39.3%, followed by disaster communication and a transportation guarantee with a weight ratio of 25.3% (both of which directly affected equipment capabilities), disaster funding (weighted at 20.2%), and finally, disaster medical rescue supplies (weighted at 15.2% (see Figure 4c). For education and exercise (A4), disaster management training directly affected education and exercise capability, with a weight ratio of 49%, followed by a disaster management exercise at 28.5% and public emergency safety education at 22.5% (see Figure 4d).

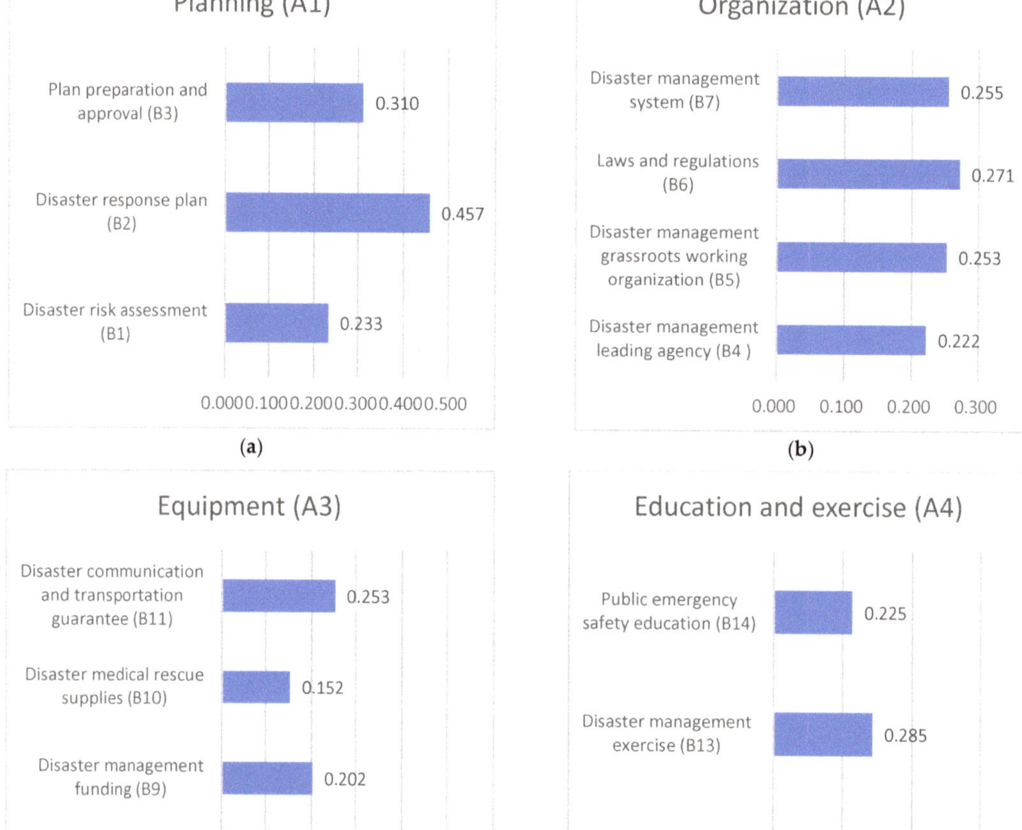

Figure 4. (**a**) The weight ranking of the third level (B1~B3) to the second level A1. (**b**) The weight ranking of the third level (B4~B7) to the second level A2. (**c**) The weight ranking of the third level (B8~B11) to the second level A3. (**d**) The weight ranking of the third level (B12~B14) to the second level A4.

4. Results

4.1. Empirical Analysis

The abovementioned evaluation method was used to assess the disaster preparedness of Zhengzhou city for this "7·20" rainstorm by combining the disaster preparedness evaluation index system and the actual disaster preparedness in response to the "7·20" rainstorm. Twenty experts engaged in government disaster management or staff of relevant government departments were selected as the subjects, and 19 valid questionnaires were finally collected, with a valid return rate of 95%. The questionnaire sought to determine contents of the three-level indicators, which were divided into quantitative and qualitative indicators according to the form of the basic data obtained from the statistical indicators. Therefore, excluding the first part of basic personal information, the questionnaire subjects were divided into two main blocks based on quantitative and qualitative content. Quantitative indicators can be judged by specific numerical values, such as the number of personnel, ambulance supplies, and shelters. Qualitative indicators were values of indicators that cannot be expressed by specific numbers, and questionnaire participants often provide descriptive data based on intuition or experience. For the convenience of calculation, a five-point Likert scale was used to convert the graded values into statistically significant indicators. We used Excel and SPSS software 26 to organize, enter, and calculate the obtained data, and descriptive statistical analysis was performed, resulting in scores of qualitative and quantitative indicators, respectively. Next, the qualitative evaluation and quantitative evaluation scores were added together and divided by 2 to obtain the comprehensive scores. A reliability test was conducted based on the results and the Cronbach's alpha was 0.993, indicating the scientific credibility of the findings.

Figure 5 shows the qualitative evaluation scores, quantitative evaluation scores, and comprehensive scores of B1~B14. After the collation and calculation, in terms of qualitative evaluation scores, the disaster management leading agency (B4) had the highest score, followed by the disaster management grassroots working organization (B5), and laws and regulations (B6); plan preparation and approval (B3) had the lowest score. The qualitative assessment of disaster preparedness revealed more recognized scores of disaster response organization, while the disaster management leading agency and grassroots working organization were more recognized; the related ability of disaster plan preparation and public emergency safety education was somewhat inadequate. The quantitative evaluation scores for B1 to B14 were 3.42, 3.32, 3.26, 3.58, 3.58, 3.42, 3.37, 3.26, 3.00, 3.21, 3.11, 2.88, 2.58 and 3.24, respectively, with the disaster management leading agency (B4) and disaster management grassroots working organization (B5) having the highest scores. This is consistent with the qualitative assessment score, followed by disaster risk assessment (B1) and laws and regulations (B6); disaster management exercise (B13) had the lowest score. Thus, in the quantitative assessment of disaster preparedness, the performance of disaster response organization and disaster response regime was more recognized, and the related ability of disaster management exercise and disaster management training was somewhat lacking.

Comprehensive scores of the individual index reveal that disaster management leading agency (B4) had the highest score in both qualitative and quantitative assessments, followed by the disaster management grassroots working organization (B5), and laws and regulations (B6). The last three rankings were disaster management exercise (B13), disaster management training (B12), and disaster management funding (B9), indicating that Zhengzhou city needs to pay more attention to disaster training, exercise, and disaster management funding in the future.

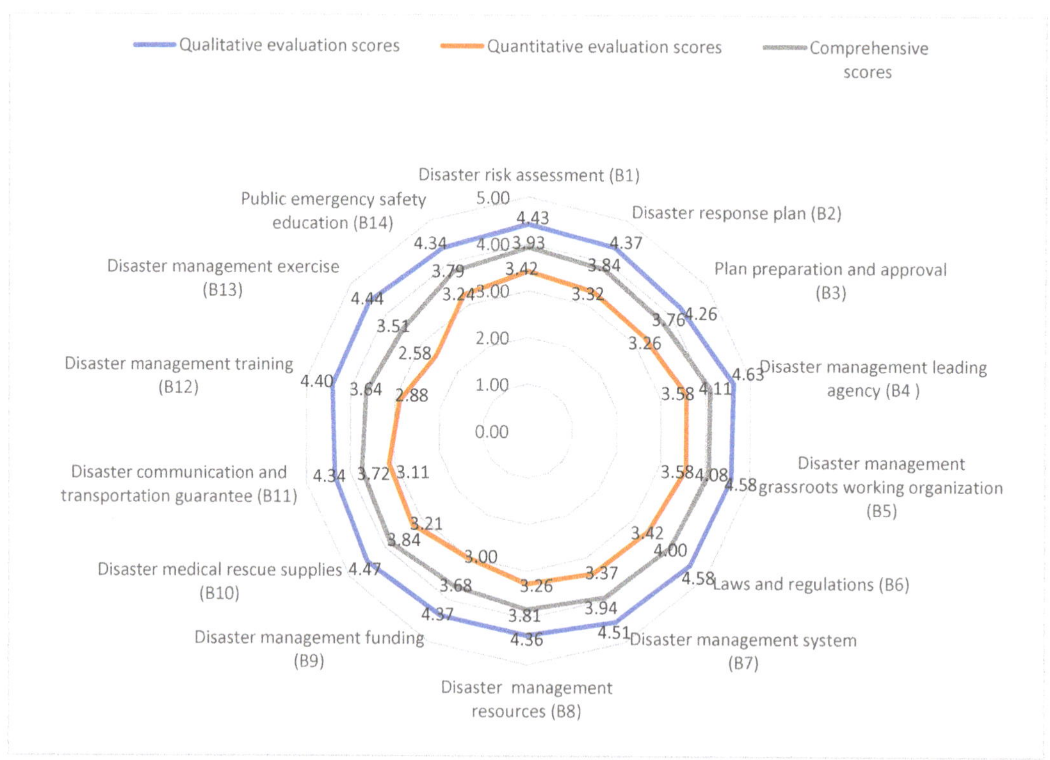

Figure 5. Radar chat of qualitative evaluation scores, quantitative evaluation scores, and comprehensive scores of B1~B14.

4.2. Comprehensive Evaluation Results

Overall, the combined assessment scores for the second-level indicators were 3.84, 4.03, 3.77, and 3.65 (out of 5) (see Table 4). The highest score for organization (A2) indicated that Zhengzhou had a more complete disaster management organization and system, followed by planning (A1), and finally, equipment (A3), and education and exercise (A4), indicating a major lack of daily disaster management training, exercise, and public emergency safety education areas in Zhengzhou. Specifically, the qualitative assessment scores of the secondary indicators were higher than the quantitative assessment scores, indicating that the Zhengzhou government had a clear understanding of the content and objectives of the work needed to improve disaster preparedness but was not sufficiently concerned about the implementation of tasks. The relevant authorities should be urged to strengthen the supervision and management of the implementation of the entire disaster preparedness process.

Table 4. Comprehensive scores of second- and third-level indicators.

Second-Level(A)	Third-Level(B)	Qualitative Evaluation Scores	Quantitative Evaluation Scores	Comprehensive Scores	
	Disaster risk assessment (B1)	4.43	3.42	3.93	
Planning (A1)	Disaster response plan (B2)	4.37	3.32	3.84	3.84
	Plan preparation and approval (B3)	4.26	3.26	3.76	

Table 4. *Cont.*

Second-Level(A)	Third-Level(B)	Qualitative Evaluation Scores	Quantitative Evaluation Scores	Comprehensive Scores	
Organization (A2)	Disaster management leading agency (B4)	4.63	3.58	4.11	
	Disaster management grassroots working organization (B5)	4.58	3.58	4.08	4.03
	Laws and regulations (B6)	4.58	3.42	4.00	
	Disaster management system (B7)	4.51	3.37	3.94	
Equipment (A3)	Disaster management resources (B8)	4.36	3.26	3.81	
	Disaster management funding (B9)	4.37	3.00	3.68	
	Disaster medical rescue supplies (B10)	4.47	3.21	3.84	3.77
	Disaster communication and transportation guarantee (B11)	4.34	3.11	3.72	
Education and exercise (A4)	Disaster management training (B12)	4.40	2.88	3.64	
	Disaster management exercise (B13)	4.44	2.58	3.51	3.65
	Public emergency safety education (B14)	4.34	3.24	3.79	

In conclusion, there were still some shortcomings in Zhengzhou City's preparation and response to the extraordinarily heavy "7·20" rainstorm. First, there was insufficient awareness of major hazard and threat information and poor awareness of disaster risk; the person in charge had a subjective sense of judgment, lacked sensitivity and alertness to major hazard signals, and ignored the forecast information made by the meteorological department. Second, there was an obvious disconnect between emergency operations and forecast information dissemination and no quick or timely alert announcement information to the society. Third, the formulation, evaluation, and revision of the plan were not refined, and the practice was not strengthened. In many disaster-prone areas, local and national governments and NGOs have worked to provide disaster education programs and emergency training to raise awareness and promote self-reliance and family preparedness [53]. Thus, the process of responding to this extraordinarily heavy rainstorm revealed that the dissemination of disaster warning information was not timely or adequate and that safety awareness and disaster prevention and avoidance capabilities were weak. The disaster education knowledge of leaders at all levels, disaster management capability training, and safety knowledge education for the public should all be improved.

5. Discussion

Based on the scores of the comprehensive evaluation indicators and the level of disaster risk in Zhengzhou, the following suggestions are provided for disaster preparedness in Zhengzhou: In terms of planning (A1), an important aspect of disaster planning is to convey to the public the nature of the risk and make appropriate adaptation strategies so that people have a clear perception of risk and know what to do and what not to do before and after a disaster [54]. We assume that all disasters are local and that the primary responsibility for managing disasters and emergencies, including informing and alerting the public, belongs to local governments [55]. An effective system requires that early warning information [1] and risk reduction be mainstreamed into the policy process and that government agencies have the capability to design and implement effective policies. Effective early warning system policy processes also require the involvement of local communities to ensure that the at-risk public is adequately informed and alerted [56]. Communities and residents are responsible for taking their own measures to prepare for disasters, and the final decision on disaster preparedness measures rests with individuals [57]. The public takes disasters more seriously when they have a large amount of information and credible disaster warnings [58].Thus, information collection on major hazards and threats should be increased, the sensitivity of major hazard signals should be maintained, the upgrading of the monitoring and warning information platform and release system should be accelerated, and multi-source information should be combined and processed quickly and efficiently to ensure the timely release of warning information to the community at the first instance of an accident. Simultaneously, to further improve the system, we should pay attention to updating and enhancing the disaster management plan, which should be filed with relevant superior departments or agencies, in order to enhance the integrity, coordination, and effectiveness of the system.

In terms of the organization (A2), we should continue to improve the disaster management system and institutional set-up to ensure disaster preparedness in an effective and organized manner. We should also enhance the efficiency of disaster management departments, while improving the disaster command and coordination mechanism to ensure coordination and linkage among functional departments and different administrative regions in an orderly and efficient manner. Additionally, the construction of a comprehensive disaster rescue system, disaster management leadership, and cooperation between the leading agencies of disaster management and local working agencies should be enhanced.

In terms of the equipment (A3), resources are stable assets that can be used to deal with a variety of situations, including those related to health, income, and social support. Having disaster resources is critical to proactively respond to disasters and crises [59]. It is necessary to focus on strengthening the provision and maintenance of disaster relief equipment and materials. Furthermore, we should increase the construction of emergency shelters, open up qualified gymnasiums, parks, and other places that can serve as emergency shelters, equip them with supporting facilities, strengthen emergency material and fund reserves, and regularly update the facilities and equipment required for living.

In terms of education and exercise (A4), education can trigger a learning process that improves disaster preparedness. In the future, managers will need to be educated on disaster preparedness planning and should work with local agencies to provide disaster training for other managers, teachers, and staff to ensure that appropriate actions are taken to minimize loss of life and property [60]. Disaster education increases coping potential, thereby reducing the adverse effects of anxiety on disaster preparedness [61]. Regular training activities for emergency management-related practitioners as well as general personnel, educational activities, and various forms of emergency plan exercises, will help improve the proportion of practitioners who meet the qualification requirements and the professionalism of rescue teams. Their focus should be on making citizens aware of their own important role and disaster preparedness actions while responding to a disaster or crisis. Due to the limited resources and capabilities of the government, it is impossible to rescue each victim in time, which makes the mutual and self-rescue of citizens very important in the early stage of disasters. Therefore, multi-form and multi-content public safety culture improvement activities should be conducted to enhance their emergency preparedness capability. Additionally in this process, the role of active opinion leaders in disaster preparedness is valued. Opinion leaders can actively organize training on disaster prevention and mitigation knowledge, strengthen public disaster risk awareness and guide the public to properly respond to natural disasters [62].

6. Limitations

The existing literature does not provide an official definition of disaster preparedness capability, and an official disaster preparedness assessment system that is applicable to various countries and governments at all levels has not yet been developed. The influencing indicators of capability are relatively complex. Therefore, the division of disaster preparedness tasks and capability described in this paper may be lacking in certain forms, as the construction of indicators is incomplete and the influencing factors are interconnected and mutually restrictive, as has been repeatedly demonstrated in combination with practical applications.

7. Conclusions

In this study, the evaluation model of disaster preparedness capability was proposed, the scores and grades of the indicators were summarized and classified to comprehensively analyze and evaluate the disaster preparedness capability of Zhengzhou city, then some suggestions were put forward. The results revealed that Zhengzhou city residents had a clear understanding of the content and objectives of the work needed to improve disaster preparedness and had a relatively complete disaster management organization and system. However, there was an overall lack of attention to the planning and implementation of tasks, which can further encourage poor management activities [63]. The research results can help find weak links in disaster preparedness in the future and improve the disaster preparedness capability of various disaster response subjects. Disaster drills and training involve many agencies and resources. There is a need to strengthen partnerships with civil society organizations, community volunteers, and local chambers of commerce and industry to build a network of organizational partners in the public, private, and non-profit sectors in disaster management [64]. The resources of these organizations and groups can improve local government weaknesses [65]. Simultaneously, effective cognitive and psychomotor skills should be developed through organized and planned training to deal effectively with disaster situations [66], and relevant departments need to be encouraged to strengthen the supervision and management of the implementation of the whole process of disaster preparedness. Regarding the disaster preparedness assessment framework proposed and constructed in this paper, with the continuous strengthening of the overall level of disaster response work, there may be some changes in the indicators in different countries and scenarios, and research will be conducted in consideration of the actual situation. Therefore, a multi-country comparative analysis will be carried out in the future and will be further improved in follow-up research.

Author Contributions: Conceptualization, L.Z. and J.E.L.; methodology, L.Z.; software, L.Z.; validation, J.E.L.; formal analysis, L.Z.; investigation, L.Z.; resources, L.Z.; data curation L.Z.; writing—original draft, L.Z.; writing—review and editing, L.Z. and J.E.L.; visualization, L.Z.; supervision, J.E.L. All authors have read and agreed to the published version of the manuscript.

Funding: This research was funded by the Ministry of Education of the Republic of Korea and the National Research Foundation of Korea (NRF-2020S1A5B8103910).

Institutional Review Board Statement: Not applicable.

Informed Consent Statement: Informed consent was obtained from all participants involved in the study.

Data Availability Statement: A copy of the questionnaire and datasets for this study are available from the corresponding author upon well-founded request.

Acknowledgments: This work was supported by the Ministry of Education of the Republic of Korea and the National Research Foundation of Korea (NRF-2020S1A5B8103910).

Conflicts of Interest: The authors declare no conflict of interest.

References

1. Kapucu, N.; Berman, E.; Wang, X. Emergency information management and public disaster preparedness: Lessons from the 2004 Florida hurricane season. *Int. J. Mass Emerg. Disasters* **2008**, *26*, 169–197. [CrossRef]
2. McEntire, D.A. Issues in disaster relief: Progress, perpetual problems and prospective solutions. *Disaster Prev. Manag.* **1999**, *8*, 351–361. [CrossRef]
3. Disaster Investigation Team of the State Council. Investigation Report on "7.20" Heavy Rainstorm Disaster in Zhengzhou, Henan. 2022. Available online: https://www.mem.gov.cn/gk/sgcc/tbzdsgdcbg/202201/P020220121639049697767.pdf (accessed on 15 May 2022).

4. Wolensky, R.P.; Wolensky, K.C. Local government's problem with disaster management: A literature review and structural analysis. *Rev. Policy Res.* **1990**, *9*, 703–725. [CrossRef]
5. Kunz, N.; Reiner, G.; Gold, S. Investing in disaster management capabilities versus pre-positioning inventory: A new approach to disaster preparedness. *Int. J. Prod. Econ.* **2014**, *157*, 261–272. [CrossRef]
6. Baker, L.R.; Baker, M.D. Disaster preparedness among families of children with special health care needs. *Disaster Med. Public Health Prep.* **2010**, *4*, 240–245. [CrossRef] [PubMed]
7. Ablah, E.; Konda, K.; Kelley, C.L. Factors predicting individual emergency preparedness: A multi-state analysis of 2006 BRFSS data. *Biosecur. Bioterror.* **2009**, *7*, 317–330. [CrossRef] [PubMed]
8. Kirschenbaum, A. Families and disaster behavior: A reassessment of family preparedness. *Int. J. Mass. Emerg. Disasters* **2006**, *24*, 111–143. [CrossRef]
9. Kapucu, N. Culture of preparedness: Household disaster preparedness. *Disaster Prevent. Manag.* **2008**, *17*, 526–535. [CrossRef]
10. Bogdan, E.E.A.; Roszko, A.M.; Beckie, M.A.; Conway, A. We're ready! Effectiveness of community disaster preparedness workshops across different community groups in Alberta, Canada. *Int. J. Disaster Risk Reduct.* **2021**, *55*, 102060. [CrossRef]
11. Suryadi, T.; Zulfan, Z.; Kulsum, K. The relationship between knowledge and attitudes about community disaster preparedness in Lambung Village, Banda Aceh. *Int. J. Disaster Manag.* **2021**, *4*, 1–10. [CrossRef]
12. Campasano, N. Community Preparedness: Creating a Model for Change. Master's Thesis, Naval Postgraduate School, Monterey, CA, USA, 2010.
13. Kohn, S.; Eaton, J.L.; Feroz, S.; Bainbridge, A.A.; Hoolachan, J.; Barnett, D.J. Personal disaster preparedness: An integrative review of the literature. *Disaster Med. Public Health Prep.* **2012**, *6*, 217–231. [CrossRef]
14. Donahue, A.K.; Eckel, C.C.; Wilson, R.K. Ready or not? How citizens and public officials perceive risk and preparedness. *Am. Rev. Public Adm.* **2014**, *44*, 89S–111S. [CrossRef]
15. Simpson, D.M. Disaster preparedness measures: A test case development and application. *Disaster Prev. Manag.* **2008**, *17*, 645–661. [CrossRef]
16. Sutton, J.; Tierney, K. Disaster preparedness: Concepts, guidance, and research. *Colo. Univ. Colo.* **2006**, *3*, 1–41.
17. Van Wassenhove, L. Humanitarian aid logistics: Supply chain management in high gear. *J. Oper. Res. Soc.* **2006**, *57*, 475–489. [CrossRef]
18. Haddow, G.; Bullock, J.A.; Coppola, D.P. *Introduction to Emergency Management*, 5th ed.; Butterworth-Heinemann: Oxford, UK, 2013.
19. Paton, D.; McClure, J. *Preparing for Disaster: Building Household and Community Capacity*; Charles C Thomas Publisher: Springfield, IL, USA, 2013.
20. National Research Council. *Facing Hazards and Disasters: Understanding Human Dimensions*; National Academies Press: Washington, DC, USA, 2006.
21. Grothmann, T.; Reusswig, F. People at risk of flooding: Why some residents take precautionary action while others do not. *Nat. Hazards* **2006**, *38*, 101–120. [CrossRef]
22. Siegrist, M.; Gutscher, H. Natural hazards and motivation for mitigation behavior: People cannot predict the affect evoked by a severe flood. *Risk Anal.* **2008**, *28*, 771–778. [CrossRef]
23. Thomalla, F.; Downing, T.; Spanger-Siegfried, E.; Han, G.; Rockström, J. Reducing hazard vulnerability: Towards a common approach between disaster risk reduction and climate adaptation. *Disasters* **2006**, *30*, 39–48. [CrossRef]
24. Berkes, F. Understanding uncertainty and reducing vulnerability: Lessons from resilience thinking. *Nat. Hazards* **2007**, *41*, 283–295. [CrossRef]
25. Norris, F.H.; Stevens, S.P.; Pfefferbaum, B.; Wyche, K.F.; Pfefferbaum, R.L. Community resilience as a metaphor, theory, set of capacities, and strategy for disaster readiness. *Am. J. Community Psychol.* **2008**, *41*, 127–150. [CrossRef]
26. Prior, T.; Eriksen, C. Wildfire preparedness, community cohesion and social–ecological systems. *Glob. Environ. Chang.* **2013**, *23*, 1575–1586. [CrossRef]
27. Russell, L.A.; Goltz, J.D.; Bourque, L.B. Preparedness and hazard mitigation actions before and after two earthquakes. *Environ. Behav.* **1995**, *27*, 744–770. [CrossRef]
28. Levac, J.; Toal-Sullivan, D.; O'Sullivan, T.L. Household emergency preparedness: A literature review. *J. Community Health* **2012**, *37*, 725–733. [CrossRef] [PubMed]
29. Karanci, A.N.; Aksit, B.; Dirik, G. Impact of a community disaster awareness training program in Turkey: Does it influence hazard-related cognitions and preparedness behaviors. *Soc. Behav. Pers.* **2005**, *33*, 243–258. [CrossRef]
30. Kirschenbaum, A.A.; Rapaport, C.; Canetti, D. The impact of information sources on earthquake preparedness. *Int. J. Disaster Risk Reduct.* **2017**, *21*, 99–109. [CrossRef]
31. Atreya, A.; Czajkowski, J.; Botzen, W.; Bustamante, G.; Campbell, K.; Collier, B.; Ianni, F.; Kunreuther, H.; Michel-Kerjan, E.; Montgomery, M. Adoption of flood preparedness actions: A household level study in rural communities in Tabasco, Mexico. *Int. J. Disaster Risk Reduct.* **2017**, *24*, 428–438. [CrossRef]
32. Xu, D.; Peng, L.; Liu, S.; Wang, X. Influences of risk perception and sense of place on landslide disaster preparedness in Southwestern China. *Int. J. Disaster Risk Sci.* **2018**, *9*, 167–180. [CrossRef]
33. Lindell, M.K.; Arlikatti, S.; Prater, C.S. Why people do what they do to protect against earthquake risk: Perceptions of hazard adjustment attributes. *Risk Anal.* **2009**, *29*, 1072–1088. [CrossRef]

34. Ebru, I.N.A.L.; Altintas, K.H.; Dogan, N. The development of a General Disaster Preparedness Belief Scale using the health belief model as a theoretical framework. *Int. J. Assess. Tools Educ.* **2018**, *5*, 146–158.

35. Onuma, H.; Shin, K.J.; Managi, S. Household preparedness for natural disasters: Impact of disaster experience and implications for future disaster risks in Japan. *Int. J. Disaster Risk Reduct.* **2017**, *21*, 148–158. [CrossRef]

36. Uprety, P.; Poudel, A. Earthquake risk perception among citizens in Kathmandu, Nepal. *Australas. J. Disaster Trauma Stud.* **2012**, *1*, 3–10.

37. Mileti, D.S.; Darlington, J.D. The role of searching in shaping reactions to earthquake risk information. *Soc. Probl.* **1997**, *44*, 89–103. [CrossRef]

38. Basolo, V.; Steinberg, L.J.; Burby, R.J.; Levine, J.; Cruz, A.M.; Huang, C. The effects of confidence in government and information on perceived and actual preparedness for disasters. *Environ. Behav.* **2009**, *41*, 338–364. [CrossRef]

39. Kusumasari, B.; Alam, Q.; Siddiqui, K. Resource capability for local government in managing disaster. *Disaster Prev. Manag.* **2010**, *19*, 438–451. [CrossRef]

40. Cigler, B.A. The "big questions" of Katrina and the 2005 great flood of New Orleans. *Public Adm. Rev.* **2007**, *67*, 64–76. [CrossRef]

41. Douglas, P.; Duncan, J. Developing disaster management capability: An assessment centre approach. *Disaster Prev. Manag.* **2002**, *11*, 115–122.

42. Turner, B.A. The organizational and interorganizational development of disasters. *Adm. Sci. Q.* **1976**, *21*, 378–397. [CrossRef]

43. Coppola, D.P. *Introduction to International Disaster Management*; Elsevier: Abingdon, UK; Butterworth-Heineman: Oxford, UK, 2015.

44. Saaty, R.W. The analytic hierarchy process—What it is and how it is used. *Math Model.* **1987**, *9*, 161–176. [CrossRef]

45. Saaty, T.L. *Fundamentals of Decision Making and Priority Theory with the Analytic Hierarchy Process*; RWS Publications: Pittsburgh, PA, USA, 1994; p. 477.

46. Schmoldt, D.L.; Kangas, J.; Mendoza, G.A. Basic principles of decision making in natural resources and the environment. In *The Analytic Hierarchy Process in Natural Resource and Environmental Decision Making*; Schmoldt, D., Kangas, J., Mendoza, G., Pesonen, M., Eds.; Springer: Dordrecht, The Netherlands, 2001; pp. 1–13.

47. Carreño, M.L.; Cardona, O.D.; Barbat, A.H.A. Disaster risk management performance index. *Nat. Hazards* **2007**, *41*, 1–20. [CrossRef]

48. Chen, G.-H.; Tao, L.; Zhang, H.-W. Study on the methodology for evaluating urban and regional disasters carrying capacity and its application. *Saf. Sci.* **2009**, *47*, 50–58. [CrossRef]

49. Orencio, P.M.; Fujii, M. A localized disaster-resilience index to assess coastal communities based on an Analytic Hierarchy Process (AHP). *Int. J. Disaster Risk Reduct.* **2013**, *3*, 62–75. [CrossRef]

50. Cardona, O.D.; Carreño, M.L. Updating the indicators of disaster risk and risk management for the Americas. *J. Integr. Disaster Risk Manag.* **2011**, *1*, 27–47. [CrossRef]

51. Saaty, T.L. *Multi-Criteria Decision Making: The Analytic Hierarchy Process, Planning, Priority Setting, Resource Allocation*; RWS Publications: Pittsburgh, PA, USA, 1990.

52. Saaty, T.L.; Kearns, K.P. *Analytical Planning: The Organization of System*; Pergamon Press, Inc.: New York, NY, USA, 1985.

53. Hoffmann, R.; Muttarak, R. Learn from the past, prepare for the future: Impacts of education and experience on disaster preparedness in the Philippines and Thailand. *World Dev.* **2017**, *96*, 32–51. [CrossRef]

54. Slovic, P. Informing and educating the public about risk. *Risk Anal.* **1986**, *6*, 403–415. [CrossRef] [PubMed]

55. MacManus, S.A.; Caruson, K. Code red: Florida city and county officials rate threat information sources and the homeland security advisory system. *State Local Gov. Rev.* **2006**, *38*, 12–22. [CrossRef]

56. Collins, M.L.; Kapucu, N. Early warning systems and disaster preparedness and response in local government. *Disaster Prev. Manag.* **2008**, *17*, 587–600. [CrossRef]

57. Adu-Gyamfi, B.; Shaw, R. Risk awareness and impediments to disaster preparedness of foreign residents in the Tokyo Metropolitan Area, Japan. *Int. J. Environ. Res. Public. Health* **2022**, *19*, 11469. [CrossRef]

58. Clarke, L. *Worst Cases: Terror and Catastrophe in the Popular Imagination*; University of Chicago Press: Chicago, IL, USA, 2006.

59. Hobfoll, S.E.; Lerman, M. Predicting receipt of social support: A longitudinal study of parents' reaction to their child's illness. *Health Psychol.* **1989**, *8*, 61–77. [CrossRef]

60. Burling, W.K.; Hyle, A.E. Disaster preparedness planning: Policy and leadership issues. *Disaster Prev. Manag.* **1997**, *6*, 234–244. [CrossRef]

61. Mishra, S.; Suar, D. Effects of anxiety, disaster education, and resources on disaster preparedness behavior. *J. Appl. Soc. Psychol.* **2012**, *42*, 1069–1087. [CrossRef]

62. He, J.; Zhuang, L.; Deng, X.; Xu, D. Peer effects in disaster preparedness: Whether opinion leaders make a difference. *Nat. Hazards* **2022**, *115*, 187–213. [CrossRef]

63. Quarantelli, E.L. Disaster crisis management: A summary of research findings. *J. Manag. Stud.* **1988**, *25*, 373–385. [CrossRef]

64. Kapucu, N.; Demiroz, F. Interorganizational networks in disaster management. In *Social Network Analysis of Disaster Response, Recovery, and Adaptation*; Jones, E.C., Faas, A.J., Eds.; Butterworth-Heinemann: Oxford, UK, 2017; pp. 25–39.

65. Dariagan, J.D.; Atando, R.B.; Asis, J.L.B. Disaster preparedness of local governments in Panay Island, Philippines. *Nat Hazards* **2021**, *105*, 1923–1944. [CrossRef]

66. Walz, B.J. Disaster training exercises: An educationally-based hierarchy. *Prehosp. Disaster Med.* **1992**, *7*, 386–388. [CrossRef]

International Journal of
***Environmental Research
and Public Health***

Article

Climate Change Adaptation Strategies at a Local Scale: The Portuguese Case Study

Margarida Ramalho [1,*], José Carlos Ferreira [1,2] and Catarina Jóia Santos [1,2]

[1] Department of Environmental Sciences and Engineering, NOVA School of Science and Technology, NOVA University Lisbon, Campus da Caparica, 2829-516 Caparica, Portugal

[2] ARNET—Aquatic Research Network/MARE—Marine and Environmental Sciences Centre, Campus da Caparica, 2829-516 Caparica, Portugal

* Correspondence: mf.ramalho@campus.fct.unl.pt

Abstract: Coastal areas are home to more than 2 billion people around the globe and, as such, are especially vulnerable to climate change consequences. Climate change adaptation has proven to be more effective on a local scale, contributing to a bottom-up approach to the problems related to the changing climate. Portugal has approximately 2000 km of coastline, with 75% of the population living along the coast. Therefore, this research had the main objective of understanding adaptation processes at a local scale, using Portuguese coastal municipalities as a case study. To achieve this goal, document analysis and a questionnaire to coastal municipalities were applied, and the existence of measures rooted in nature-based solutions, green infrastructures, and community-based adaptation was adopted as variable. The main conclusion from this research is that 87% of the municipalities that answered the questionnaire have climate change adaptation strategies implemented or in development. Moreover, it was possible to conclude that 90% of the municipalities are familiar with the concept of nature-based solutions and all the municipalities with adaptation strategies include green infrastructure. However, it was also possible to infer that community-based adaptation is a concept that most municipalities do not know about or undervalue.

Keywords: coastal areas; community-based adaptation; nature-based solutions; green infrastructure

Citation: Ramalho, M.; Ferreira, J.C.; Jóia Santos, C. Climate Change Adaptation Strategies at a Local Scale: The Portuguese Case Study. *Int. J. Environ. Res. Public Health* **2022**, *19*, 16687. https://doi.org/10.3390/ijerph192416687

Academic Editors: Mikio Ishiwatari and Daisuke Sasaki

Received: 25 October 2022
Accepted: 4 December 2022
Published: 12 December 2022

Publisher's Note: MDPI stays neutral with regard to jurisdictional claims in published maps and institutional affiliations.

1. Introduction

In the last few decades, climate change has become a major topic of discussion all around the world. Climate change will have long-term effects on human lives and beings, due to the drastic changes in the ecosystem's patterns and processes. In some cases, this has already led to changes in some economic activities [1–3].

To address climate change and the consequences that derive from it, various authors have produced the concept of climate change adaptation, varying in their approach to the concept, which has resulted in various definitions [4–8]. Schmidt-Thomé (2017) [4] explores the concept of climate change adaptation, mentioning that the concept reflects the context in which it is applied and giving the example of the definition by the United Framework on Climate Change: "practical steps to protect countries and communities from the likely disruption and damage that will result from effects of climate change".

Moreover, Smit and Pilifosova (2001) [6] define adaptation as "changes in processes, practices, or structures to moderate or offset potential damages or to take advantage of opportunities associated with changes in climate".

However, the more consensual definition is the one by the Intergovernmental Panel on Climate Change [8], which specifies that climate change adaptation can be defined as "the process of adjustment to actual or expected climate and its effects in order to moderate harm or take advantage of beneficial opportunities (…)".

Since the effects of climate change and global warming have become more severe in the last decade, many countries have decided to implement national and local strategies for climate change adaptation, to better prepare themselves for the adverse consequences that can be projected for each area [9,10]. It has also been found that climate change impacts are mainly experienced at a local level, which helped to prompt the mentioned strategies [11].

Climate change adaptation has more success when applied at a local scale, contributing to bottom-up approaches instead of top-down approaches, and the potential transferability of the best adaptation processes [12]. As such, climate change policies, including adaptation strategies, must be local, mainly because of the different contexts provided by a variety of community stakeholders, which contribute to the proximity to the challenges and an understanding of bigger problems at a local scale [11,13]. The implementation of climate change adaptation at a local level is illustrated in studies undertaken by Roberts (2008), Laukkonen et al. (2009), and Rauken et al. (2014) [14–16].

It is important to note that larger cities in developed countries have easier access to resources for climate change adaptation planning—benefiting from the presence of research institutions, conferences, and other entities. Moreover, smaller cities generally have fewer resources and have more difficulty being engaged in national and international networks [17]. Therefore, it is crucial to evaluate case studies of implemented climate change adaptation strategies to identify current good practices and, simultaneously, overcome discrepancies such as the ones mentioned.

Climate change adaptation options can include measures derived from the concepts of nature-based solutions (NBS), green infrastructure (GI), or community-based adaptation (CBA). Some uses of NBS include green building–integrated systems and technologies—green roofs and walls [18,19], green parks [20], permeable pavements, and stormwater ponds [21,22].

According to several authors [23–25], GI can have several benefits that contribute to the adaptation and mitigation of climate change, such as air quality improvement, carbon storage increase, urban noise reduction, and stormwater management.

Jarillo & Barnett (2021) and Regmi et al. (2015) mention that CBA tackles adaptation by learning from the experiences of local communities with climate change effects, identifying the problems, and coming up with activities that capacitate the people [26,27].

This research aims to draw conclusions on adaptation processes at a local scale, using the Portuguese study case, and assess how to replicate it in a global context, since it is understood that local adaptation planning is the way most communities can adapt to climate change effects.

Additionally, this research also intends to make a national survey of the local adaptation strategies of coastal municipalities in Portugal, and their inclusion of nature-based solutions, green infrastructure, and community-based adaptation. This is particularly relevant considering the new Portuguese Climate Law (Lei de Bases do Clima) [28], which mandates municipalities to have local adaptation strategies by the end of 2023.

2. Research Framework

2.1. Climate Change Adaptation in the European Union

The first European Union (EU) adaptation strategy was presented in April 2013. This document had three main objectives: promoting action by the Member States, promoting better-informed decision-making, and promoting adaptation in key vulnerable sectors [16].

The implementation of the strategy was based on eight actions [29]:

1. Encourage all Member States to adopt comprehensive adaptation strategies by providing guidelines and identifying key indicators to measure Member States' level of readiness;
2. Provide LIFE funding to support capacity building and step-up adaptation action in Europe (2014–2020);
3. Introduce adaptation in the Covenant of Mayors framework (2013/2014), especially to support adaptation in cities;
4. Bridge the knowledge gap by working with the Member States and stakeholders to identify tools and methodologies that can address adaptation knowledge gaps;

5. Further develop Climate-ADAPT as the "one-stop shop" for adaptation information in Europe through the improvement of access to information between this portal and other relevant platforms;
6. Facilitate the climate-proofing of the Common Agricultural Policy (CAP), the Cohesion Policy, and the Common Fisheries Policy (CFP) by integrating climate change adaptation measures in these documents, and by capacitating relevant stakeholders in this process;
7. Ensuring more resilient infrastructure by mandating European standardization organizations to map industry-relevant standards in the areas of energy, transport, and buildings and to identify standards that can better include adaptation considerations. The Commission also provided guidelines to climate-proof vulnerable investments and additional guidance for its Communication on Green Infrastructure;
8. Promote insurance and other financial products for resilient investment and business decisions.

In 2018, the 2013 Strategy was evaluated to understand the progress the Member States had made since its implementation. The main conclusions from the evaluation drew on the need for a greater focus on adaptation efforts at the EU level, specifically on water and drought, climate change adaptation at a local and urban level, agriculture policy, climate finance, insurance, and business. It is also concluded that the EU Strategy has likely enhanced the political focus of Member States on adaptation issues, in addition to the Paris Agreement [30].

In 2021, the European Commission released a new EU Strategy for Adaptation to Climate Change. This document has three main objectives, building on the 2013 strategy and its evaluation [31]:

- Make adaptation smarter by improving knowledge and availability of data;
- Make adaptation more systemic through the support of policy development;
- Speed up adaptation across the board by accelerating the development and rollout of adaptation solutions.

With these goals in mind, the strategy presents the long-term vision for the EU in terms of adaptative capacity to minimize vulnerability to the effects of climate change while being in synergy with other Green Deal policies. It is also noted by the Commission that there is an urgent need to develop effective and inclusive governance mechanisms that can connect policymakers and scientists [31].

Through this document, the importance of the Climate-ADAPT Platform is also reinforced, which the EU aims to make the authoritative European platform for climate change adaptation. Climate-ADAPT is a partnership between the European Commission and the European Environment Agency (EEA), which aims to share data and information about adaptation, the national adaptation strategies and actions of the Member States, case studies, and tools that support adaptation planning [32].

2.2. Adaptation in Portugal

Portugal has had a National Strategy for Climate Change Adaptation (ENAAC) since 2010, introduced by resolution no. 24/2010 of the Council of Ministers of 1 April 2010 [33]. The main objectives of this strategy were the following:

- Information and knowledge by developing a scientific a technical basis of information;
- Reduce vulnerability and increase the capacity to respond by identifying and defining priorities in terms of climate change adaptation measures;
- Participate, raise awareness, and publicize through the contribution of stakeholders;
- International cooperation by approaching the national responsibilities in terms of cooperating with international adaptation policies.

In 2015, the same strategy was renewed, originating ENAAC 2020, which was supposed to run until 2020 but was extended to 2025 due to the approval of the National Plan of Energy and Climate (PNEC 2030). According to resolution no. 53/2020 of the Council of

Ministers of 10 July 2020 [34], this strategy is also accompanied by an action plan, which defines concrete lines of action for adaptation to climate change.

ENAAC 2020 was approved in resolution no. 56/2015 of the Council of Ministers of 30 July 2015 [35], renewing ENAAC 2010–2013 and with the following main objectives:

- Better the knowledge level about climate change by updating and developing information about climate change, while assessing its risks, impacts, and consequences. This material must be exposed through communication platforms and awareness campaigns;
- Implement adaptation measures through two options: consulting stakeholders from the sectorial working groups and through the collection of information regarding best practices, both on a national and international level (especially south European countries);
- Promote the integration of adaptation in sectorial policies.

The strategy includes six thematic areas, which were selected through the knowledge obtained in the ENAAC 2010–2013, which are seen in most activity sectors. The areas are research and development (R&D), financing, international cooperation, communication, territory planning, and water resources management.

Moreover, the nine updated priority sectors are the basis of ENAAC 2020 and are as follows: agriculture, biodiversity, economics, energy, forests, health, security of people and goods, transport and communications, and coastal areas.

For the mentioned sectors, and as it happened in the first ENAAC, a working group was defined by the competent public authority.

According to this strategy, most scenarios for 2080–2100 for Portugal project the following climate change effects:

- A general increase in the average yearly temperature in every region of the country;
- An increase of up to 3 °C for the highest temperature in summer, for the coastal areas, and an increase of up to 7 °C for the countryside. For the Madeira and Azores islands an increase between 1 °C and 3 °C is projected;
- A reduction in frost days and an increase in hotter days and tropical nights;
- An increase in forest fire risk, change of land use, and implications for water resources;
- A significant change in the precipitation cycle, which may include a reduction in precipitation during spring, summer, and autumn in mainland Portugal. There is also a possibility of a decrease in yearly precipitation and an increase in winter rainfall, due to the rise in the number of days with stronger rain.

Due to the mentioned projections, as well as other factors, the Portuguese Government presented the new Portuguese Climate Law of 31 December 2021 [28]. This law establishes several objectives related to the environment and the climate emergency, among them the need to "reinforce resilience and the national capacity to adapt to climate change".

Climate change causes more relevant impacts at a local scale and as a result, there has been a continuous effort to shift policies from national and regional levels to the local one [36]. In view of the new Climate Law [28], each Portuguese municipality must have adaptation strategies and action plans implemented by the end of 2023.

As mentioned earlier in the chapter, for the present paper, ClimAdaPT.Local assumes relevance in the present paper. This program was created in 2016, and its main goal is to encourage local climate change adaptation in Portugal, through the following steps [32]:

- Facilitate experience exchange between municipalities;
- Promote knowledge exchange between municipalities, universities, and research centers, as well as local companies;
- Promote international cooperation relations;
- Promote the empowerment of municipalities.

2.3. Natured-Based Solutions

Due to the more frequent effects and consequences of extreme phenomena caused by climate change, there has been a focus on implementing local-scale climate change adaptation actions that help reduce the vulnerability of the receiving environment [36,37].

These interventions make up the concept of nature-based solutions (NBS), which can be defined, broadly, as solutions that are inspired and supported by nature and are cost-effective, while simultaneously providing environmental, social, and economic benefits and building resilience while bringing more nature and diverse natural features and processes into urban spaces through resource-efficient and systemic interventions [21,38–40].

Moreover, NBS are interventions that are supported by nature, which can evolve and change. This implies the active management of these systems to ensure that their services are provided [41].

2.4. Green Infrastructure

Green infrastructure (GI) is a relatively recent concept that originated in the 1990s, associated with green spaces [42]. One of its different definitions is that it is an interconnected network of green space that conserves natural ecosystem values and functions and provides associated benefits to human beings. However, GI has both an ecological and engineering approach to it and is becoming a priority for decision-makers [42,43].

Additionally, GI can aid in adaptation to climate change in three main aspects: urban heat island effect—by regulating temperature in urban spaces, especially in population-dense locations [44–46]; flood risk management—through green cover that can reduce water runoff [44,47]; and ecosystem resilience—by preventing ecosystem fragmentation while increasing the number of protected areas and maintaining habitat connectivity [44,48].

Furthermore, the more widely accepted definition of GI comes from the European Commission (2013) and can be defined as "a strategically planned network of natural and semi-natural areas with other environmental features designed and managed to deliver a wide range of ecosystem services [29]".

On land, GI is present in rural and urban settings, incorporating green spaces (or blue, if aquatic ecosystems are concerned) and other physical features in terrestrial areas [29]. GI can have an ecological approach, for instance in the form of a natural system composed of national parks, parkways, forests, community gardens, and green corridors, among others [49]. Because of this, GI can respond to a wide range of environmental, social, and economic challenges, including climate change adaptation [42].

NBS and GI are closely related concepts, being complementary to a certain degree. NBS entails a more holistic perspective, aiming to support the implementation of solutions that approach biodiversity conservation, ecosystem service protection, and green infrastructure [50,51]. As such, GI is an important component of NBS.

2.5. Community-Based Adaptation

The community-based adaptation (CBA) concept has evolved in recent years, due to the more frequent impacts of climate change, especially in coastal communities, namely the rise in the sea level. These accumulated effects and consequences have become the drive needed for the development of community-based approaches in terms of climate change adaptation [52,53].

CBA happens in local communities vulnerable to the impacts of climate change by identifying, assisting, and implementing activities that will help strengthen the adaptive capacity of these populations [54]. CBA is a bottom-up approach to climate change adaptation with the aim of enhancing adaptive capacity to climate change [55]. As such, the activities mentioned are usually related to participatory processes, connecting local stakeholders and the local communities in the reduction of the risk associated with the rising effects of climate change [56,57].

Since CBA is also a place-based approach, the planning process for the communities must consider the social and ecological dynamics and priorities. According to Basel et al. [58], the integration of the CBA process "is achieved through a high level of trust and community engagement and input, over a long duration to establish meaningful relationships and understand community priorities and drivers".

Nevertheless, this process is not without criticism. One recurrent criticism of CBA is that there is a need to make the process more relevant to risks and policies outside of communities, that is, on a bigger scale—upscaling [54]. Another criticism is that CBA's techniques can be difficult to replicate, due to their adaptation to the specifics of the community they are applied to. Additionally, there are concerns about how to make this process a mainstream approach, in a way that makes decision-makers adapt it through policies [54].

Forsyth (2017) also draws on another CBA challenge: its capacity for representing local people fairly, and the simplistic way the term "community" can be treated. Community implies people in a certain locale act as a unit, which rarely happens—with communities having internal divisions. This can also contribute to the difficulties in implementing CBA, namely in its participatory component [59].

3. Material and Methods

3.1. Methodology

The methodology presented in this section has four main steps, as described in Figure 1. Each of these steps will be explained in depth later in this chapter.

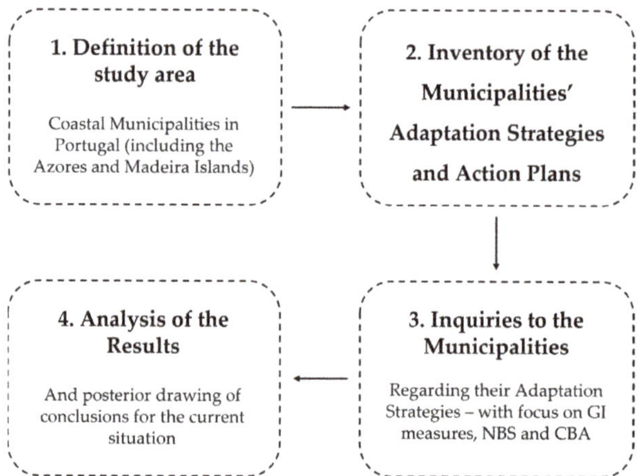

Figure 1. Methodological Framework.

3.1.1. Step 1—Definition of the Study Area

The study area defined for the purpose of this research is the coastal municipalities in Portugal, which also include the coastal municipalities of the islands of Madeira and the Azores, as seen in Figure 2.

The coastline of Portugal, including the Azores and Madeira islands, is approximately 2000 km long and, as such, 75% of the Portuguese population is concentrated on the coast. This area also generates roughly 80% of the Portuguese Gross Domestic Product (GDP), which proves its importance on a national level [60].

Due to the length of its coast, as well as its exposure to the sea waves from the North Atlantic, Portugal is one of the countries most affected by coastal erosion in Europe. With the increasing frequency of events caused by climate change, it is expected that coastal erosion will worsen on the Portuguese coast [61].

Therefore, it is necessary to establish and implement measures and solutions to adapt to coastal cities and towns of the country. Such measures and solutions are one of the bases of the current study.

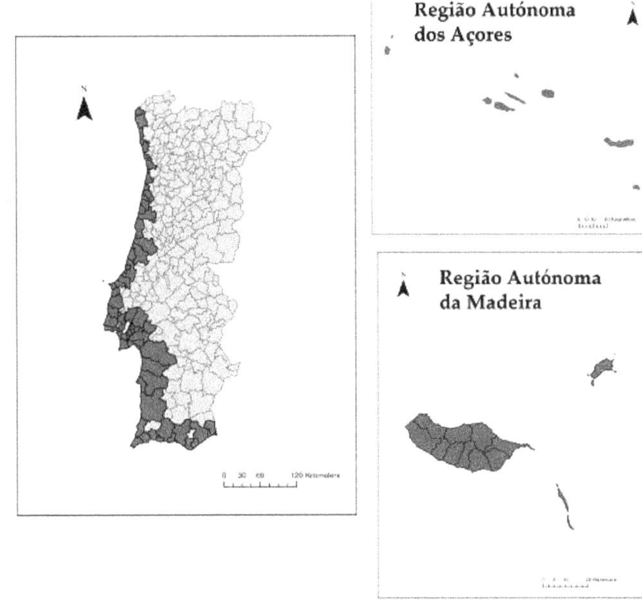

Figure 2. NUTS II and study area (coastal municipalities), where AML is Lisbon Metropolitan Area (Data from: CAOP 2021).

3.1.2. Step 2—Documentary Analysis

One of the first actions related to this research was the making of an inventory of all the Portuguese coastal municipalities and their work towards climate change adaptation so far.

To complete this inventory, it was necessary to read and analyze the adaptation strategies and plans made by each municipality. It is important to note that not all coastal municipalities have implemented climate change adaptation strategies. This will be presented in the following chapter, which is dedicated to the results of this study.

In this step of the methodological framework, the authors of this research aimed to answer the following questions to categorize the municipalities' adaptation strategies and plans:

1. Is the municipality a participant in the ClimAdaPT program?
2. What was the methodology used for the elaboration of the strategy?
3. What are the main climate change projections for the municipality?
4. Does the adaptation strategy mention green infrastructure or infrastructure measures?
5. How many adaptation options are mentioned in the strategy?
6. What is the priority of green infrastructure among the other adaptation options?
7. Does the adaptation strategy plan how to integrate the strategy into the territory management instruments (IGT)?

It is important to note that questions 5 and 6 were only asked if the municipality was a part of the ClimAdaPT program. This happened because all ClimAdaPT's participants are required to follow a similar structure for their adaptation strategy structure.

3.1.3. Step 3—Inquiries to the Coastal Municipalities

To achieve the purpose of this research, there was a need to undertake a survey of the adaptation strategies of coastal municipalities in Portugal.

The initial survey was directed to all 92 municipalities on the Portuguese coast, including Madeira and the Azores islands. For these municipalities, the questions asked focused on the adaptation strategies and their respective inclusion of green infrastructure, nature-based solutions, and community-based adaptation.

The inquiries were made on Google Forms, since this tool also allows the export of data and, consequently, better data treatment.

The inquiries were sent to all 92 municipalities via the author's institutional e-mail, with the period of the responses being between 3 June and 7 October.

The inquiries were divided into the following sections:

- General information about the strategy of the municipality;
- Green infrastructure and nature-based solutions and their inclusion in the strategy;
- Community-based adaptation;

The last two sections of the inquiry were only presented if the municipality answered that they had implemented a climate change adaptation strategy.

3.1.4. Step 4—Data Treatment

After obtaining the responses to the inquiry, the data were explored further using Microsoft Excel.

As such, a group of parameters was explored, namely:

1. General data concerning the inquiry;
2. Data from municipalities that answered they do not have an adaptation strategy;
3. Data from municipalities that have an adaptation strategy in development.
4. Data from municipalities that answered they have an adaptation strategy.

For the first point in the list above, the results obtained through the data from the questionnaire were the characterization of the municipalities, in terms of their answers to the inquiries and if said municipalities have a climate change adaptation strategy.

As for the second point, the data acquired allowed us to explore the reasons why municipalities do not have climate change adaptation strategies implemented.

Moreover, the third point reflects on the municipalities that have a strategy in development, and at which stage of development it is.

Lastly, the final point allowed for wider data treatment and more conclusive results, due to extensive questions asked to municipalities with adaptation strategies. As such, it was possible to divide this point into four sub-sections of results: general data, nature-based solutions, green infrastructure, and community-based adaptation.

4. Results

4.1. Results from the Documentary Analysis

From the documentary research, it was possible to observe that out of all 92 municipalities, only 18 have an individual climate change adaptation strategy available for consultation online. In the following table, the mentioned municipalities are presented, along with their Nomenclature of Territorial Units for Statistical Purposes (NUTS II).

Table 1 shows that Norte is the region with the most individual adaptation strategies, with a total of five, which corresponds to 56% of the coastal municipalities in this NUTS II.

From the municipalities that do not have a climate change adaptation strategy available for consultation online, it was possible to understand that the regions with the fewest strategies are the Azores and Madeira islands, with 95% and 91%, respectively, of their municipalities, not having the documents.

Another parameter observed was the participation of municipalities in the ClimAdaPT project. Approximately 60% of municipalities are participants in this project, meaning that 11 municipalities have elaborated their strategy in compliance with this program and its methodology.

Additionally, the documentary research showed that 44 municipalities are included in either inter-municipal or metropolitan plans, such as the Lisbon and Porto Metropolitan Areas adaptation strategies, the plans from the Madeira and Azores islands, the Algarve and the Oeste inter-municipal adaptation plans, and the inter-municipal plan of the region CIM of Coimbra.

Table 1. Municipalities with individual adaptation strategies, by NUTS II.

NUTS II	Municipalities
Norte	Viana do Castelo Esposende Vila do Conde Porto Espinho
Centro	Aveiro Ílhavo Leiria Torres Vedras
Lisbon Metropolitan Area	Mafra Barreiro Cascais
Alentejo	Benavente Odemira
Algarve	Loulé Faro
Região Autónoma da Madeira	Funchal
Região Autónoma da Açores	Vila Franca do Campo

However, if only individual municipal strategies are considered, then the total of municipalities that do not have an adaptation strategy document is 74. This means that 80.4% of Portuguese municipalities do not have an individual adaptation strategy in place.

Another factor that was examined in the climate change adaptation strategies was their inclusion of green infrastructures. It was possible to verify that, of the 18 individual strategies, 15 mentioned and included adaptation options regarding green infrastructure. However, three municipalities did not consider GI in their documents. The three municipalities that did not include GI are all from the Norte: Espinho, Vila do Conde, and Esposende.

4.2. Results from the Questionnaire

For this inquiry, all 92 coastal municipalities of Portugal were contacted and invited to answer a questionnaire about adaptation strategies.

4.2.1. General Data

Firstly, it is important to acknowledge that 50% of the Portuguese coastal municipalities answered the questionnaire. Therefore, the results of this evaluation are not completely representative of all coastal municipalities in Portugal and only represent 46 municipalities.

Considering the NUTS II, the NUTS II with the highest percentage of responses to the questionnaire was the Norte, with 66.7% of the municipalities surveyed answering. However, the NUTS II with the smallest percentage of responses was Alentejo, with only 16.7% of the surveyed municipalities of this region answering.

Due to the higher number of municipalities contacted, it was expected that the Metropolitan Area of Lisbon would have a higher percentage response. Nevertheless, this NUTS II had 56.3% of responses out of the 16 municipalities contacted.

Regarding the municipalities from the islands, Azores had responses from 52.6% of its municipalities, while Madeira had 27.3% of responses.

Concerning only the municipalities that answered the inquiry, and if they have a climate change adaptation strategy implemented or in development, the results are described in Figure 3.

Does the municipality have an adaptation strategy?

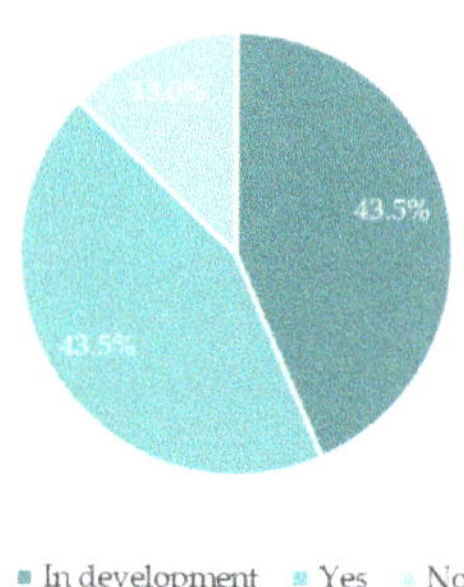

Figure 3. Percentage of surveyed municipalities that have climate change adaptation strategies.

From this graphic, it is possible to understand that 13.0% of the municipalities do not have an adaptation strategy while 43.5% do, and the remaining 43.5% have adaptation strategies in development.

Considering the existence of the climate change adaptation strategies, the NUTS II with the fewest strategies is Madeira—with one of the three municipalities not having a strategy implemented.

Regarding strategies in development, it is possible to verify that the NUTS II with the most climate change adaptation strategies in development is the Azores. Out of the 11 answers, 7 municipalities are developing strategies.

As for the region with the most adaptation strategies, the Norte, all the municipalities that answered the questionnaire have strategies in place. After the Norte, the Lisbon Metropolitan Area is the NUTS II with the most municipal adaptation strategies implemented, as six out of the nine municipalities that have answered the inquiry have these documents put into effect.

4.2.2. Municipalities That Do Not Have a Climate Change Adaptation Strategy

Out of the 46 municipalities that answered the inquiry, six of them do not have a climate change adaptation strategy. In the table below Table 2, the municipalities that do not have individual strategies and their corresponding NUTS II are presented.

Table 2. Municipalities that do not have strategies and their corresponding NUTS II.

NUTS II	Municipalities
Centro	Marinha Grande Vagos
Algarve	Olhão
Região Autónoma dos Açores	Horta Santa Cruz da Graciosa
Região Autónoma da Madeira	Câmara de Lobos

Olhão Municipality, although part of the inter-municipal plan for climate change adaptation of the Algarve Metropolitan Area (PIAAC AMAL), stated they did not have a climate change adaptation strategy. The specific reason Olhão indicated for not having an individual adaptation strategy is that it did not have an environment/sustainability department until recently.

The same happened with Câmara de Lobos Municipality and Horta and Santa Cruz, which are included in the regional climate change adaptation plans for Madeira and Azores, respectively. These municipalities claimed the following reasons for not having individual adaptation strategies:

- Lack of decision by the executive and lack of human resources (Câmara de Lobos);
- The adaptation strategy is not planned (Santa Cruz da Graciosa);
- Lack of interest by the executive (Horta).
- Regarding Marinha Grande and Vagos, when asked about the lack of development of the adaptation strategies, the following reasons were given:
- Lack of human resources;
- Lack of funding for the elaboration of the document.

4.2.3. Municipalities That Have a Climate Change Adaptation Strategy in Development

To better explore the results obtained through the inquiry, municipalities that have strategies in development were also accounted for. In total, 20 municipalities have climate change adaptation strategies in development, which are presented in the following table alongside the corresponding NUTS II.

Table 3 shows that, apart from the municipalities from the Norte region, all of the NUTS II regions have municipalities with strategies in development. The Norte region is not represented because all the municipalities that responded to the inquiry, and that are from this NUTS II, have strategies implemented.

To study how far along the climate change adaptation strategies are, results showing their stages of development are presented in Figure 4.

From the graphic in Figure 4, it is possible to observe that 10 municipalities have their strategy documents in elaboration, while 7 municipalities have the strategy in preparation, and 2 municipalities admit that the process is delayed. In this last parameter, it was not possible to verify why the process is delayed.

Moreover, these results show a clear interest from these municipalities in developing an adaptation strategy to comply with the New Climate Law and to implement measures that can adapt these municipalities and avoid a higher risk.

Table 3. Municipalities that have strategies in development and their corresponding NUTS II.

NUTS II	Municipalities
Região Autónoma dos Açores	Praia da Vitória Ponta Delgada Angra do Heroísmo Povoação Velas Madalena Santa Cruz das Flores
Alentejo	Santiago do Cacém
Algarve	Tavira Albufeira Castro Marim Silves
Centro	Ovar Caldas da Rainha Nazaré
Lisbon Metropolitan Area	Oeiras Seixal Sesimbra
Região Autónoma da Madeira	Calheta Ribeira Brava

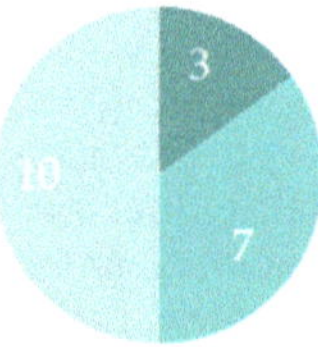

- The elaboration of the document is delayed

- The document is still in preparation, however its elaboration hasn't started

- The document is being elaborated

Figure 4. Municipalities with strategies in development.

4.2.4. Municipalities That Have a Climate Change Adaptation Strategy

As already mentioned, 20 municipalities out of the 46 municipalities that answered the inquiry have a climate change adaptation strategy implemented. These municipalities are presented in Table 4, as well as their corresponding NUTS II.

From Table 4, it is possible to observe that, according to the answers to the inquiry, no municipality from Alentejo and Madeira islands has a climate change adaptation strategy.

Regarding the remaining answers, the results presented show that they are well distributed within the NUTS II, with representation in five out of the seven NUTS II regions.

Through the questionnaire, it was possible to obtain the dates when the strategies of the municipalities above were published/elaborated, with the results being shown in Figure 5.

Table 4. Municipalities that have strategies and their corresponding NUTS II.

NUTS II	Municipalities
Norte	Esposende Viana do Castelo Vila do Conde Matosinhos Vila Nova de Gaia Porto
Centro	Ílhavo Figueira da Foz Leiria Óbidos Torres Vedras
Área Metropolitana de Lisboa	Almada Cascais Lisboa Loures Setúbal Sintra
Algarve	Loulé Faro
Região Autónoma dos Açores	Ribeira Grande

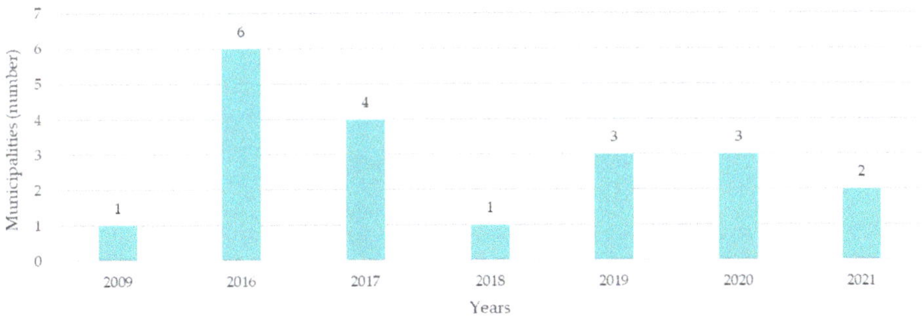

Figure 5. Climate change adaptation strategies approved by year.

From this graphic, it is shown that 2016 was the year when most adaptation strategies were launched, followed by 2017, 2019, and 2020.

The responses to the inquiries also allowed us to ascertain the scale of the strategy—whether it acts at a municipal or inter-municipal scale. Figure 6 displays the results.

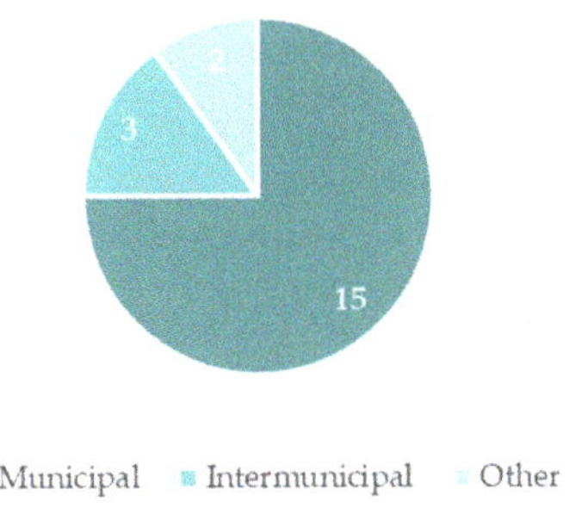

Figure 6. Scale of action of the adaptation strategies.

Through Figure 6, it is possible to verify that most adaptation strategies act on a municipal level. However, three municipalities have strategies that act on an inter-municipal scale, and the scale of the remaining two municipalities is metropolitan.

The responses to the questionnaire also made it possible to understand at what phase the adaptation strategies from the municipalities are. As such, 11 municipalities have concluded their document, while the remaining 7 are in the process of executing their strategies.

As stated earlier in this section, through the inquiry, the municipalities were questioned about their inclusion of nature-based solutions in their climate change adaptation strategies. The following graphic presents the number of municipalities that utilized this measure in their documents.

As presented in Figure 7, 18 municipalities consider nature-based solutions in their strategies, and 2 municipalities do not. As per this distribution in NUTS II, it was verified that the two municipalities that do not include NBS in their strategies are from the Norte region, namely, Vila do Conde and Matosinhos. The remaining 18 municipalities are distributed through all the Portuguese NUTS II, except the Madeira islands and Alentejo.

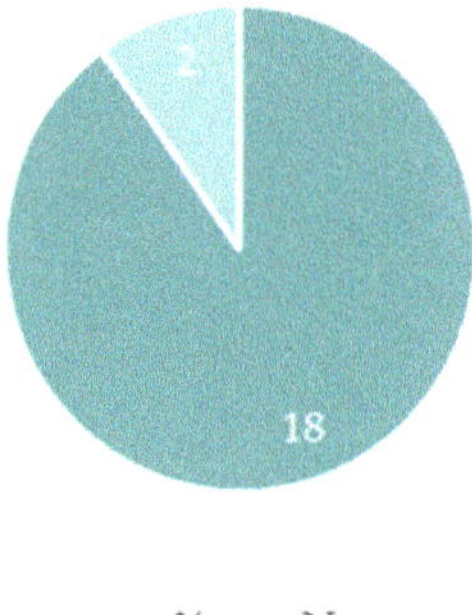

Figure 7. Nature-based solutions in adaptation strategies.

The motives stated by Vila do Conde and Matosinhos for the lack of NBS in their strategies were the following: even though NBS were not explicit in the document, they are implicitly integrated into the adaptation strategy, and the NBS were already contemplated in another action plan from the municipality.

Regarding specific NBS measures adopted in the climate change adaptation strategies, the results are presented in Figure 8.

Figure 8 shows that 89% of the municipalities that have adaptation strategies expect to recover and restore water lines in their territory, 78% want to restore the ecosystems in their municipality, and 72% hope to accomplish more urban green spaces. As for the other NBS measures mentioned by two municipalities, they encompass sustainable drainage systems.

NBS are well-known to municipalities, as it is possible to see from the above results. In most municipalities, adaptation options consider NBS, whether in its green infrastructure component or in the recovery of ecosystems and water lines.

GI data were also obtained because of the inquiry sent to the coastal municipalities, specifically the ones that answered they have a strategy in motion. All the municipalities that have a climate change adaptation strategy mention and plan to implement GI. As such, Figure 9 presents the results obtained regarding the types of GI the municipalities expect to execute through their adaptation strategies.

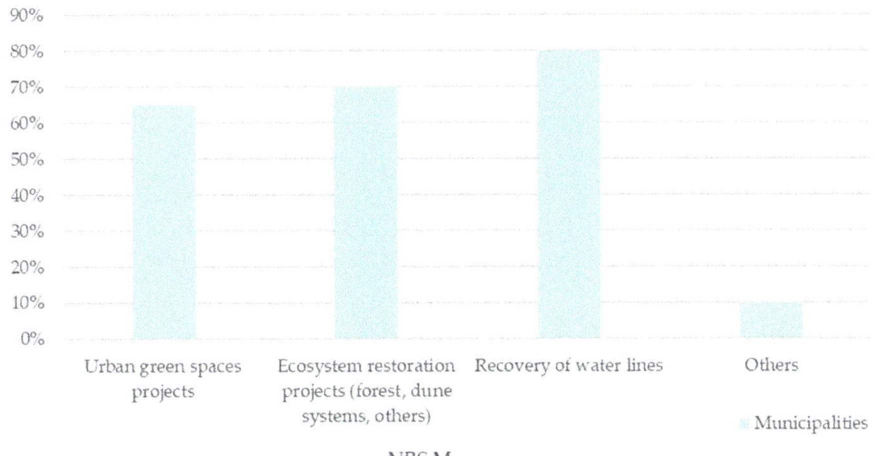

Figure 8. Main NBS measures in the adaptation strategies.

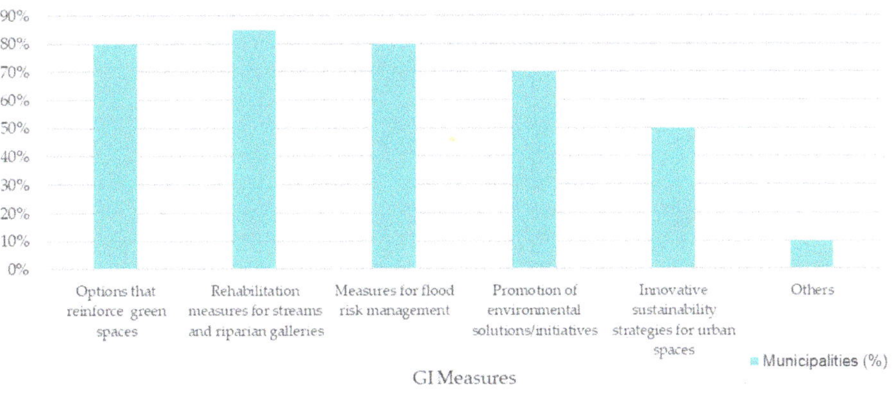

Figure 9. GI measures mentioned in the adaptation strategies.

Figure 9 shows that the top GI measure for the surveyed municipalities is related to the rehabilitation measures for streams and associated riparian galleries, with 85% of the municipalities having this GI option in their strategy. Moreover, 80% of the municipalities consider options that reinforce green spaces in their territory and specific measures for flood risk management. Most municipalities also regard the promotion of sustainable solutions and initiatives in their strategies and half of the municipalities have innovative sustainability strategies for urban spaces. Lastly, 10% of the municipalities also refer to other GI options, such as the promotion of sustainable agriculture practices and the adaptation of more resilient species in the management of green infrastructure in the territory.

Through the questionnaire, it was possible to verify that 15 municipalities do not include community-based adaptation in their strategies, while 5 do. These five municipalities are Ribeira Grande, Ílhavo, Matosinhos, Loures, and Cascais from the Azores, Centro, Norte, and Lisbon Metropolitan area NUTS II, respectively.

In order to know in which way the municipalities are utilizing the CBA concept, a question was asked to the municipalities about the measures used that can be encompassed by CBA. The results are presented in Figure 10.

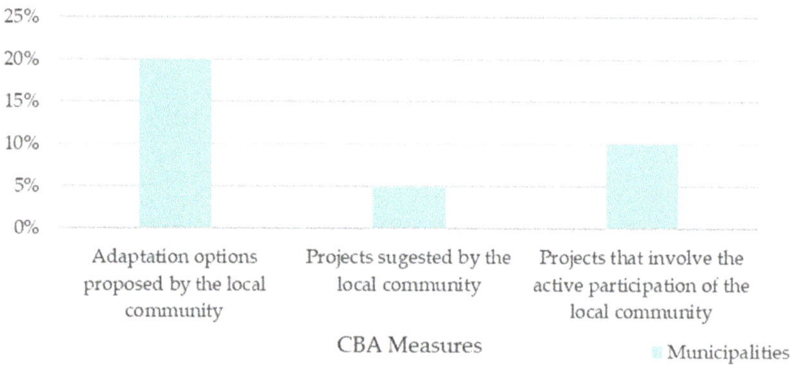

Figure 10. CBA measures mentioned in the adaptation strategies.

Out of all the municipalities that have a strategy, 10% have implemented CBA in their strategies through adaptation options proposed by the local community, 5% have applied projects suggested by the local population, and 20% have projects that require active participation by the community.

Concerning the municipalities that do not include CBA in their strategies, the main reason why municipalities did not include CBA in their adaptation strategies is that they did not understand the need for it. Two municipalities also answered they were unaware of the utility of this concept and, finally, six municipalities referred to other motives. Three municipalities did not answer this question.

The motives included in "others" were the following:

- The creation of a stakeholders commission already covers the active participation of the local community in strategies such as the climate change adaptation strategy;
- The municipality followed the ClimAdaPT methodology, which did not include CBA at that date;
- The municipality is contemplating the creation of a climate change adaptation local network.

5. Discussion

5.1. Documentary Analysis

Through the results presented in the previous section of the documentary analysis, it was possible to verify that the number of adaptation strategies available for consultation is still small, compared to the number of coastal municipalities in Portugal. Most of the municipalities are included in regional or inter-municipal strategies, which have a tendency to treat the municipalities as a unit. This can result in a lack of detail for the municipalities, on an individual level, or can lead to some municipalities being overlooked in favor of others.

Approximately 60% of municipalities that have climate change adaptation strategies are participants in the ClimAdaPT project. This means that this project was an important factor in the elaboration of the strategies.

To get a better geographical distribution of the participating municipalities, ClimAdaPT invited several municipalities throughout the Portuguese territory, including those from NUTS II that usually do not have the resources (either human resources or financial resources) to elaborate on these documents.

As such, this choice can contribute to the sharing of know-how about local climate change adaptation with the neighboring municipalities.

It was also possible to verify that strategies that were elaborated in the scope of the ClimAdaPT project present in detail their adaptation options, even ranking these options by priority order. However, the other strategies made by the municipalities that were not participants in this project were not so clear in their adaptation measures, with their descriptions being made very superficially.

5.2. Questionnaires to the Coastal Municipalities

Through the results from the questionnaire, it was also possible to conclude that half of the coastal municipalities answered it. The lack of response to this questionnaire might signify underlying problems, namely the number of public administration employees of each municipality. As is understandable, bigger municipalities, or municipalities in certain areas of influence, have more employees and, as such, deploy more human resources to either answer the inquiry or participate in the team that will work on the strategies.

Regarding the answers to the questionnaire, it was possible to expect some of them, due to the previous documentary analysis. These answers, even if expected, were also able to help the authors identify the knowledge of municipalities towards the concepts in the study and if the concepts were included in their respective adaptation strategies. Some answers also allowed the authors to identify inconsistencies, which will be explored further.

Moreover, it was possible to verify that almost half of the municipalities that responded to the inquiry have an adaptation strategy and approximately 40% have their own strategy in development. For municipalities that do not have adaptation strategies, four out of the six municipalities are included in regional or inter-municipal climate change adaptation plans. Several reasons were given by the municipalities for the lack of individual strategies, such as a lack of human resources and lack of funding. As mentioned before, these responses can relate to the number of employees in the local administration of municipalities.

Data collected from PorData show that, as of 2020, there were only 135,125 people working in local administration in Portugal. The Azores Islands have 2878 local administration employees, while the Madeira islands have 3133. Moreover, the Centro region has 29,242 employees, and the Algarve has 9733. This can contribute to the lack of implemented adaptation strategies in municipalities from this NUTS II.

By exploring the data of the municipalities that do not have strategies implemented on an individual scale, it is possible to see the number of employees in local administration in these areas (Table 5).

Table 5. Number of local public administration employees (source: PorDATA).

Municipality	Number of Local Public Administration Employees
Marinha Grande	266
Vagos	247
Olhão	593
Horta	54
Santa Cruz da Graciosa	166
Câmara de Lobos	248

For the municipalities in the islands, the number of public employees is justifiable. However, for the mainland municipalities, the data shows there could be enough human resources for the development of an adaptation strategy.

Another reason these municipalities give for not having an adaptation strategy is that it is not planned, or that there is no interest in it. This may reveal a lack of interest in the executive regarding climate laws, namely the New Climate Law of 31 December 2021 which requires municipalities to have a municipal or regional adaptation strategy implemented.

Results from the municipalities that have adaptation strategies in development allowed us to verify a clear interest from municipalities in developing an adaptation strategy to comply with the New Climate Law and to implement measures that can adapt these municipalities and avoid a higher risk in face of climate change.

As for the municipalities that do have adaptation strategies, most of the documents were approved in 2016 and 2017. This also derives from the ClimAdaPT program, which took place between January 2015 and December 2016. Of the six strategies approved in 2016, four were from municipalities that participated in this program, and from the three strategies approved in 2017, two were also from participating municipalities.

The fact that some municipalities have strategies that are not on a municipal/individual level is concerning, mainly because it means these documents are not as specific as they ought to be since the role of the municipality is not considered with such dept when integrated into an inter-municipal community or in a metropolitan region. As such, it is necessary that these municipalities develop adaptation strategies on a municipal scale, making them more specific and in accordance with local problems and with common solutions led by the different stakeholders that are relevant to the municipality, while simultaneously ensuring active participation throughout the whole process.

The information from the results allows us to conclude that these municipalities have an overall knowledge of the concept of NBS and GI, and adaptation options that fall into the latter concept are included in all strategies. This will ensure that there is less risk of damage in the municipalities, either from a social, environmental, or economic perspective—which can also be interconnected.

Crossing data from the results of the documentary analysis with the results of the questionnaire, regarding NBS and GI, it is possible to verify that two municipalities such as Vila do Conde and Esposende answered that they have contemplated GI and NBS in their strategies. However, analyzing the strategies from these two municipalities, GI and NBS are not mentioned (Vila do Conde)in any of them and are implicit in in one of the municipalities, the adaptation options (Esposende) are implicit. It is important to note that these two municipalities did not participate in the ClimAdaPT project and, as such, followed a different structure for their respective documents.

However, the results concerning CBA are not optimistic. A large percentage of municipalities have no knowledge of this concept and did not include it in the adaptation strategy.

The participation of the community is a relevant part of the success of climate change adaptation, and it is acknowledged by the municipalities in their strategies and in the answers to the questionnaire. The members of the community can contribute with different knowledge and that can contribute to better solutions to the problems. Nevertheless, the authors also recognize that the participation of the community has a number of constraints associated: the lack of involvement of people and internal divisions in the community.

Therefore, there is a knowledge gap in the Portuguese coastal municipalities regarding CBA. It must be addressed for the adaptation strategies to succeed and include adaptation measures that come from it.

6. Conclusions

This research had the main purpose of addressing climate change adaptation strategies at a local scale, using Portuguese coastal municipalities as a case study. Moreover, this research aimed to understand the inclusion of the local communities in the elaboration of these strategies.

From questionnaires addressed to the Portuguese coastal municipalities, it was possible to understand that most municipalities have climate change adaptation strategies implemented or in development (86.7% of the inquired municipalities). For the municipalities that do not have strategies, the main reason given was the lack of human resources to work on this document. This also connects to the evaluation made by the European Commission in 2018, regarding the Member States' adaptation preparedness. For Portugal's fiche, it was assessed that few adaptation plans were moving onto the implementation stage—with this being the weakest part of the process, along with monitoring and evaluating policies [30]. It was also concluded that there is a deep knowledge of nature-based solutions and green infrastructure among the municipalities, with most adaptation strategies having adaptation options and measures that reference these notions.

However, a concerning conclusion was made through the obtained results: most coastal municipalities in Portugal have little knowledge of the concept of community-based adaptation. In fact, because it is such a recent concept, there are still not many scientific articles published on this topic, most of them being studies regarding the indigenous population and their knowledge of nature [62–64].

The engagement of the community in local processes is essential for the success of public policies. This is also applied to climate change adaptation since the different local stakeholders have knowledge of different themes, which can contribute to solutions that better serve the community. As such, it is important to invest in the education of the local public administration regarding CBA, to ensure the engagement of people in the decision-making process. In this process, it would be relevant to include stakeholders such as R&D institutions and universities, as well as companies and the community [36]. This can be accomplished through short-term mandatory courses or seminars aimed mainly at employees that work directly with these policies or, ultimately, through law enforcement—as a more serious measure required by the executive power.

One example of the engagement of the local community in climate change adaptation in Portugal is the PLAAC-Arrábida Project. In this project, three municipalities, with the coordination of one energy agency and the collaboration of two universities and their local stakeholders, have successfully developed three local climate change adaptation strategies. Over the course of 15 months, all parties continuously collaborated in three different meetings and five different workshops to discuss which local community members should also be considered for this cooperation, to identify the main areas at risk, to elaborate the specific actions and measures for climate change adaptation and, finally, to choose the most relevant ones for their specific territory.

This project, and the results obtained throughout this research, exhibit the importance of community participation for the success of climate change adaptation at a local level, and should, therefore, be a driver for all coastal communities worldwide. Thus, it is crucial that more CBA studies are undertaken to consolidate a framework for its use in different places and communities globally.

Author Contributions: Conceptualization, M.R., J.C.F. and C.J.S.; methodology, M.R. and J.C.F.; formal analysis, M.R., J.C.F. and C.J.S.; investigation, M.R.; resources, M.R., J.C.F. and C.J.S.; data curation, M.R.; writing—original draft preparation, M.R.; writing—review and editing, J.C.F. and C.J.S.; visualization, M.R.; supervision, J.C.F. and C.J.S.; funding acquisition, J.C.F. All authors have read and agreed to the published version of the manuscript.

Funding: This work had the support of national funds through Fundação para a Ciência e a Tecnologia (FCT I.P., Portugal), under the strategic projects UIDB/04292/2020 and UIDP/04292/2020 granted to MARE—Marine and Environmental Sciences Centre, and the project LA/P/0069/2020 granted to the Associate Laboratory ARNET—Aquatic Research Network.

Institutional Review Board Statement: Not applicable.

Informed Consent Statement: Informed consent was obtained from all subjects involved in the study.

Acknowledgments: The authors would like to thank all the municipalities that have collaborated in this research by answering the questionnaire about climate change adaptation strategies.

Conflicts of Interest: The authors declare no conflict of interest.

References

1. Teixeira, C.P.; Fernandes, C.O.; Ahern, J. Adaptive planting design and management framework for urban climate change adaptation and mitigation. *Urban For. Urban Green.* **2022**, *70*, 127548. [CrossRef]
2. Hoegh-Guldberg, O.; Jacob, D.; Bindi, M.; Brown, S.; Camilloni, I.; Diedhiou, A.; Djalante, R.; Ebi, K.; Engelbrecht, F.; Guiot, J.; et al. Impacts of 1.5 °C Global Warming on Natural and Human Systems. In *Global Warming of 1.5 °C: An IPCC Special Report*; Masson-Delmotte, V., Zhai, P., Pörtner, H.O., Roberts, D., Skea, J., Shukla, P.R., Pirani, A., Moufouma-Okia, W., Péan, C., Pidcock, R., et al., Eds.; IPCC Secretariat: Geneva, Switzerland, 2018; pp. 175–311.
3. Loucks, D.P. Impacts of climate change on economies, ecosystems, energy, environments, and human equity: A systems perspective. In *The Impacts of Climate Change*; Elsevier: Amsterdam, The Netherlands, 2021; pp. 19–50. [CrossRef]
4. Schmidt-Thomé, P. Climate Change Adaptation. *Oxf. Res. Encycl. Clim. Sci.* **2017**, 1–26. [CrossRef]
5. Fankhauser, S. Adaptation to Climate Change. *Annu. Rev. Resour. Econ.* **2017**, *9*, 209–230. [CrossRef]

6. Smit, B.; Pilifosova, O. Adaptation to climate change in the context of sustainable development and equity. *Sustain. Dev.* **2003**, *8*, 9.
7. Parry, M.L.; Canziani, O.; Palutikof, J.; Van der Linden, P.; Hanson, C. *Climate Change 2007-Impacts, Adaptation and Vulnerability: Working Group II Contribution to the Fourth Assessment Report of the IPCC*; Cambridge University Press: Cambridge, UK, 2007.
8. Pörtner, H.O.; Roberts, D.C.; Adams, H.; Adler, C.; Aldunce, P.; Ali, E.; Begum, R.A.; Betts, R.; Kerr, R.B.; Biesbroek, R.; et al. *Climate Change 2022: Impacts, Adaptation and Vulnerability. IPCC Sixth Assessment Report*; Cambridge University Press: Cambridge, UK, 2022.
9. Nalau, J.; Cobb, G. The strengths and weaknesses of future visioning approaches for climate change adaptation: A review. *Glob. Environ. Chang.* **2022**, *74*, 102527. [CrossRef]
10. Hossen, M.A.; Netherton, C.; Benson, D.; Rahman, M.R.; Salehin, M. A governance perspective for climate change adaptation: Conceptualizing the policy-community interface in Bangladesh. *Environ. Sci. Policy* **2022**, *137*, 174–184. [CrossRef]
11. Aguiar, F.C.; Bentz, J.; Silva, J.M.N.; Fonseca, A.L.; Swart, R.; Santos, F.D.; Penha-Lopes, G. Adaptation to climate change at local level in Europe: An overview. *Environ. Sci. Policy* **2018**, *86*, 38–63. [CrossRef]
12. Schlingmann, A.; Graham, S.; Benyei, P.; Corbera, E.; Sanesteban, I.M.; Marelle, A.; Soleymani-Fard, R.; Reyes-García, V. Global patterns of adaptation to climate change by Indigenous Peoples and local communities. A systematic review. *Curr. Opin. Environ. Sustain.* **2021**, *51*, 55–64. [CrossRef]
13. Kirby, A. Analyzing climate change adaptation policies in the context of the local state. *Environ. Sci. Policy* **2021**, *123*, 160–168. [CrossRef]
14. Roberts, D. Prioritizing climate change adaptation and local level resilience in Durban, South Africa. *Environ. Urban.* **2010**, *22*, 397–413. [CrossRef]
15. Laukkonen, J.; Blanco, P.K.; Lenhart, J.; Keiner, M.; Cavric, B.; Kinuthia-Njenga, C. Combining climate change adaptation and mitigation measures at the local level. *Habitat Int.* **2009**, *33*, 287–292. [CrossRef]
16. Rauken, T.; Mydske, P.K.; Winsvold, M. Mainstreaming climate change adaptation at the local level. *Local Environ.* **2013**, *20*, 408–423. [CrossRef]
17. Lioubimtseva, E.; da Cunha, C. Local climate change adaptation plans in the US and France: Comparison and lessons learned in 2007–2017. *Urban Clim.* **2020**, *31*, 100577. [CrossRef]
18. Lei nº 98/2021, de 31 de Dezembro. Diário da República, 1.ª Série-Nº 253. Available online: https://dre.pt/dre/detalhe/lei/98-2021-176907481 (accessed on 29 August 2022).
19. Basu, A.S.; Pilla, F.; Sannigrahi, S.; Gengembre, R.; Guilland, A.; Basu, B. Theoretical Framework to Assess Green Roof Performance in Mitigating Urban Flooding as a Potential Nature-Based Solution. *Sustainability* **2021**, *13*, 13231. [CrossRef]
20. Pineda-Martos, R.; Calheiros, C.S.C. Nature-Based Solutions in Cities—Contribution of the Portuguese National Association of Green Roofs to Urban Circularity. *Circ. Econ. Sustain.* **2021**, *1*, 1019–1035. [CrossRef]
21. Sekulova, F.; Anguelovski, I. The Governance and Politics of Nature-Based Solutions. Naturvation Deliverable. 2017. Available online: https://naturvation.eu/sites/default/files/news/files/naturvation_the_governance_and_politics_of_nature-based_solutions.pdf (accessed on 21 September 2022).
22. Le Coent, P.; Graveline, N.; Altamirano, M.A.; Arfaoui, N.; Benitez-Avila, C.; Biffin, T.; Calatrava, J.; Dartee, K.; Douai, A.; Gnonlonfin, A.; et al. Is-it worth investing in NBS aiming at reducing water risks? Insights from the economic assessment of three European case studies. *Nat.-Based Solut.* **2021**, *1*, 100002. [CrossRef]
23. Ferreira, J.; Monteiro, R.; Silva, V. Planning a Green Infrastructure Network from Theory to Practice: The Case Study of Setúbal, Portugal. *Sustainability* **2021**, *13*, 8432. [CrossRef]
24. Liberalesso, T.; Cruz, C.O.; Silva, C.M.; Manso, M. Green infrastructure and public policies: An international review of green roofs and green walls incentives. *Land Use Policy* **2020**, *96*, 104693. [CrossRef]
25. Asl, S.R.; Pearsall, H. How Do Different Modes of Governance Support Ecosystem Services/Disservices in Small-Scale Urban Green Infrastructure? A Systematic Review. *Land* **2022**, *11*, 1247. [CrossRef]
26. Jarillo, S.; Barnett, J. Contingent communality and community-based adaptation to climate change: Insights from a Pacific rural atoll. *J. Rural. Stud.* **2021**, *87*, 137–145. [CrossRef]
27. Regmi, B.R.; Star, C.; Filho, W.L. An overview of the opportunities and challenges of promoting climate change adaptation at the local level: A case study from a community adaptation planning in Nepal. *Clim. Chang.* **2016**, *138*, 537–550. [CrossRef]
28. European Commission. Strategy on Adaptation to Climate Change. Communication from the Commission to the European Parliament, the Council, the European Economic and Social Committee and the Committee of the Regions. 2013. Available online: https://eur-lex.europa.eu/legal-content/EN/TXT/PDF/?uri=CELEX:52013SC0137&from=EN (accessed on 4 August 2022).
29. European Commission. Study to Support the Evaluation of the EU Adaptation Strategy Final Report. 2018. Available online: https://ec.europa.eu/clima/system/files/2018-11/adapt_strat_eval_report_en.pdf (accessed on 18 August 2022).
30. European Commission. Forging a Climate-Resilient Europe—The New EU Strategy on Adaptation to Climate Change. 2021. Available online: https://ec.europa.eu/jrc/en/peseta-iv/economic-impacts (accessed on 18 August 2022).
31. Rede de Municípios para a Adaptação Local às Alterações Climáticas. Carta de Compromisso. Available online: https://www.adapt-local.pt/ (accessed on 19 August 2022).
32. Presidência do Conselho de Ministros. Resolução do Conselho de Ministros n.º 24/2010 de 1 de Abril que Aprova a Estratégia Nacional de Adaptação às Alterações Climáticas. Diário Da República, 1.ª Série, 1090–1106. 2010. Available online: https://files.dre.pt/1s/2010/04/06400/0109001106.pdf (accessed on 19 August 2022).

33. Presidência do Conselho de Ministros. Resolução do Conselho de Ministros n.º 56/2015 de 30 de julho que Aprova o Quadro Estratégico para a Política Climática, o Programa Nacional para as Alterações Climáticas e a Estratégia Nacional de Adaptação às Alterações Climáticas, Determina os Valores de Redução das Emissões de Gases Com Efeito de Estufa para 2020 e 2030 e Cria a Comissão Interministerial do Ar e das Alterações Climáticas. Diário Da República, 1.ª série, 5114–5168. 2015. Available online: https://dre.pt/dre/detalhe/resolucao-conselho-ministros/56-2015-69905665 (accessed on 18 August 2022).
34. Presidência do Conselho de Ministros. Resolução do Conselho de Ministros n.º 53/2020 de 10 de Julho que Aprova o Plano Nacional Energia e Clima 2030 (PNEC 2030). Diário Da República, 1.ª Série, 2–158. 2020. Available online: https://dre.pt/dre/detalhe/resolucao-conselho-ministros/53-2020-137618093 (accessed on 19 August 2022).
35. Baker, I.; Peterson, A.; Brown, G.; McAlpine, C. Local government response to the impacts of climate change: An evaluation of local climate adaptation plans. *Landsc. Urban Plan.* **2012**, *107*, 127–136. [CrossRef]
36. Holden, P.B.; Rebelo, A.J.; Wolski, P.; Odoulami, R.C.; Lawal, K.A.; Kimutai, J.; Nkemelang, T.; New, M.G. Nature-based solutions in mountain catchments reduce impact of anthropogenic climate change on drought streamflow. *Commun. Earth Environ.* **2022**, *3*, 51. [CrossRef]
37. Moosavi, S. Design experimentation for Nature-based Solutions: Towards a definition and taxonomy. *Environ. Sci. Policy* **2022**, *138*, 149–161. [CrossRef]
38. Ascenso, A.; Augusto, B.; Silveira, C.; Rafael, S.; Coelho, S.; Monteiro, A.; Ferreira, J.; Menezes, I.; Roebeling, P.; Miranda, A.I. Impacts of nature-based solutions on the urban atmospheric environment: A case study for Eindhoven, The Netherlands. *Urban For. Urban Green.* **2020**, *57*, 126870. [CrossRef]
39. Tyllianakis, E.; Martin-Ortega, J.; Banwart, S.A. An approach to assess the world's potential for disaster risk reduction through nature-based solutions. *Environ. Sci. Policy* **2022**, *136*, 599–608. [CrossRef]
40. Raška, P.; Bezak, N.; Ferreira, C.S.; Kalantari, Z.; Banasik, K.; Bertola, M.; Bourke, M.; Cerdà, A.; Davids, P.; de Brito, M.M.; et al. Identifying barriers for nature-based solutions in flood risk management: An interdisciplinary overview using expert community approach. *J. Environ. Manag.* **2022**, *310*, 114725. [CrossRef]
41. Fernandes, J.P.; Guiomar, N. Nature-based solutions: The need to increase the knowledge on their potentialities and limits. *Land Degrad. Dev.* **2018**, *29*, 1925–1939. [CrossRef]
42. Monteiro, R.; Ferreira, J.C.; Antunes, P. Green Infrastructure Planning Principles: Identification of Priorities Using Analytic Hierarchy Process. *Sustainability* **2022**, *14*, 5170. [CrossRef]
43. Washbourne, C.-L. Environmental policy narratives and urban green infrastructure: Reflections from five major cities in South Africa and the UK. *Environ. Sci. Policy* **2021**, *129*, 96–106. [CrossRef]
44. Sussams, L.; Sheate, W.; Eales, R. Green infrastructure as a climate change adaptation policy intervention: Muddying the waters or clearing a path to a more secure future? *J. Environ. Manag.* **2014**, *147*, 184–193. [CrossRef] [PubMed]
45. Zhang, D.; Zhou, C.; Zhou, Y.; Zikirya, B. Spatiotemporal relationship characteristic of climate comfort of urban human settlement environment and population density in China. *Front. Ecol. Evol.* **2022**, *10*, 495. [CrossRef]
46. Shen, Z.-J.; Zhang, B.-H.; Xin, R.-H.; Liu, J.-Y. Examining supply and demand of cooling effect of blue and green spaces in mitigating urban heat island effects: A case study of the Fujian Delta urban agglomeration (FDUA), China. *Ecol. Indic.* **2022**, *142*, 109187. [CrossRef]
47. Green, D.; O'Donnell, E.; Johnson, M., Slater, L.; Thorne, C.; Zheng, S.; Stirling, R.; Chan, F. K. S.; Li, L.; Boothroyd, R. J. Green infrastructure: The future of urban flood risk management? *Wiley Interdiscip. Rev. Water* **2021**, *8*, 1–18. [CrossRef]
48. Wu, X.; Zhang, J.; Geng, X.; Wang, T.; Wang, K.; Liu, S. Increasing green infrastructure-based ecological resilience in urban systems: A perspective from locating ecological and disturbance sources in a resource-based city. *Sustain. Cities Soc.* **2020**, *61*, 102354. [CrossRef]
49. Nieuwenhuijsen, M.J. Green Infrastructure and Health. *Annu. Rev. Public Health* **2021**, *42*, 317–328. [CrossRef]
50. Green Infrastructure and Nature-Based Solutions for Urban Biodiversity with Case Studies Position Paper. 2020. Available online: www.ceeweb.org (accessed on 19 November 2022).
51. Ferrari, B.; Quatrini, V.; Barbati, A.; Corona, P.; Masini, E.; Russo, D. Conservation and enhancement of the green infrastructure as a nature-based solution for Rome's sustainable development. *Urban Ecosyst.* **2019**, *22*, 865–878. [CrossRef]
52. McNamara, K.E.; Buggy, L. Community-based climate change adaptation: A review of academic literature. *Local Environ.* **2016**, *22*, 443–460. [CrossRef]
53. Phong, N.T.; Quang, N.H.; Van Sang, T. Shoreline change and community-based climate change adaptation: Lessons learnt from Brebes Regency, Indonesia. *Ocean Coast. Manag.* **2022**, *218*, 106037. [CrossRef]
54. Forsyth, T. Community-based adaptation: A review of past and future challenges. *WIREs Clim. Chang.* **2013**, *4*, 439–446. [CrossRef]
55. Ayers, J.; Forsyth, T. Community-Based Adaptation to Climate Change. *Environ. Sci. Policy Sustain. Dev.* **2009**, *51*, 22–31. [CrossRef]
56. Picketts, I.M.; Werner, A.T.; Murdock, T.Q.; Curry, J.; Déry, S.J.; Dyer, D. Planning for climate change adaptation: Lessons learned from a community-based workshop. *Environ. Sci. Policy* **2012**, *17*, 82–93. [CrossRef]
57. Duus, E.; Montag, D. Protecting women's health in a changing climate: The role of community-based adaptation. *J. Clim. Chang. Health* **2022**, *6*, 100120. [CrossRef]

58. Basel, B.; Goby, G.; Johnson, J. Community-based adaptation to climate change in villages of Western Province, Solomon Islands. *Mar. Pollut. Bull.* **2020**, *156*, 111266. [CrossRef] [PubMed]
59. Forsyth, T. Community-Based Adaptation to Climate Change. *Oxf. Res. Encycl. Clim. Sci.* **2017**, *29*, 1–30. [CrossRef]
60. Santos, F.; Moniz, G.; Lopes, A.; Ramos, L.; Taborda, R. Relatório do Grupo de Trabalho do Litoral. In Proceedings of the VIII Congresso sobre Planeamento e Gestão das Zonas Costeiras dos Países de Expressão Portuguesa & 1ª Conferência Internacional "Turismo em Zonas Costeiras—Oportunidades e Desafios", Aveiro, Portugal, 16 October 2015.
61. Schmidt, L.; Mourato, J. Políticas públicas costeiras e adaptação às alterações climáticas: Que limites de implementação? In Proceedings of the VIII Congresso sobre Planeamento e Gestão das Zonas Costeiras dos Países de Expressão Portuguesa & 1ª Conferência Internacional "Turismo em Zonas Costeiras—Oportunidades e Desafios", Aveiro, Portugal, 16 October 2015.
62. Ford, J.D.; Sherman, M.; Berrang-Ford, L.; Llanos, A.; Carcamo, C.; Harper, S.; Lwasa, S.; Namanya, D.; Marcello, T.; Maillet, M.; et al. Preparing for the health impacts of climate change in Indigenous communities: The role of community-based adaptation. *Glob. Environ. Chang.* **2018**, *49*, 129–139. [CrossRef]
63. Nalau, J.; Becken, S.; Schliephack, J.; Parsons, M.; Brown, C.; Mackey, B. The Role of Indigenous and Traditional Knowledge in Ecosystem-Based Adaptation: A Review of the Literature and Case Studies from the Pacific Islands. *Weather. Clim. Soc.* **2018**, *10*, 851–865. [CrossRef]
64. Bronen, R.; Pollock, D.; Overbeck, J.; Stevens, D.; Natali, S.; Maio, C. Usteq: Integrating indigenous knowledge and social and physical sciences to coproduce knowledge and support community-based adaptation. *Polar Geogr.* **2019**, *43*, 188–205. [CrossRef]

MDPI

St. Alban-Anlage 66

4052 Basel

Switzerland

Tel. +41 61 683 77 34

Fax +41 61 302 89 18

www.mdpi.com

International Journal of Environmental Research and Public Health Editorial Office

E-mail: ijerph@mdpi.com

www.mdpi.com/journal/ijerph